Electromagnetic Modeling in Power Electronics

Electromagnetic Modeling in Power Electronics

Editors

Ivica Stevanovic
Bernhard Wunsch

MDPI • Basel • Beijing • Wuhan • Barcelona • Belgrade • Manchester • Tokyo • Cluj • Tianjin

Editors
Ivica Stevanovic
Radio Technology
Swiss Federal Office of Communications OFCOM
Biel/Bienne
Switzerland

Bernhard Wunsch
Corporate Research
ABB Switzerland
Baden-Dättwil
Switzerland

Editorial Office
MDPI
St. Alban-Anlage 66
4052 Basel, Switzerland

This is a reprint of articles from the Special Issue published online in the open access journal *Energies* (ISSN 1996-1073) (available at: www.mdpi.com/journal/energies/special_issues/EM_PE).

For citation purposes, cite each article independently as indicated on the article page online and as indicated below:

LastName, A.A.; LastName, B.B.; LastName, C.C. Article Title. *Journal Name* **Year**, *Volume Number*, Page Range.

ISBN 978-3-0365-1987-6 (Hbk)
ISBN 978-3-0365-1986-9 (PDF)

Contents

About the Editors

Ivica Stevanovic

Ivica Stevanovic received his PhD degree in Electrical Engineering from the École Polytechnique Fédérale de Lausanne, Switzerland, in 2005. He worked in Freescale Semiconductor and ABB Corporate Research before joining the Swiss Federal Office of Communications in 2013. His research interests include computational and applied electromagnetics, spectrum engineering, radio propagation, and wireless power transfer. Dr. Stevanovic represents Switzerland in international committees on spectrum engineering and standardization in wireless communications. In the Electronic Communications Committee, he is Vice Chairman of the Working Group Spectrum Engineering and Chairman of the Project Team SE45. In ITU-R, he is Chairman of Working Groups 3K-1 and 3M-3. He was awarded an NSF fellowship by the California Institute of Technology in 2000 and has been an IEEE Senior Member since 2012. The Swiss Academy of Sciences elected him to the URSI National Committee in 2018.

Bernhard Wunsch

Dr. Bernhard Wunsch received his PhD in condensed matter physics from the University of Hamburg in 2006. Later he worked as postdoc on graphene nanostructures at the Complutense University of Madrid, and the Materials Science Institute of Madrid. Then he joined the condensed matter theory group at Harvard University and worked on superconductivity, magnetism and optical lattices. In 2011 Dr. Wunsch joined ABB Corporate Research, Switzerland, where his research topics include electromagnetic compatibility of power electronic converters, computational and applied electromagnetics, design of magnetic components and inductive power transfer.

Preface to "Electromagnetic Modeling in Power Electronics"

Electromagnetic effects can play a major role in the operation of power electronic devices and systems. They may either be the primary underlying operating mechanism or they may need to be mitigated to allow for operation free of electromagnetic interference and noise. In both cases, there is a substantial benefit in having accurate and efficient methods and techniques to model these effects. This allows the performance of devices and systems to be predicted, evaluated, and optimized through simulations at an early stage of design, prior to prototyping, thereby reducing cost and minimizing subsequent redesigns.

This book contains new and original research highlighting advances in methods and techniques for electromagnetic modeling of power electronic components and systems.

We are grateful to all the contributing authors for agreeing to have their research published in this book.

Special thanks go to the authors of the invited papers: Prof. Carl Ho (University of Manitoba, Canada), Prof. Jean-Luc Schanen (University Grenoble Alpes, France), Dr. Eckart Hoene (Fraunhofer IZM, Germany), Prof. Giulio Antonini (Università degli Studi dell'Aquila, Italy), Prof. Jonas Ekman (LuleåUniversity of Technology, Sweden), and Prof. Flavia Grassi (Politecnico di Milano, Italy).

Ivica Stevanovic, Bernhard Wunsch
Editors

energies

Article

Electromagnetic Field Tests of a 1-MW Wireless Power Transfer System for Light Rail Transit

Gunbok Lee [1], Myung Yong Kim [1], Changmu Lee [1], Donguk Jang [2], Byung-Song Lee [1] and Jae Hee Kim [3,*]

1 Smart Electrical & Signaling Division, Korea Railroad Research Institute (KRRI), Uiwang 16105, Korea; gunbok@krri.re.kr (G.L.); mykim@krri.re.kr (M.Y.K.); cmlee@krri.re.kr (C.L.); bslee@krri.re.kr (B.-S.L.)
2 Advanced Railroad Technology Planning Department, Korea Railroad Research Institute (KRRI), Uiwang 16105, Korea; dujang@krri.re.kr
3 School of Electrical, Electronics and Communication Engineering, Korea University of Technology and Education (KOREATECH), Cheonan 31253, Korea
* Correspondence: jaehee@koreatech.ac.kr; Tel.: +82-41-560-1172

Abstract: The high-power wireless power transfer (WPT) system in railways does not require physical contact to transfer electrical power, is electrically safe, and reduces maintenance costs from wear and tear. However, a high-power system generates a strong magnetic field that can result in problems of electromagnetic field (EMF) exposure and electromagnetic interference (EMI). In this study, EMF and EMI were measured at various positions under in-motion environment conditions for a 1-MW WPT light rail transit system. The measured maximum EMF was 2.41 μT, which is lower than the international guideline of 6.25 μT for the various locations with a potential presence of passengers. The measured EMI also satisfied international standards in the frequency range of 150 kHz–1 GHz.

Keywords: wireless power transfer; light rail; electromagnetic field; electromagnetic interference; radiated emission

Citation: Lee, G.; Kim, M.Y.; Lee, C.; Jang, D.; Lee, B.-S.; Kim, J.H. Electromagnetic Field Tests of a 1-MW Wireless Power Transfer System for Light Rail Transit. *Energies* **2021**, *14*, 1171. https://doi.org/10.3390/en14041171

Academic Editors: Ivica Stevanovic and Bernhard Wunsch

Received: 10 February 2021
Accepted: 18 February 2021
Published: 22 February 2021

1. Introduction

The wireless power transfer (WPT) system transfers electrical power using magnetic fields without any contact with electrical cables, avoiding the electrical [1] and mechanical [2] issues in conventional contact systems. The WPT technology has already been utilized in small electronic devices and home appliances such as mobile phones, and with the rapid development in battery technology, the WPT system can be applied to high-power systems such as electric vehicles (EVs) and railways. WiTricity entered the WPT system market for EVs in 2018 [3]. WPT systems for stationary applications have a power of 11 kW and an efficiency of 90–93%. Moreover, the dynamic WPT system has been tested at a frequency of 85 kHz, and it could attain a power of 20 kW. However, a WPT system with a higher power has been applied on an electric bus. In 2013, Korea Advanced Institute of Science and Technology applied a 100-kW WPT system on an electric bus, which was plied on an actual route [4]. The frequency was 20 kHz, and the air gap was approximately 300 mm. In 2019, Toshiba applied a 44-kW wireless charging technology on an electric bus [5]. An efficiency of approximately 85% was achieved with a frequency of 85 kHz. The transmission distance between the charging pads was 100–130 mm.

In addition, high-power WPT systems of more than 100 kW have been studied for applications on railways. In 2013, Bombardier Inc. demonstrated the application of a 200-kW WPT technology for light rail transit [6–8]. The notable feature of this technology was the use of a three-phase current. In 2015, the Korea Railroad Research Institute applied a 1-MW WPT system where the frequency was 60 kHz and the air gap was 50 mm on a high-speed railway system and demonstrated its successful pilot operation [9]. The same year, the Railway Technical Research Institute (RTRI) applied a 50-kW WPT system with a frequency of 10 kHz, an air gap of 75 mm, and the primary current at 400 A for application in railways [10].

Although there is an insufficient number of studies or reports regarding WPT systems with a power of more than 500 kW, a number of major research institutes have plans to perform tests by increasing the power transfer capacity. In addition, active R&D is underway for dynamic charging or distributed charging at stations rather than a simple system of stationary charging. In particular, dynamic charging is preferred for high-power WPT systems because the charging duration becomes longer with stationary charging.

One of the challenges for commercial applications of high-power WPT technology arises from the electromagnetic field (EMF) and electromagnetic interference (EMI) [11–14]. The levels of EMF and EMI exposure to humans during the operation of the WPT system must comply with the international standards. At the WPT frequency of 60 kHz, the standard limit of EMF is 6.25 μT. The EMI standard is defined from 150 kHz up to 1 GHz, and the limit is different for each frequency. Although the EMF and EMI of WPT systems have been discussed in certain previous studies, test results for various conditions still need to be presented. The technology for EMF reduction in the operation of the WPT system has been mostly investigated on a laboratory scale or using simulations, where the major techniques modified the transmitter/receiver coil design and the shape of the shielding material.

A technique using two loops in the transmission line has been proposed for reducing EMI in WPT-powered buses [15,16]. Oak Ridge National Laboratory (ORNL) proposed a method based on magnetic shielding for reducing EMF in a 100-kW wireless EV charging system [17]. In addition, the National Renewable Energy Laboratory (NREL) measured the EMF and touch current at various locations for a 25-kW inductive power transfer system applied to a minibus and presented the measurement data [18]. Moreover, a method has been proposed for reducing EMF by using a three-phase current in the transmitter coil of a WPT system [19]. Furthermore, a technique has been developed to reduce EMF by attaching additional coils to the outside of the transmitter/receiver coils, and fixedly attached or switch-controlled passive and active coils have been proposed as well [20–22]. All previously presented EMF measurements and leakage reduction techniques have been performed at the laboratory level or tested at medium power less than 100 kW. The EMF and EMI measurement data in the actual high-power WPT system have not been presented.

In this study, we present the EMF and EMI levels measured based on stationary and dynamic charging application of a WPT system for light rail transit. Recently, we applied a 1-MW WPT system on the light rail transit of Korea-automated guideway transit (K-AGT) and measured the EMF and EMI levels in various environments during a test operation. The EMF and EMI standards applicable to WPT systems are briefly outlined in Section 2. Thereafter, Section 3 briefly introduces the newly developed WPT system for light rail transit. The EMF and EMI values measured during operation on the test route are presented in Section 4, and the obtained results are concluded in Section 5.

2. International Guidelines on EMF and EMI

2.1. EMF Guidelines

In this study, the high-power WPT system for light rail transit used a frequency of 60 kHz. Therefore, the applicable regulations for this frequency and that for EMF are reviewed in this section. The International Commission on Non-Ionizing Radiation Protection (ICNIRP) 1998 [23] presented reference levels for general public exposure to frequency bands of 1 Hz–300 GHz, whereas ICNIRP 2010 [24] provides reference levels for frequency bands from 1 Hz to 10 MHz. In particular, ICNIRP 1998 represents a more rigorous standard than ICNIRP 2010 for the frequency band of 60 kHz. Thus, the current study analyzed the measurements by applying the ICNIRP 1998 standard, which is an adopted standard for human protection against electromagnetic fields in South Korea. Table 1 depicts the standards of electromagnetic fields for the general public, as presented in the ICNIRP guidelines.

Table 1. Reference levels for general public exposure to time varying electric and magnetic fields.

	Frequency Range	H-Field Strength (A/m)	B-Field (μT)
ICNIRP 1998	3–150 kHz	5	6.25
ICNIRP 2010	3 kHz–10 MHz	21	27

The EMF measurement methods for the railway sector along with the measurement locations of the rolling stock are specified in IEC62597 [25], according to which the measurement location is divided for the rolling stock and the surrounding infrastructure. The measurements inside the rolling stock are specified to be taken at three points: 0.3 m, 0.9 m, and 1.5 m from the floor. In addition, a distance of at least 0.3 m needs to be maintained from the wall. For measurements outside the rolling stock, the points must be horizontally separated from the vehicle by at least 0.3 m, and the measurements are to be acquired at 0.5, 1.5, and 2.5 m in terms of height. In addition, the measurement points on the platform need to be separated at least 0.3 m from the end of the platform at specified heights of 0.9 m and 1.5 m. Moreover, the power supply system installed in the infrastructure ought to have a minimum separation distance of 0.3 m, and its EMF has to be measured at 0.3, 0.9, and 1.5 m in height.

2.2. EMI Guidelines

Radiated emission (RE) for light rail systems is specified in IEC 62236-2: 2018 [26] and IEC 62236-3-1: 2018 [27]. According to IEC standards, the reference value of RE varies with the type of rolling stock, and the strictest rules have been applied for light rail systems plying in urban areas. In addition, the H-field is specified to be measured in the frequency band 150 kHz–30 MHz, whereas the E-field is measured in the 30 MHz–1 GHz frequency band. The limits of the fields for the light rail system defined in IEC standards can be formulated with the following equations:

$$H_{max}(f) = 60 - 21.7 * log_{10}\left(\frac{f}{150,000}\right) \text{dB}\mu\text{A/m for 150 kHz–30 MHz}$$
$$(60 \text{ dB}\mu\text{A/m at 150 kHz}),$$

$$E_{max}(f) = 85 - 16.4 * log_{10}\left(\frac{f}{30,000,000}\right) \text{dB}\mu\text{V/m for 30 MHz–1 GHz}$$
$$(85 \text{ dB}\mu\text{V/m at 30 MHz}),$$

where f is the frequency in Hz, $H_{max}(f)$ is maximum magnetic field intensity, and $E_{max}(f)$ is maximum electric field intensity. However, the fields at 60 kHz are not defined. Thus, it is necessary to examine whether or not the RE of the harmonic frequencies of the operating frequency of WPT system exceeds the standard level. The magnetic field was measured at 10 m away from the center of the track with a loop antenna placed at a height of 1–2 m. The electric field was measured at a distance of 10 m from the center of the track using a log-periodic antenna at a height of approximately 2.5–3.5 m.

3. WPT System for Light Rail Transit

3.1. Overview of WPT System for Light Rail Transit

The WPT system manufactured for the K-AGT light rail transit is presented in Figure 1; it was designed with 1 MW, and the detailed specifications are listed in Table 2. It used four pick-ups and generated 250 kW of rated power output per pick-up. The output voltage was derived at 750 V DC to match the supply power of the existing light rail transit. The air-gap distance between the transmission line and the pick-up was designed to be 60 mm. In addition, the batteries of the stated system could be charged when both stationary and in motion. Both driving energy and battery-charged energy was supplied to the WPT rolling stock in sections with a transmission line, but the sections without a power line could utilize only the battery energy for running, and a power supply system using a third rail was not required. As the rolling stock used the maximum power in the accelerating section post departure, the WPT transmission line was installed in this section to ensure sufficient power supply for charging and driving.

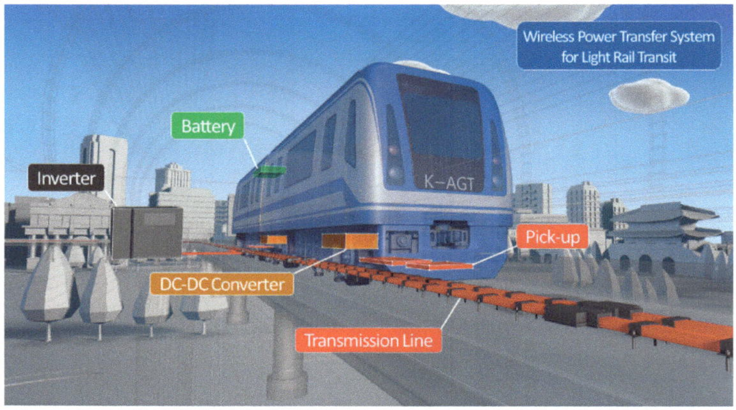

Figure 1. WPT system for light rail transit.

Table 2. Specifications of WPT system for light rail transit.

Category		Description
Frequency		60 kHz
Air gap		60 mm
Maximum efficiency		90.1%
Inverter	Maximum current	500 A
	Maximum output power	1.2 MW
Transmission line	Rated current	500 A
	Length	201 m
Pick-up	Output voltage	$750V_{DC}$
	Output power	250 kW/unit

3.2. Structure of the Power Line and Pick-Up

As depicted in Figure 2, the transmission line was set with wires carrying a maximum current of 500 A wrapped around the track, where two cables of 35 mm diameter were connected in parallel to achieve this purpose. The high-frequency loss in the cable was minimized with a Litz wire comprising 25,200 strands, and the inside of one cable was approximately 160 mm from that of the opposite cable. The width of the cable was set to be narrow so that the magnetic field formed in the center part of the rolling stock can be reduced at points away from the power line without using core shielding. In addition, the ferrite core was removed to reduce the cost of the transmission line as compared with the existing method of using the core therein. Moreover, the length of the transmission line was set at 200 m in consideration of the distance traveled by the K-AGT rolling stock till accelerating up to the maximum speed.

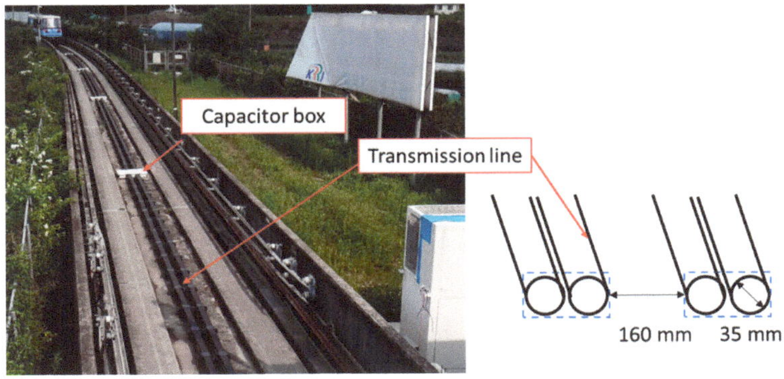

Figure 2. Structure of power line.

The structure of the pick-up is illustrated in Figure 3, where the pick-up contains a receiving coil that receives the magnetic field, a rectifier that converts AC into DC, and a regulator. The coil was formed by winding a 20-mm-diameter cable four times, and the regulator was designed for a constant output at 750 V by rectifying the voltage of the pick-up. In particular, the rated power per pick-up was 250 kW, and the maximum current flow in the pick-up was set to 392 A. In addition, a ferrite core and an aluminum plate were used above the cable. The ferrite core served to increase the pick-up efficiency and reduce the magnetic field leakage from the pick-up, whereas the aluminum plate served to prevent the magnetic field generated from the lower side of the rolling stock to transmit into the vehicle.

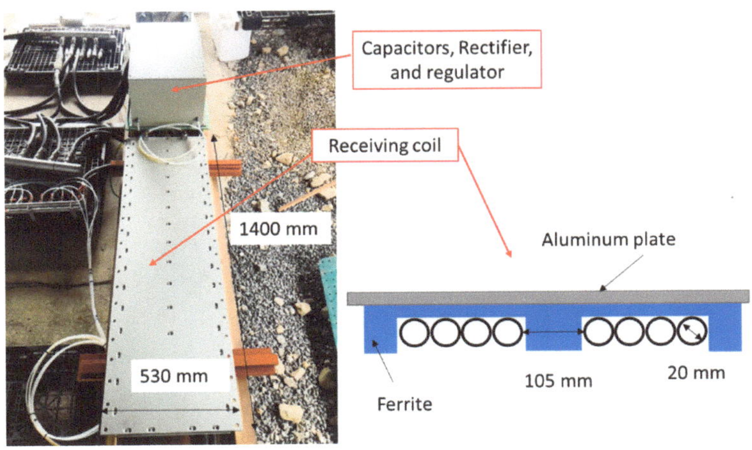

Figure 3. Structure of the pick-up.

3.3. Analysis of Electric Field by Simulation

The magnetic field generated by the WPT system was analyzed under simulation performed using the designed and manufactured transmission and pick-up structures, as depicted in Figure 4. A 3D electromagnetic analysis tool, Ansys HFSS, was used for the full-wave simulation. The cables used for the transmission line and pick-up were assumed as copper with a conductivity of 5.8×10^7 S/m. The top of the transmission line and the bottom of the pick-up was separated by 60 mm to maintain the primary and pick-up coils as close as possible, and allow a small margin of around 50 mm as the minimum distance between the track surface and the underside of the vehicle. In addition, the current flowing through the transmission line during the actual operation of the light rail transit was set to 450 A. Moreover, the magnetic field value around the pick-up did not alter significantly with the number of pick-ups increasing longitudinally over the transmission line; thus, only a single pick-up was used in the simulation. Furthermore, a 180-nF capacitor was connected in series with the coil to ensure resonance of the pick-up at 60 kHz.

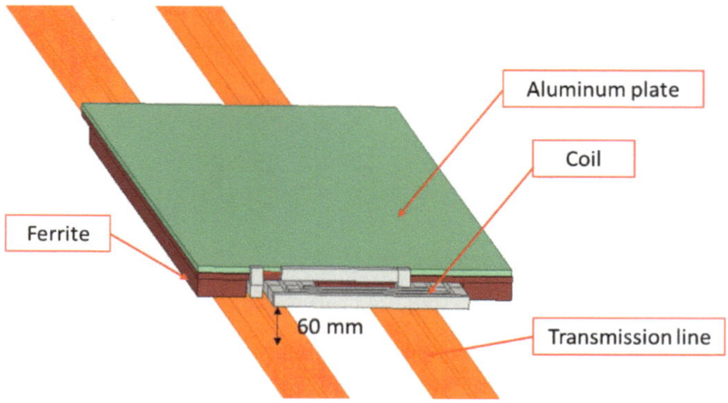

Figure 4. Structure of the power line and pick-up for EMF simulation.

The magnetic field distribution between the transmission line and the pick-up is presented in Figure 5. An impedance-matching method was applied to ensure maximum output of the pick-up with 2 Ω at its output terminal. As can be observed from Figure 5, a stronger magnetic field was formed sideways around the pick-up owing to a phase difference of 90° between the magnetic field generated by the transmission line and that produced from the current flowing in the pick-up coil—transferring power to the pick-up. Therefore, the net magnetic field was augmented by the magnetic fields generated by both the transmission line and the pick-up.

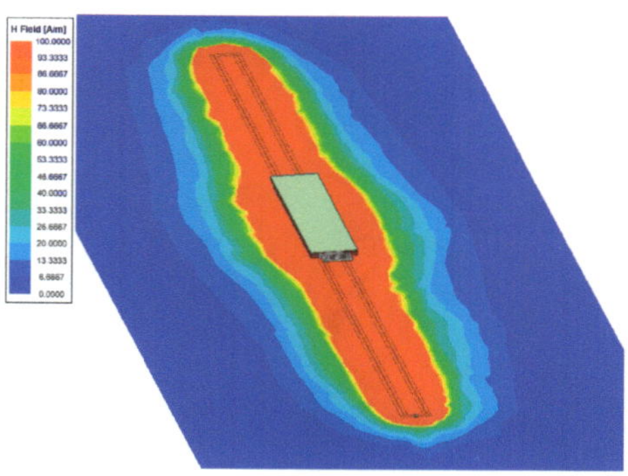

Figure 5. Magnetic field distribution between the power line and pick-up.

The pick-up of the actual light rail system was attached to the underside of the rolling stock, and an external infrastructure such as a platform was present for the train to stop. Therefore, a simulation considering the presence of a platform was performed to consider all these effects. The magnetic field distribution in Figure 6 illustrates the cross-section of the location where the pick-up was attached. Here, the maximum range of the magnetic field was derived at 5 A/m in accordance with the maximum human-exposure level. In addition, the rolling stock was assumed as a perfect conductor, and the platform was assumed of concrete with a dielectric constant of 6 and a loss tangent of 0.06. Although the actual pick-up was attached to the underside of the rolling stock, Figure 6a depicts the distribution of a magnetic field for a transmission line and pick-up without the rolling stock. On the contrary, Figure 6b presents the magnetic field distribution for the pick-up attached to the underside of the rolling stock, which serves to shield the magnetic field. Moreover, the amount and intensity of the magnetic field on the platform were observed to have evidently reduced. Therefore, the EMF levels of the designed system will not exceed the reference level in an actual environment with the rolling stock.

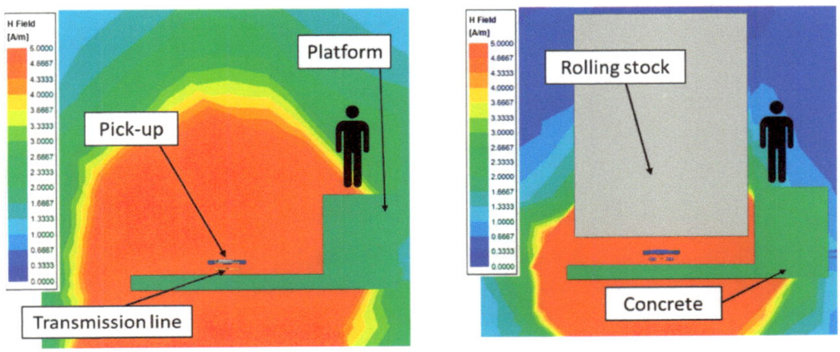

(a) (b)

Figure 6. Magnetic field distribution in a platform (**a**) without a rolling stock and (**b**) with a rolling stock.

4. Measurement Results of EMF and EMI

The EMF and EMI were measured on the application of the WPT on the K-AGT rolling stock to determine whether the manufactured WPT system satisfied the standards for electromagnetic waves. In addition, the rolling stock consumed lesser power than the design owing to the battery charging capacity and the vehicle power consumption limit at low speed. As the charging speed varied based on the actual battery level, the measured values could change as well. The following measurement results were obtained using two pick-ups (rated power: 500 kW) in the in-motion state, and 150 kW of power was transferred during charging in the stationary state. Moreover, up to 400 kW of power was transferred during dynamic charging. The measurement values stated in the results correspond to the above conditions.

The various EMF and EMI measurement points are illustrated in Figure 7. The measurements were performed when the rolling stock was stationary on the platform and when it started moving with the use of adequate energy. The EMF was measured at two locations inside the rolling stock, one on the platform, one outside the rolling stock, and one around an inverter that generated high-frequency power. In addition, the only RE for EMI was measured in this paper. The RE measurement of rolling stock was performed for three cases: stationary state, low-speed state, and high-speed state. For the stationary and low-speed states, the measurements were obtained at point A, whereas the measurements for the high-speed state were acquired at point B, as the maximum speed can be achieved only after operating a sufficient distance.

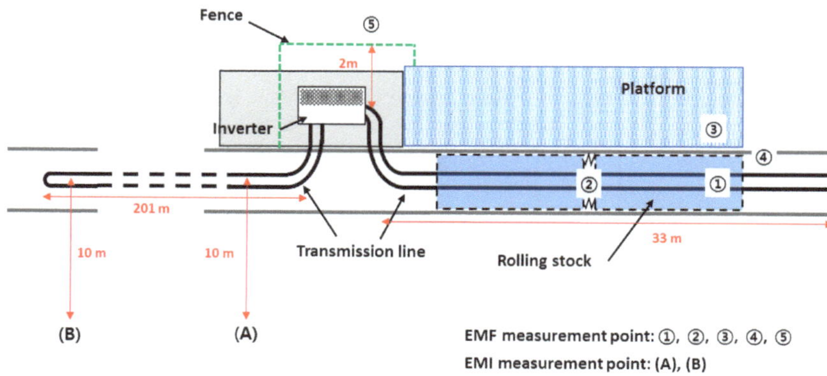

Figure 7. Measurement points for EMF and EMI.

4.1. EMF Measurement Results

When the rolling stock is in dynamic charging the battery with the maximum power collection state, Figure 8 depicts the EMF measurement points inside the rolling stock, and Table 3 outlines the measurement results. Three magnetic field testers (FT3470-50, Hioki) were used to measure magnetic fields at each point simultaneously. The width of the rolling stock was 2.4 m, and the measurement inside the rolling stock was acquired at the midpoint—the closest position to the external transmission line, i.e., 1.2 m away from the end of the rolling stock. The measurement location from the exit side and that at the gangway point connecting the vehicles are portrayed in Figure 8a,b, respectively. The magnetic field strength measured on the gangway side was four times larger than that on the exit side owing to the gaps on the gangway floor. In addition, the sidewalls were covered with a non-metallic material such as rubber that provided insufficient shielding from the magnetic fields. Nonetheless, all the measured magnetic field strengths satisfied the reference levels as per the ICNIRP 1998 standard, because the inside of the rolling stock is 1.2 m away from the transmission line, and the bottom of the gangway is made of metal.

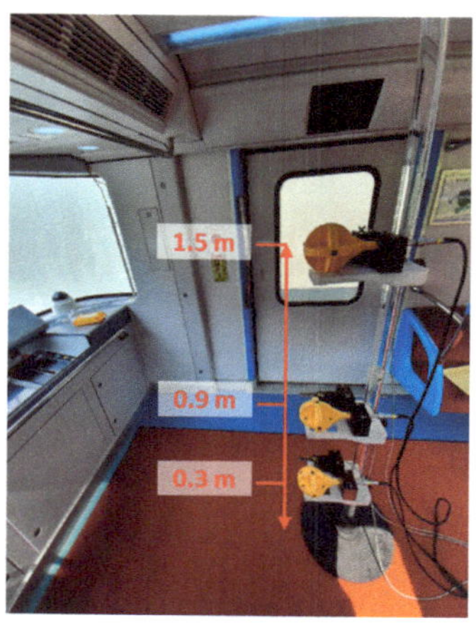

(a)

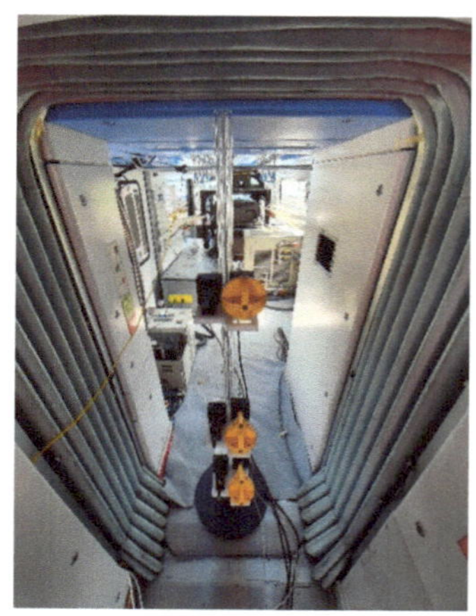

(b)

Figure 8. EMF measurement points inside the rolling stock.

Table 3. Magnetic field measurements inside the rolling stock.

Measurement Points	Measurement Height	Measured B-Field (μT) (Intensity Limit: 6.25)
Inside the rolling stock	0.3 m	0.49
	0.9 m	0.45
	1.5 m	0.33
Gangway	0.3 m	2.41
	0.9 m	1.90
	1.5 m	1.78

The three EMF measurement points—outside the rolling stock, on the platform, and on the inverter—are presented in Figure 9. First, the measurements outside the rolling stock were obtained on the track separated by 0.3 m from the front of the light-rail rolling stock. As the transmission line, in this case, was installed at the center of the track, the measurement could not be acquired at this point. Therefore, a point 1 m away from the center of the line was considered for the measurement. The second measurement point was located on the platform, where the passengers usually wait for the light rail transit, and the exit where the passengers stand in most cases. Therefore, the EMF measurement location was selected at a point 0.3 m away from the end of the platform, which is the closest point to the transmission line for the exit location. Although the high-frequency power of 60 kHz was generated by the inverter, the amount of EMF leakage was small because the inverter was shielded with a metal housing. However, a strong magnetic field could be generated near the cable connection, as the transmitter cable was buried in the floor. Therefore, a fence was installed at a distance of 2 m to restrict public access in the vicinity of the inverter, and the magnetic field was measured at a point 0.3 m away from the fence.

(a) (b) (c)

Figure 9. EMF measurement points: (**a**) outside the rolling stock, (**b**) platform, (**c**) near the inverter.

The measurement results for the instant (departure while charging) when the largest amount of power was transferred to the light rail transit are listed in Table 4. All the measurements in the platform were within the standard value, because the measurement points were located at a sufficient height—almost 1 m higher than the transmission line. In addition, the magnetic field value at a point 0.5 m away from the ground and in front of the rolling stock was measured at 14.5 μT, which exceeds the standard value. However, this does not cause a practical problem because the transmission line is operated with the flow of current only when the train enters, and the measurement point was located in an area where human access to the track side is strictly restricted and the rolling stock is in motion. Thus, the measurement is not posing any risk of human exposure. In addition, the reference level in the revised ICNIRP 2010 is increased to 27 μT; therefore, the measured value satisfies this standard value. Although the magnetic field strength in the vicinity of the inverter with the buried transmission cable is higher than that on the platform, the measurement was within the reference value range. Moreover, the inverter is a high-voltage electric device, and a safety fence can be installed around the inverter to restrict public access. In this case, the risk of exposure to the magnetic field would be significantly reduced.

Table 4. Magnetic fields measured outside the rolling stock and surrounding infrastructure.

Measurement Points	Measurement Height	Measured B-Field (μT) (Intensity Limit: 6.25)	Simulated B-Field (μT)
Front of the rolling stock	0.5 m	14.50	13.88
	1.5 m	4.09	6.62
	2.5 m	1.37	3.77
Platform	0.5 m	1.93	1.17
	0.9 m	1.48	1.01
	1.5 m	1.08	0.73
Near the inverter	0.3 m	4.03	-
	0.9 m	3.98	-
	1.5 m	1.99	-

Simulation results were also presented for the platform environment and outside of the vehicle in Table 4. It can be seen that the trend of the simulation and measurement result are similar. The numerical differences are due to the simplification of the vehicle and platform structures in the simulation. Only EMF data were presented inside the vehicle and near the inverter due to the difficulty of simulation.

4.2. EMI Measurement Results

The measurement location of the RE is shown in Figure 10, where a loop antenna and a log-periodic antenna of vertical and horizontal polarization was used at a point 10 m vertically away from the center of the transmission line. The measurements were obtained according to the train operation mode under the condition of WPT, and the measurements were acquired for three categories: stationary, low-speed motion (0–35 km/h), and high-speed motion (60 km/h). The measurements for the stationary and low-speed states were obtained around the power inverter (point A in Figure 7) and those for the high-speed state were measured at the end of the transmission line (point B in Figure 7). The EMI measurement data are presented in Figure 11. Although the frequency band of 0–150 kHz was not defined in the standard, these values represented the magnetic field measurements for reference purposes. As the RE of the WPT system required to be measured, the EMI was initially measured when the light rail system was not under operation, as illustrated in Figure 11a. Moreover, certain values exceeding the standards corresponded to the broadcasting and communication bands. Figure 11b presents the EMI measurements during wireless charging in the stationary state. At frequencies beyond 150 kHz, all the measurements were under the standard levels, and at the wireless charging frequency of 60 kHz, the magnetic field was measured at 64 dBμA/m. The EMI measurements in low- and high-speed states of the rolling stock during dynamic charging are presented in Figure 11c,d, where the measurement data can be observed to be similar. During the same wireless charging, the noise level in low-speed operations was slightly higher than that of high-speed operations, because the rolling stock used energy with maximum acceleration at low speed, whereas less energy was used for the coastdown at high speeds. In addition, the noise level was observed to rise corresponding to various frequencies and results in a stationary state. In particular, noise from the rolling stock operations was generated at a frequency of around 180 kHz, which increased the wireless power energy. Furthermore, the noise was presumably generated from the vehicle itself, such as during switching of propulsion inverters of the rolling stock.

Figure 10. Radiation emission measurement location.

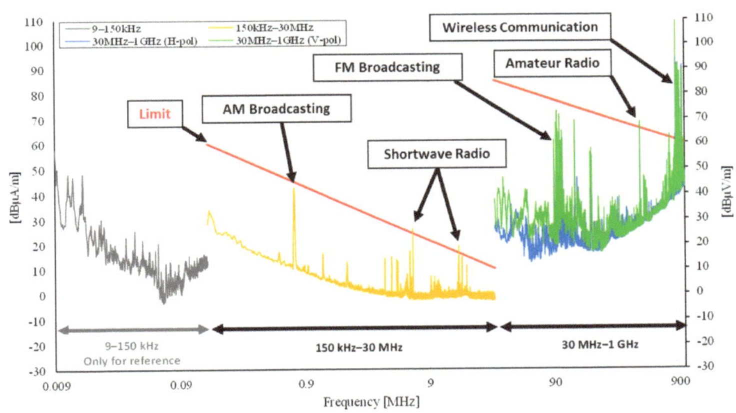

(**a**)

Figure 11. *Cont.*

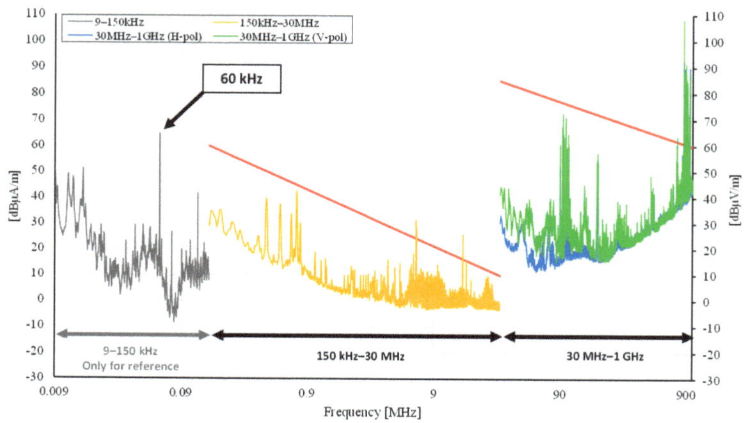

(b)

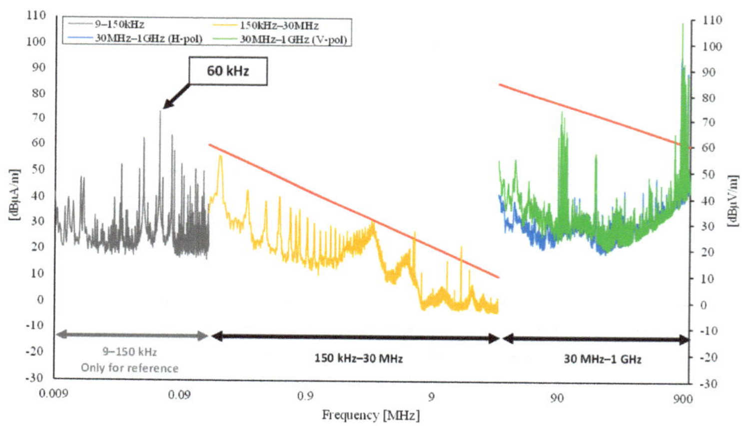

(c)

Figure 11. *Cont.*

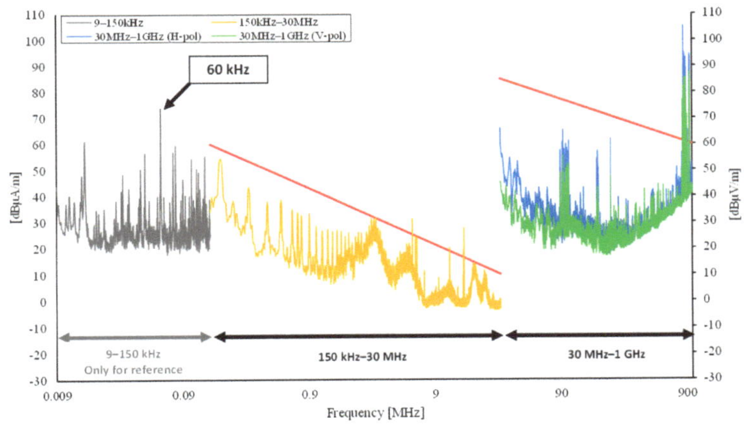

(d)

Figure 11. RE measurement results for each scenario: (**a**) without WPT, (**b**) stationary charging, (**c**) dynamic charging: low speed, (**d**) dynamic charging: high speed.

5. Conclusions

The current study presented the results of EMF and EMI measurements acquired inside and around the rolling stock in a light rail transit system operating in a real environment using a WPT system. This study is the first to present EMF and EMI data in accordance with the railway standard for the 1 MW WPT system. The EMF simulations shows that the EMF varied according to the platform environment of the rolling stock. In addition, the measured EMF and EMI were below the reference level, thus satisfying the safety standards. Moreover, the EMF inside the rolling stock was weak, because the magnetic field was shielded by the vehicle itself. However, the magnetic field values in the gangway were relatively higher than those inside the rolling stock owing to the presence of several gaps in the vehicle body at this location. Therefore, magnetic field reduction method at the gangway needs to be considered for using a strong magnetic field in the future. As the height of the platform in the railway environment is adequate to be well separated from the transmission line, the magnetic field strength on the platform satisfied the standard reference level. However, a strong magnetic field can be produced near the buried transmission cable in the high-frequency power inverter. Therefore, the standard level at this location can be satisfied by maintaining a safe distance with a fence around the inverter. In addition, the installation of additional shielding structures must be considered before the transmission line reaches the track. The current study verified that all the measurements of the magnetic field obtained from recommended locations for a high-power WPT system satisfied the reference levels in the ICNIRP 1998 standard, thereby indicating the commercial potential of the WPT system.

Author Contributions: Conceptualization, G.L., M.Y.K. and J.H.K.; methodology, G.L. and J.H.K.; software, G.L., C.L. and D.J.; validation, G.L. and M.Y.K.; formal analysis, G.L. and M.Y.K.; investigation, G.L. and J.H.K.; resources, G.L., M.Y.K., C.L. and D.J.; data curation, C.L. and D.J.; writing—original draft preparation, J.H.K. and G.L.; writing—review and editing, G.L. and J.H.K.; visualization, G.L. and J.H.K.; supervision, B.-S.L.; project administration, B.-S.L., M.Y.K. and G.L.; funding acquisition, B.-S.L. and M.Y.K. All authors have read and agreed to the published version of the manuscript.

Funding: This research was supported by a grant from R&D Program of the Korea Railroad Research Institute, Republic of Korea, and by the Railroad Technology Development Program funded by the

Ministry of Land, Infrastructure and Transport (MOLIT) of the Korean government under Grant 19RTRP-B097048-05.

Institutional Review Board Statement: Not applicable.

Informed Consent Statement: Not applicable.

Conflicts of Interest: The authors declare no conflict of interest.

References

1. Midya, S.; Bormann, D.; Schutte, T.; Thottappillil, R. Pantograph arcing in electrified railways—Mechanism and influence of various parameters—Part I: With DC traction power supply. *IEEE Trans. Power Del.* **2009**, *24*, 1931–1939. [CrossRef]
2. Song, Y.; Liu, Z.; Rønnquist, A.; Nåvik, P.; Liu, Z. Contact wire irregularity stochastics and effect on high-speed railway pantograph–catenary interactions. *IEEE Trans. Instrum. Meas.* **2020**, *69*, 8196–8206. [CrossRef]
3. Zhang, B.; Carlson, R.B.; Smart, J.G.; Dufek, E.J.; Liaw, B. Challenges of future high-power wireless power transfer for light-duty electric vehicles—technology and risk management. *eTransportation* **2019**, *6*, 100088. [CrossRef]
4. Song, K.; Koh, K.E.; Zhu, C.; Jiang, J.; Wang, C.; Huang, X. A Review of dynamic wireless power transfer for in motion electric vehicles. *Wirel. Power Transf.* **2016**, 109–128. [CrossRef]
5. Obayashi, S.; Shijo, T.; Suzuki, M.; Moritsuka, F.; Ogawa, K.; Ogura, K.; Kanekiyo, Y.; Ishida, M.; Takanaka, T.; Tada, N.; et al. 85 kHz band 44 kW wireless rapid charging system for field test and public road operation of electric bus. *World Electr. Veh. J.* **2019**, *10*, 26. [CrossRef]
6. Elliott, G.A.J.; Covic, G.A.; Kacprzak, D.; Boys, J.T. A new concept: Asymmetrical pick-ups for inductively coupled power transfer monorail systems. *IEEE Trans. Magn.* **2006**, *42*, 3389–3391. [CrossRef]
7. Brecher, A.; David, A. *Review and Evaluation of Wireless Power Transfer (WPT) for Electric Transit Applications*; No. FTA Report No. 0060; John, A., Ed.; Volpe National Transportation Systems Center: Cambridge, MA, USA, 2014.
8. Winter, J.; Mayer, S.; Kaimer, S. Inductive power supply for heavy rail vehicles. In Proceedings of the 3rd International Electric Drives Production Conference (EDPC), Nuremberg, Germany, 29–30 October 2013; pp. 1–9. [CrossRef]
9. Kim, J.H.; Lee, B.-S.; Lee, J.-H.; Lee, S.-H.; Park, C.-B.; Jung, S.-M.; Lee, S.-G.; Yi, K.-P.; Baek, J. Development of 1-MW inductive power transfer system for a high-speed train. *IEEE Trans. Ind. Electron.* **2015**, *62*, 6242–6250. [CrossRef]
10. Ukita, K.; Kashiwagi, T.; Sakamoto, Y.; Sasakawa, T. Evaluation of a non-contact power supply system with a figure-of-eight coil for railway vehicles. In Proceedings of the 2015 IEEE PELS Workshop on Emerging Technologies: Wireless Power (2015 WoW), Daejeon, Korea, 5–6 June 2015. [CrossRef]
11. Campi, T.; Cruciani, S.; Santis, V.D.; Maradei, F.; Feliziani, M. EMC and EMF safety issues in wireless charging system for an electric vehicle (EV). In Proceedings of the International Conference of Electrical and Electronic Technologies for Automotive, Torino, Italy, 15–16 June 2017. [CrossRef]
12. Jeschke, S.; Maarleveld, M.; Baerenfaenger, J.; Schmuelling, B.; Burkert, A. Challenges in EMC testing of EV and EVSE equipment for inductive charging. In Proceedings of the International Symposium on Electromagnetic Compatibility (EMC EUROPE), Amsterdam, The Netherlands, 27–30 August 2018. [CrossRef]
13. Zhang, W.; White, J.C.; Malhan, R.K.; Mi, C.C. Loosely coupled transformer coil design to minimize EMF radiation in concerned areas. *IEEE Trans. Veh. Technol.* **2016**, *65*, 4779–4789. [CrossRef]
14. Yashima, Y.; Omori, H.; Morizane, T.; Kimura, N.; Nakaoka, M. Leakage magnetic field reduction from wireless power transfer system embedding new eddy current-based shielding method. In Proceedings of the International Conference on Electrical Drives and Power Electronics (EDPE), The High Tatras, Slovakia, 21–23 September 2015. [CrossRef]
15. Shijo, T.; Ogawa, K.; Suzuki, M.; Kanekiyo, Y.; Ishida, M.; Obayashi, S. EMI reduction technology in 85 kHz band 44 kW wireless power transfer system for rapid contactless charging of electric bus. In Proceedings of the IEEE Energy Conversion Congress and Exposition (ECCE), Milwaukee, WI, USA, 18–22 September 2016. [CrossRef]
16. Song, B.; Dong, S.; Gao, X.; Li, Y.; Cui, S. A tripolar wireless power transfer system with low leakage magnetic field for railway vehicles. In Proceedings of the 21st European Conference on Power Electronics and Applications (EPE′19 ECCE Europe), Genova, Italy, 2–5 September 2019. [CrossRef]
17. Asa, E.; Mohammad, M.; Onar, O.C.; Pries, J.; Galigekere, V.; Su, G.-J. Review of safety and exposure limits of electromagnetic fields (EMF) in wireless electric vehicle charging (WEVC) applications. In Proceedings of the IEEE Transportation Electrification Conference & Expo (ITEC), Chicago, IL, USA, 22–26 June 2020. [CrossRef]
18. Mohamed, A.A.S.; Meintz, A.; Schrafel, P.; Calabro, A. Testing and assessment of EMFs and touch currents from 25-kW IPT system for medium-duty EVs. *IEEE Trans. Veh. Technol.* **2019**, *68*, 7477–7487. [CrossRef]
19. Kim, M.; Kim, H.; Kim, D.; Jeong, Y.; Park, H.-H.; Ahn, S. A three-phase wireless-power-transfer system for online electric vehicles with reduction of leakage magnetic fields. *IEEE Trans. Microw. Theory Tech.* **2015**, *63*, 3806–3813. [CrossRef]
20. Mohammad, M.; Wodajo, E.T.; Choi, S.; Elbuluk, M.E. Modeling and design of passive shield to limit EMF emission and to minimize shield loss in unipolar wireless charging system for EV. *IEEE Trans. Power Electron.* **2019**, *34*, 12235–12245. [CrossRef]
21. Choi, S.Y.; Gu, B.W.; Lee, S.W.; Lee, W.Y.; Huh, J.; Rim, C.T. Generalized active EMF cancel methods for wireless electric vehicles. *IEEE Trans. Power Electron.* **2013**, *29*, 5770–5783. [CrossRef]

22. Cruciani, S.; Campi, T.; Maradei, F.; Feliziani, M. Wireless charging in electric vehicles: EMI/EMC risk mitigation in pacemakers by active coils. In Proceedings of the IEEE PELS Workshop on Emerging Technologies: Wireless Power Transfer (WoW), London, UK, 17–27 June 2019. [CrossRef]
23. Ahlbom, A.; Bergqvist, U.; Bernhardt, J.H.; Cesarini, J.P.; Court, L.A.; Grandolfo, M.; Hietanen, M.; McKinlay, A.F.; Repacholi, M.H.; Sliney, D.H.; et al. Guidelines for limiting exposure to time-varying electric, magnetic, and electromagnetic fields (up to 300 GHz). *Health Phys.* **1998**, *74*, 494–521.
24. International Commission on Non-Ionizing Radiation Protection. Guidelines for limiting exposure to time-varying electric and magnetic fields (1 Hz to 100 kHz). *Health Phys.* **2010**, *99*, 818–836. [CrossRef]
25. International Standard. *Magnetic Field Levels Generated by Electronic and Electrical Apparatus in the Railway Environment with Respect to Human Exposure–Measurement Procedures*; IEC 62597; American National Standards Institute (ANSI): New York, NY, USA, 2019.
26. International Standard. *Railway Applications–Electromagnetic Compatibility–Part 2: Emission of the Whole Railway System to the Outside World*; IEC 62236-2; American National Standards Institute (ANSI): New York, NY, USA, 2018.
27. International Standard. *Railway Applications–Electromagnetic Compatibility–Part 3-1: Rolling Stock–Train and Complete Vehicle*; IEC 62236-3-1; American National Standards Institute (ANSI): New York, NY, USA, 2018.

energies
MDPI

Article

Generalized Behavioral Modelling Methodology of Switch-Diode Cell for Power Loss Prediction in Electromagnetic Transient Simulation

Yanming Xu [1], Carl Ngai Man Ho [1,*], Avishek Ghosh [1] and Dharshana Muthumuni [2]

[1] Department of Electrical and Computer Engineering, University of Manitoba, Winnipeg, MB R3T5V6, Canada; xuy34527@myumanitoba.ca (Y.X.); ghosha3@myumanitoba.ca (A.G.)
[2] Manitoba Hydro International, Winnipeg, MB R3P1A3, Canada; dharshana@hvdc.ca
* Correspondence: carl.ho@umanitoba.ca

Abstract: Modern wide-bandgap (WBG) devices, such as silicon carbide (SiC) or gallium nitride (GaN) based devices, have emerged and been increasingly used in power electronics (PE) applications due to their superior switching feature. The power losses of these devices become the key of system efficiency improvement, especially for high-frequency applications. In this paper, a generalized behavioral model of a switch-diode cell (SDC) is proposed for power loss estimation in the electromagnetic transient simulation. The proposed model is developed based on the circuit level switching process analysis, which considers the effects of parasitics, the operating temperature, and the interaction of diode and switch. In addition, the transient waveforms of the SDC are simulated by the proposed model using dependent voltage and current sources with passive components. Besides, the approaches of obtaining model parameters from the datasheets are given and the modelling method is applicable to various semiconductors such Si insulated-gate bipolar transistor (IGBT), Si/SiC metal–oxide–semiconductor field-effect transistor (MOSFET), and GaN devices. Further, a multi-dimensional power loss table in a wide range of operating conditions can be obtained with fast speed and reasonable accuracy. The proposed approach is implemented in PSCAD/ Electromagnetic Transients including DC, EMTDC, (v4.6, Winnipeg, MB, Canada) and further verified by the hardware setups including different daughter boards for different devices.

Keywords: semiconductor; model; power loss

Citation: Xu, Y.; Ho, C.N.M.; Ghosh, A.; Muthumuni, D. Generalized Behavioral Modelling Methodology of Switch-Diode Cell for Power Loss Prediction in Electromagnetic Transient Simulation. *Energies* **2021**, *14*, 1500. https://doi.org/10.3390/en14051500

Academic Editors: Ivica Stevanovic and Bernhard Wunsch

Received: 14 January 2021
Accepted: 4 March 2021
Published: 9 March 2021

1. Introduction

A power electronics (PE) system plays a key role in the process of efficient energy control, conversion, and management. Power semiconductor devices are the core components in a PE system and have a significant impact on system efficiency, reliability, and cost [1]. For decades, silicon-based devices, such as insulated-gate bipolar transistors (IGBTs) [2], metal-oxide-semiconductor field-effect transistors (MOSFETs) [3], are mainly and widely used in various modern PE applications (e.g., Photovoltaics (PV) [4], Power Factor Correction (PFC) [5], power supply [6], and other power converters [7,8]). However, the PE system performance and efficiency are hindered by Si-based devices due to the fundamental material limits. Recently, wide-bandgap (WBG) devices [9–11], such as silicon carbide (SiC) MOSFETs [12], enhancement-mode gallium nitride (eGaN) high-electron-mobility transistors (HEMTs) [13,14], have emerged and gained great popularity due to the superior features of fast switching speed and low switching loss. Thereby, the switching frequency can be further increased bringing the merits of size reduction for magnetic components, high power density, and high efficiency. Whereas the increased power losses of semiconductors are typically the main contributor to total loss especially for high-frequency (HF) applications, and the generated heat energy during switching transition may lead to fatigue failure and affect the reliability [15]. Hence, an accurate power loss model, which is applicable for different semiconductors and provides a deep insight into the switching process, is highly desirable for device selection and PE system optimization.

Currently, the ideal switch or two-state resistances model is typically adopted in most electromagnetic transient (EMT) simulators such as PSCAD/EMTDC and MATLAB/Simulink [16]. This simple model is mainly used to evaluate the overall system response and control strategy, and only the conduction loss is roughly considered. The conduction loss can be directly determined by the output curves in the datasheet, while the switching loss is more complicated and can be measured in the double-pulse test (DPT) [17]. Although DPT is widely used and can achieve high accuracy, it typically involves expensive probes and much peripheral bulky equipment such as a high voltage power supply. Designing a testing board with low parasitics is challenging and also significant for WBG devices due to the fast switching. Recently, several physic-based semiconductor models [18–21], such as simulation program with integrated circuit emphasis (SPICE) models [22,23], have been proposed to accurately describe the transient behaviors of the devices. However, the geometrical parameters for the model are often not available in the datasheet and thus the applicability of the model is limited. Another type of model (i.e., behavioral model [24,25]) has been developed, which focuses more on the external behaviors of the devices instead of the internal physics. As a result, the complexity is reduced and fast simulation speed can be achieved. It is adequate and widely used for system-level study, but more detailed transient concerns are needed to accurately evaluate the switching performance and estimate the power losses.

To have a better description of the switching transients, a lot of analytical loss models have been proposed [26,27]. Piecewise linearizing the switching process of the device is a commonly used method which enables simple and rapid loss estimation [28]. Whereas, the accuracy is still limited due to the ignorance of the parasitics. To improve it, more comprehensive loss models are developed considering various factors, such as temperature-dependent parameters [29,30], interactions between diode and switch [31], cross-talk issue [32], displacement current [33,34], and non-flat miller plateau [35]. Thereby, the switching loss can be obtained by solving the equivalent circuit for each switching sub-stage. Further, the entire switching process of eGaN HEMT in synchronous buck converter application is presented in [36,37] considering the third quadrant operation with the help of the 2-dimensional electron gas (2DEG). However, these methods are complicated involving huge computational burdens, not to mention the convergence issue. The measurement techniques and loss distribution including the capacitive losses for eGaN HEMT are illustrated in [38,39] and the scalable loss estimation method is further proposed based on the measurements. However, the measured data in the datasheet is typically under specific conditions, which limits the applicability and accuracy.

In a PE system, a power switch is typically paired with a diode as a switch-diode cell (SDC) to provide current commutation [40]. This basic commutation cell as shown in Figure 1a is widely used in PE applications and it consists of the active power switch (S), diode (D), equivalent circuit voltage (V_{dd}) and load current (I_L) [41]. Note that, four configurations of S, namely Si/SiC MOSFET, Si IGBT and eGaN HEMT, are taken into account in this paper and D can be a single positive-intrinsic-negative (PIN) diode, a Schottky barrier diode (SBD), the body diode of MOSFET or the equivalent diode of eGaN HEMT. During switching transition, power loss is resulted mainly from the switching and conduction losses of S as well as the conduction loss and reverse recovery loss of D. In order to estimate these power losses in a PE system, a generalized behavioral modelling method of switch-diode cell in electromagnetic transient simulation (EMT) is proposed and it is an extension of previous work [42–45]. There are three technical contributions in this paper comparing with the conventional methods.

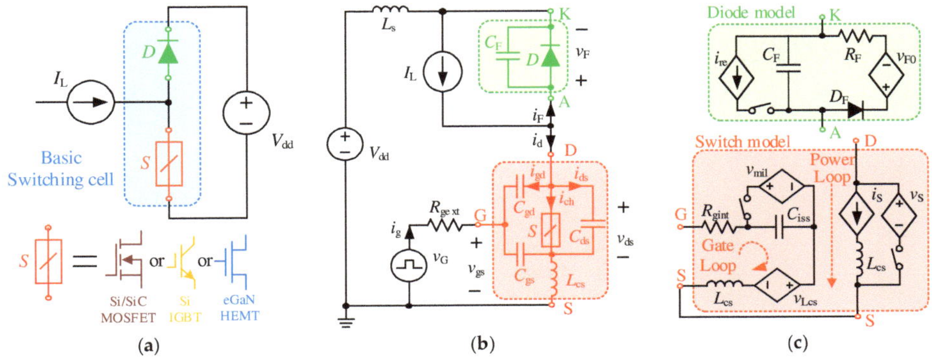

Figure 1. Schematic diagrams of (**a**) switch-diode cell circuit; (**b**) DPT circuit; (**c**) Proposed switch and diode models.

- A generalized behavioral model of SDC is proposed which is realized by dependent sources with passive components considering the impacts of parasitics, the temperature-dependent parameters, and the reverse recovery behavior of *D*. This model is not limited to a specific device and it is applicable to various devices including Si/SiC MOSFET, Si IGBT, and eGaN HEMT. In addition, most of the model parameters can be obtained from the device datasheets by the curve fitting method and no additional measurement is involved. Based on the specific requirement, the model can be modified and integrated into different simulators accordingly.
- The switching process of the switch-diode cell in a clamped inductive switching circuit is studied analytically based on the equivalent circuits for each switching sub-stage. Accordingly, the semiconductor model is developed and implemented in PSCAD/EMTDC. The switching analysis in this paper is more comprehensive considering the respective features of different devices such as the tail current of IGBT and the third quadrant operation of eGaN HEMT.
- A DPT setup was designed for experimental verification. To meets the different requirements of different semiconductors, three daughter boards were specifically designed incorporating with the main control board to characterize various devices and measure the corresponding power losses in a wide range of operating conditions. The simulated results are compared with the experimental results and show good agreements within 10% average error range.

2. Model Description

The simulation procedure of the proposed model is demonstrated in Figure 2. Initially, the device requirements for the desired PE application (e.g., voltage, current, temperature, and frequency) are determined. Based on those requirements, a specific semiconductor device is preliminarily selected for modelling and characterization. According to the device's datasheet, the key model parameters can be extracted by the curve fitting method. Afterwards, the model parameters together with the operating conditions are input to the proposed device model, and a DPT simulation using the proposed model is further carried out. Subsequently, the transient voltage and current waveforms can be obtained, and simultaneously the power loss of the device can be computed. If the simulated results, in terms of switching transient behaviors and power loss, meet the requirements within the acceptable range, then the semiconductor is eventually selected for this application. Otherwise, it is necessary to reselect another device and evaluate the performance until the design is optimized.

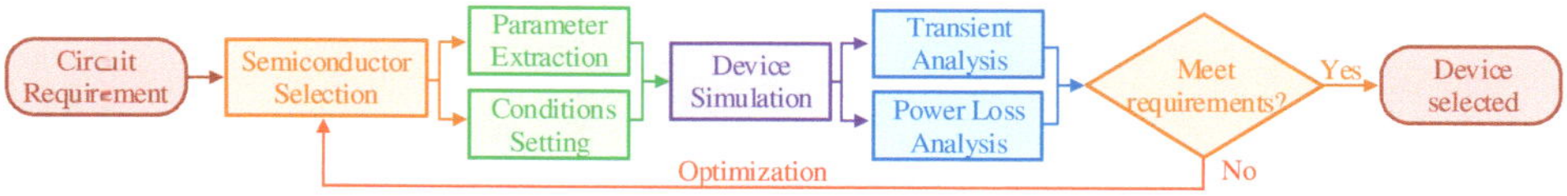

Figure 2. Flow chart of the proposed modelling procedure.

In order to understand the switching behaviors of the SDC, a diode-clamped inductive load circuit (i.e., DPT circuit) is taken as an example which is widely used for device characterization. As shown in Figure 1b, the basic commutation unit consists of two complementary switches, one operates as a freewheeling diode D and the other is an active switch S controlled by the gate drive voltage (v_{G}) through the external gate resistance R_{gext}. In a typical hard-switching PE system, S is identified by a positive drain current i_{d} (collector current i_{c} for IGBT) direction matching with the direction of I_{L}. Since the commutation time is sufficiently short, I_{L} and V_{dd} are hardly changing during switching transition and thus they are treated as constant current and voltage sources, respectively.

Note that, the crucial circuit parasitic elements are also included as shown in Figure 1b. All the stray inductances in the power loop including the printed circuit board (PCB) trace and device package inductance are lumped and represented by L_{s}, while the common source inductance (L_{cs}) of S is considered separately. In addition, the parasitic capacitances of S include gate-drain capacitance (C_{gd}), gate-source capacitance (C_{gs}) and drain-source capacitance (C_{ds}). Besides, the equivalent capacitance of D (C_{F}) denotes for junction capacitance of diode. It should be mentioned that when D is configured as the body diode of a switch rather than a single diode, C_{F} will be the corresponding parasitic capacitance of the switch. During a switching transition, I_{L} commutates between S and D. When a positive v_{G} is given, the gate-source voltage (v_{gs}) will increase to turn on S. Subsequently, i_{d} including the channel current (i_{ch}), the gate-drain current (i_{gd}) and the drain-source current (i_{ds}) starts rising, meanwhile the diode forward current (i_{F}) declines gradually. When S is fully turn on, the drain-source voltage (v_{ds}) decreases to the on-state voltage and the diode forward voltage (v_{F}) rises to V_{dd}. The behavioral models of a SDC as illustrated in Figure 1c are proposed to reproduce the switching behaviors of S and D, respectively. The details of the model descriptions including the active switch and diode model are presented as follows.

2.1. Active Switch Model

As shown in Figure 1c, the proposed active switch model consists of two parts, the gate loop and the power loop. It is noted that L_{cs} is shared by both loops and thus each loop includes one L_{cs} in order to decoupling both loops. Additionally, a dependent voltage source (v_{Lcs}) is added in the gate loop to reflect the interactive impact of the current source (i_{S}) on L_{cs} as expressed by,

$$v_{\mathrm{Lcs}} = L_{\mathrm{cs}} \cdot \mathrm{d}i_{\mathrm{S}}/\mathrm{d}t \tag{1}$$

- Gate Loop Part

The external v_{G} is typically flipped between V_{gon} (20 or 15 V) and V_{goff} (−5 or 0 V) based on the specific gate drive requirement of the switch. The device internal gate resistance (R_{gint}) is merged into R_{gext} as the total gate resistance (R_{G}). Furthermore, the gate related junction capacitances (i.e., C_{gd} and C_{gs}) are represented by the input capacitance (C_{iss}) and an additional dependent voltage source (v_{mil}). This equivalent v_{mil} becomes valid only when the miller plateau occurs on v_{gs} during switching transition and its value can be computed by

$$v_{\mathrm{mil}} = v_{\mathrm{th}} + i_{\mathrm{ch}}/g_{\mathrm{fs}} \tag{2}$$

where v_{th} and g_{fs} stand for threshold voltage and transconductance of a switch, respectively. Additionally, i_{ch} during miller plateau period typically equals to I_{L} which will be discussed in Section 2.3.

It is noted that the gate inductance is neglected here for simplicity although it can introduce a slight delay on v_{gs}. In fact, this delay is mainly resulted from L_{cs} and v_{Lcs} due to the fast change of i_d. Besides, the gate drive circuit is normally placed close to S in order to minimize the potential oscillation introduced by the gate inductance and thus this inductance is negligible.

- Power Loop Part

In this model, i_d is represented by i_S which is the sum of i_{ch}, i_{gd} and i_{ds}. Note that, most of the time, i_S is the same as i_{ch} except for the voltage transition period when a displacement current is introduced due to the process of charging and discharging of the parasitic capacitance. In order to reflect the voltage change during switching transition as well as the on-state voltage (v_{on}) of S, an equivalent dependent voltage source (v_S) is adopted here. The value of v_{on} can be determined by $R_{ds(on)}$ with I_L in (3) or the saturation voltage (v_{cesat}) for the case of IGBT.

$$v_{on} = R_{ds(on)} \cdot I_L. \tag{3}$$

Besides, L_{cs} is also included in the power loop part which is associated with L_s to influence the transient waveforms. Thereby, the gate loop and power loop parts are decoupled and their interaction is represented by the equivalent dependent sources instead of nonlinear junction capacitances resulting in a reduction of model complexity.

2.2. Diode Model

The static model of a diode typically can be represented by an ideal diode (D_F), a forward resistance (R_F) and a voltage source (v_{F0}) based on the forward characteristics in the device datasheet. Typically, v_F of the diode can be computed by,

$$v_F = R_F \cdot i_F + v_{F0}. \tag{4}$$

It should be mentioned that, for the case of eGaN HEMT as D, the diode behavior is realized by 2DEG and thus the calculation of v_F is based on the reverse conduction characteristic of the GaN device [44] which is highly affected by the gate drive voltage (v_{GF}) of GaN device as shown in (5),

$$v_F(\text{GaN}) = R_{Fr} \cdot i_F + v_{thF} - v_{GF} \tag{5}$$

where R_{Fr} and v_{thF} are on-resistance in the third quadrant and threshold voltage of a GaN device, respectively. Since negative v_{GF} is typically provided to avoid the cross-talk issue, higher v_F is thus resulted which will increase the conduction loss of D. Notice that, if a positive v_{GF} is provided enabling the channel fully on, the on-state resistance will be the same value in the first quadrant.

Moreover, the dynamic characteristic of D is described by C_F in parallel with an equivalent dependent current source (i_{re}) for the reverse recovery behavior of D. When D switches from forwarding conduction to off-state, i_F cannot be eliminated immediately and it takes a while to extinguish the excess carriers, this time is called reverse recovery time (t_{rr}). The reverse recovery process occurs as soon as i_F becomes negative, i_{re} can be expressed by [45],

$$i_{re} = \begin{cases} \mathrm{d}i_F/\mathrm{d}t \cdot t, & t < t_{rm} \\ I_{rm} \cdot \exp((t - t_{rm})/\tau_{re}), & t \geq t_{rm} \end{cases}, \tag{6}$$

where τ_{re} denotes decay time constant and i_{re} reaches the peak current (I_{rm}) at time t_{rm}. In addition, the slew rate of diode current ($\mathrm{d}i_F/\mathrm{d}t$) typically keeps the same as the turn-off slew rate of S. It needs to notice that, this reverse recovery behavior commonly exists in PIN diode and body diode of S, while the reverse recovery loss is eliminated for the case of SBD or eGaN HEMT and thus i_{re} can be neglected for simplicity. In fact, in these cases, the effect of C_F is the main concern which can introduce displacement current resulting in capacitive loss during switching transition.

2.3. Switching Transient Modelling

The switching process of the SDC in the DPT circuit is thoroughly analyzed based on the switching waveforms and the equivalent circuits as follows.

- Turn-on transition ($t_0 - t_3$)

The typical turn-on waveforms along with power loss information are illustrated in Figure 3 considering the case of PIN or body diode as well as the case of SBD or eGaN HEMT. The equivalent circuits during this period are also provided in Figure 4.

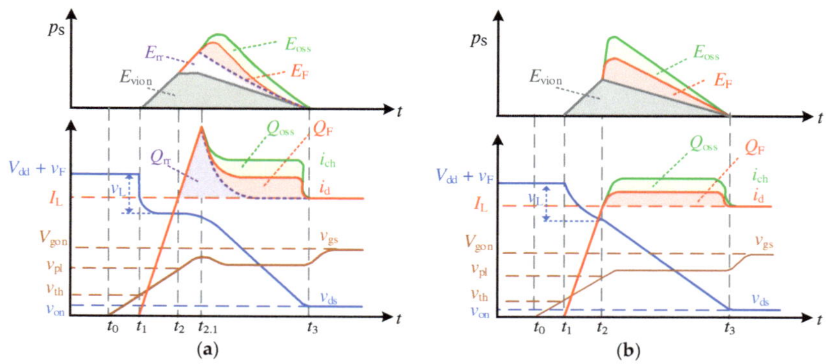

Figure 3. Typical turn on waveforms for *S* with *D* (**a**) PIN or body diode; (**b**) SBD or eGaN HEMT.

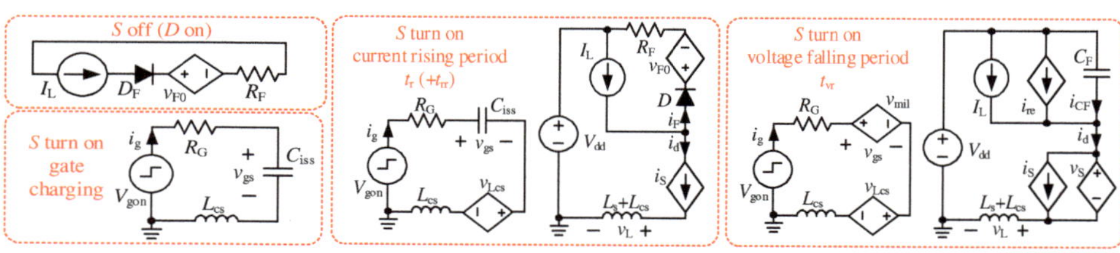

Figure 4. Simplified equivalent circuits during *S* turn on transition.

Initially, *S* is in the off-state, and all the I_L flows through *D*. The corresponding v_F can be estimated by (5). At t_0, the gate charging period begins with a positive V_{gon} applying to v_G and C_{iss} is charged up through R_G. Subsequently, v_{gs} will increase accordingly with the time constant ($\tau_{iss} = R_G \cdot C_{iss}$). Note that L_{cs} in the gate loop will prolong the turn-on time causing more power losses.

The current rising period begins when v_{gs} goes beyond v_{th}. During this interval, the conductive channel of *S* is forming and i_S starts rising from zero to I_L which can be expressed by

$$i_S = g_{fs} \cdot (v_{gs} - v_{th}). \tag{7}$$

The fast change of i_S on one hand, will introduce a negative feedback v_{Lcs} from power loop to gate loop due to L_{cs} to further delay the turn-on process. On the other hand, it will result in a total voltage drop (v_L) on L_s and L_{cs}. Simultaneously, v_{ds} decreased by v_L as shown in Figure 3.

As I_L commutates from *D* to *S*, i_S reaches I_L at t_2 and v_{gs} will be clamped at v_{mil}. At the same time, i_F decreases to zero and *D* enters into the reverse recovery as shown in Figure 3a. This additional i_{re} will add to i_S (= I_L + i_{re}) resulting a current spike and thus a bump in v_{gs} according to (2). When i_{re} reaches I_{rm}, it starts declining and the voltage falling period begins. Subsequently, v_{ds} starts decreasing which is controlled by v_S and the corresponding slew rate can be determined by,

$$dv_{ds}/dt = dv_S/dt = -(V_{gon} - v_{mil} - v_{Lcs})/\left(C_{gd} \cdot R_G\right). \tag{8}$$

As v_{ds} keeps decreasing and v_F increases simultaneously, the output capacitance (C_{oss}) of S and C_F of D is discharged and charged, respectively. Since the voltage of C_{oss} and C_F are clamped to V_{dd}, they share the same absolute value of voltage change. The resultant capacitive displacement current for C_{oss} (i_{oss}) can be expressed by (9). Additionally, this i_{oss} along with the counterpart for C_F ($i_{CF} = -C_F \cdot dv_{ds}/dt$) will affect i_d as can be seen in Figure 3. By applying Kirchhoff's law, i_d can be determined by (10) and i_S is modified accordingly to consider these displacement currents.

$$i_{oss} = i_{gd} + i_{ds} = C_{oss} \cdot dv_{ds}/dt \tag{9}$$

$$i_d = i_S = I_L + i_{re} - C_F \cdot dv_{ds}/dt \tag{10}$$

Based on Figure 1b along with the above equations, i_{ch} can be further obtained,

$$i_{ch} = i_d - i_{oss} = I_L + i_{re} - (C_{oss} + C_F) \cdot dv_{ds}/dt \tag{11}$$

Consequently, during this period, i_d includes I_L, i_{re} and i_{CF}, while the additional i_{oss} is further added to i_{ch} as shown in Figure 3. As a result, v_{mil} will also change according to (2). This period ends when v_{ds} drop to v_{on} at t_3. Thereafter, v_{gs} will continue climbing until reaches V_{gon}.

Furthermore, the turn-on waveforms for the case of SBD or eGaN HEMT are presented in Figure 3b. Since the reverse recovery behavior is neglected for these cases as mentioned previously, i_{re} keeps zero and the voltage falling period starts right after i_S reaches I_L. Apart from that, the turn-on modelling and analysis are the same as the case of the PIN diode.

- Turn-off transition ($t_4 - t_7$)

The turn-off process can be considered as the opposite of turn-on transition and the typical transient waveforms are illustrated in Figure 5. In order to turn off S, V_{gon} is replaced by a negative gate drive signal V_{goff} and thus C_{iss} is discharged through R_G resulting the reduction of v_{gs}. As v_{gs} drops to v_{mil}, C_{gd} absorbs nearly all the i_g and thus v_{ds} begins to rise which again causes a current decline of i_d. When v_{ds} reaches V_{dd}, the miller plateau disappears and i_d begins decreasing with v_{gs} which results an additional v_L on v_{ds} as shown in Figure 5a.

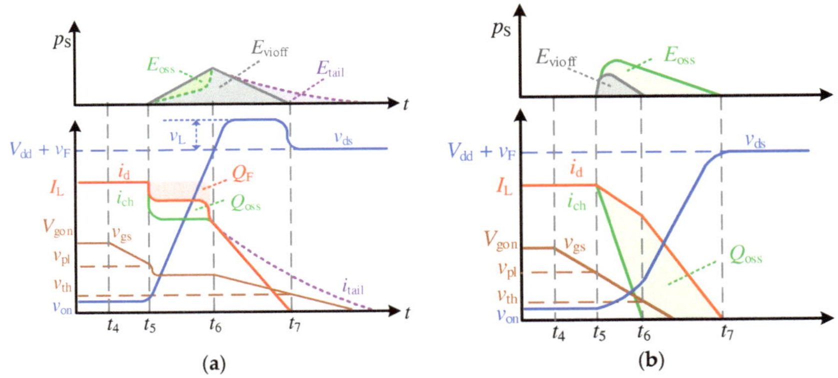

Figure 5. Typical turn off waveforms for S as (**a**) Si IGBT or Si/SiC MOSFETs; (**b**) Typical eGaN HEMT.

As v_{gs} drops below v_{th}, i_d becomes zero and S turns off completely. However, for the case of Si IGBT, the tail current (i_{tail}) is considered due to the recombination of the excess carriers. This i_{tail} will prolong the turn off time and can be modelled by the exponential function [42],

$$i_{tail} = I_{tail0} \cdot \exp(-(t - t_{tail0})/\tau_{tail}) \quad (12)$$

where τ_{tail} stands for carrier transit time and the tailing period starts at t_{tail0} with the initial value of I_{tail0}. These parameters can be estimated from the turn-off current waveform.

As can be seen from Figure 5b, a notable difference for the case of eGaN HEMT is v_{gs} does not typically plateau due to the much smaller capacitance and it keeps decreasing until reaches V_{goff}. As a result, i_{ch} quickly declines synchronized with v_{gs} based on (7), meanwhile v_{ds} rises slightly and the slew rate is limited by the relatively high C_{oss} at low v_{ds}. In fact, the channel turns off completely before v_{ds} is significantly rising. However, i_d does not follow the fast decreasing i_{ch} since its changing rate is limited by the inductances in the power loop. Additionally, C_{oss} is charged by the current difference between i_d and i_{ch} resulting a slight increase of v_{ds} which can be expressed as

$$dv_{ds}/dt = dv_S/dt = (i_d - i_{ch})/C_{oss} \quad (13)$$

Meanwhile, C_F is discharged resulting in a reduction of v_F. Additionally, i_d thus can be obtained by

$$i_d = i_S = I_L - C_F \cdot dv_F/dt \quad (14)$$

Once v_{gs} drops below v_{th}, namely i_{ch} becomes zero, the channel shuts down and I_L is shared by C_{oss} and C_F. During this period, v_{ds} keeps rising according to (14). When v_{ds} rises to V_{dd}, I_L starts commuting to D and the S turn-off transition finishes. Note that if very high R_G is used for eGaN HEMT, the turn-off analysis will be the same as the case of MOSFET as shown in Figure 5a.

Based on the above analysis v_S can be considered as an open circuit except for voltage rising/falling periods and S on-state. During the voltage transition period, v_s is modelled as a dependent voltage source with a voltage slew rate as mentioned previously. In addition, the key expressions of i_S for different conditions can be summarized in Table 1. It is noted that when v_{gs} is less than v_{th}, the conduction channel is not established and theoretically, no current is flowing through the device. As a result, i_S is modelled with zero ampere under this condition in PSCAD/EMTDC which can be considered as open circuit. In this paper, all analytical equations for v_s and i_s are implemented and programmed with conditions in the custom programming modules in PSCAD. In this way, it is feasible and convenient to make any modifications as necessary.

Table 1. Key expressions for i_S in the proposed model.

	Condition	$v_{gs}<v_{th}$	$i_d<I_L$	$i_d>I_L$	Tail Period for IGBT	Turn-off for eGaN HEMT
i_S	Expression	0	(7)	(10)	(12)	(14)

3. Power Loss Analysis and Parameter Extraction

In general, the power losses of SDC mainly include conduction loss and switching loss. Typically the conduction losses of S and D can be calculated directly as the product of operating current (i.e., I_L) and the on-state voltage drop based on (3–5). In addition, the reverse recovery loss of D can be estimated based on the reverse recovery charge (Q_{rr}) and v_F from the device datasheet and the switching loss of S is analyzed as follows.

3.1. Turn on Loss(E_{on})

The instantaneous power of S (p_S) along with E_{on} are presented in Figure 3. Basically, E_{on} consists of the turn-on V-I overlap loss (E_{vion}), the reverse recovery related loss (E_{rr}) and the capacitive losses (E_{oss} and E_F) for C_{oss} and C_F, respectively. E_{vion} graphically can be divided into two parts, the i_d rising period and v_{ds} decreasing period. Hence, E_{vion} can be expressed as

$$E_{vion} = \int_{t_1}^{t_2} v_{ds} \cdot i_d dt + I_L \cdot \int_{t_2}^{t_3} v_{ds} d \tag{15}$$

Since reverse recovery behavior of D and the displacement current of C_{oss} already have been considered in the modelling of i_S, therefore the sum of E_{vion}, E_{rr} and E_{oss}, which is actually the measured turn-on loss (E_{onm}), can be directly obtained by integrating i_d and v_{ds}. This significantly reduces the complexity comparing with the analytical loss model by computing the switching time for each sub-stages. Moreover, according to (11), both the discharging current of C_{oss} and charging current of C_F are flowing through the channel of S and thus these capacitive energy losses (i.e., E_{oss}, E_F) are dissipated into the channel. Based on the capacitance curves, E_{oss} can be expressed as

$$E_{oss} = \int_0^{V_{dd}} v_{ds} \cdot C_{oss} dv_{ds} \tag{16}$$

Since the charging current of C_F is provided by V_{dd} and part of the energy is stored in C_F, thus the energy loss dissipated in the channel (i.e., E_F) can be derived based on the charge of C_F (Q_F),

$$E_F = V_{dd} \cdot Q_F - \int_0^{V_{dd}} v_F \cdot C_F dv_F = \int_0^{V_{dd}} (V_{dd} - v_F) \cdot C_F dv_F \tag{17}$$

Consequently, E_F theoretically should also be included in E_{on} which can be expressed as,

$$E_{on} = E_{onm} + E_F = \int_{t_1}^{t_3} i_d \cdot v_{ds} dt + E_F \tag{18}$$

3.2. Turn off Loss(E_{off})

Generally, the power loss during turn-off transition occurs from t_5 to t_7 which includes the turn-off V-I overlap loss (E_{vioff}), E_{oss} and the tailing loss for the case of IGBT (E_{tail}). The analysis of E_{vioff} is significantly different for the slow-switching scenario in Figure 5a and the typical fast-switching for eGaN HEMT in Figure 5b. As for the former case, v_{gs} is fixed at v_{mil} and thus i_d is relatively constant throughout the voltage rising period. Thereafter, i_d decreases significantly meanwhile v_{ds} keeps relatively constant. Hence, E_{vioff} for this case can be graphically calculated as

$$E_{vioff} = \int_{t_5}^{t_6} v_{ds} \cdot i_d dt + \int_{t_6}^{t_7} (V_{dd} + v_L) \cdot i_d d \tag{19}$$

Similarly, the sum of E_{vioff} and E_{tail}, namely the measured turn-off loss (E_{offm}), is typically an integral of i_d and v_{ds}. It is noticed that, C_F is discharged and the energy is transferred to the inductive load during the voltage rising period resulting in a reduction of i_d, while C_{oss} is charged and the corresponding energy (i.e., E_{oss}) is stored which will be dissipated in the next turn-on transition. Therefore, E_{oss} should be theoretically excluded from E_{off} which can be expressed as

$$E_{off} = E_{offm} - E_{oss} = \int_{t_5}^{t_7} v_{ds} \cdot i_d dt - E_{oss} \tag{20}$$

As for the typical eGaN HEMT scenario, v_{gs} skips the plateau and the channel turns off quickly before v_{ds} rises significantly as discussed previously. Afterwards, the energy is commutating between the inductive load and the two capacitances (i.e., C_F and C_{oss}) which is almost lossless. Since the resistive overlap loss only occurs as long as the channel is on, it is significantly reduced for this case due to the relatively low v_{ds} during this time. Nevertheless, E_{off} still can be calculated by (20).

3.3. Parameter Extraction

The key model parameters can be directly extracted from the corresponding curves provided in the device datasheet by the curve fitting method [46–48], to avoid the supplementary experiments which are usually not practical. This method is adopted in this paper since it is applicable to different semiconductor devices and provides a relatively balanced tradeoff between accuracy and practicability. As an example, different types of semiconductors from different manufacturers as listed in Table 2 are selected for modelling and validation of the proposed method. The extracting sequence is discussed in detail as follows.

Table 2. Semiconductor devices selected for modelling and validation.

Device Type	Si IGBT	SiC MOSFET	Si MOSFET	eGaN HEMT	SiC SBD
Part Number	IKW40T120	SCT2080KE	NVHL072N65S3	GS66506T	SCS220KG
Manufacturer	Infineon	Rohm	On Semiconductor	GaN System	ROHM

- Static characteristic

In order to reproduce the switching behavior of S, two key parameters (i.e., v_{th} and g_{fs}) are considered first. Since v_{th} typically is a temperature-dependent parameter rather than a constant value, it can be fitted by the 2nd order polynomial of junction temperature (T_j) from the corresponding curve in the datasheet. Likewise, the transfer characteristic of S can be fitted by the quadratic function of v_{gs} and subsequently g_{fs} can be further determined based on v_{th} and (7) as

$$g_{fs} = i_s \cdot \sqrt{k_{ga} / \left(i_s - k_{gb}\right)} \cdot k_{gTa} \cdot \left(T_j / T_a\right)^{k_{gTb}} \quad (21)$$

where k_{ga}, k_{gb}, k_{gTa}, and k_{gTb} are fitting constants. T_a is room temperature which is considered as 25 °C here. In this way, g_{fs} under the given T_j in datasheet can be obtained. The fitted results of transfer characteristics are compared and shown in good agreement with the datasheet in Figure 6. It is also found that there is a positive correlation between T_j and g_{fs} for Si IGBT and Si/SiC MOSFET while it shows a negative correlation for the case of eGaN HEMT.

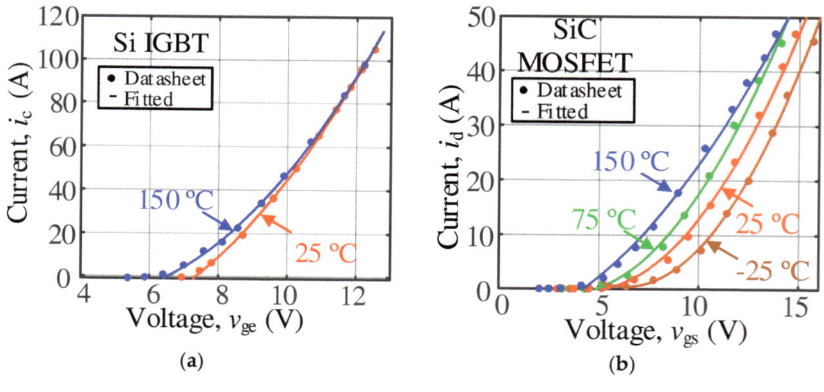

Figure 6. *Cont.*

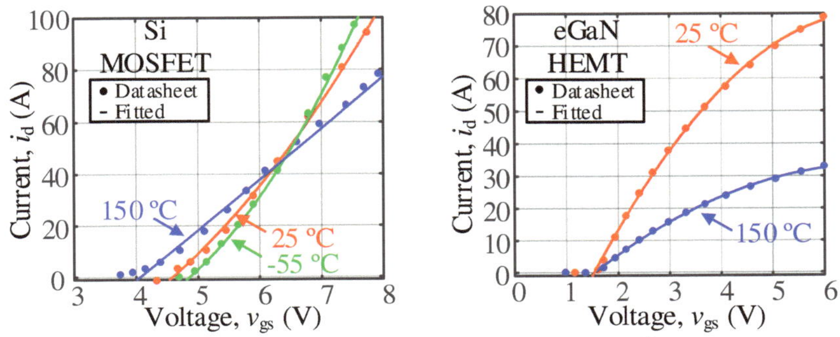

(c) (d)

Figure 6. Fitted results of transfer curves for (**a**) Si IGBT; (**b**) SiC MOSFET; (**c**) Si MOSFET; (**d**) eGaN HEMT.

Furthermore, in order to represent the on-state characteristic of S as expressed in (3), the parameter v_{cesat} for IGBT or $R_{\text{ds(on)}}$ for other cases is needed to be extracted. Typically, both of v_{cesat} and $R_{\text{ds(on)}}$ are affected by T_j and i_S according to the curves in the datasheet. Therefore, v_{cesat} can be obtained by the following Equation (22).

$$v_{\text{cesat}} = (k_{\text{cea}} + r_{\text{cea}} \cdot i_S) + (k_{\text{ceb}} + r_{\text{ceb}} \cdot i_S) \cdot (T_j - T_a) \tag{22}$$

where k_{cea}, r_{cea}, k_{ceb} and r_{ceb} are fitting coefficients. Note that the gate voltage is assumed as constant in the parameter extraction for simplicity.

Likewise, $R_{\text{ds(on)}}$ for the cases of MOSFET and eGaN HEMT can be extracted by

$$R_{\text{ds(on)}} = (k_{\text{ona}} + r_{\text{ona}} \cdot i_S) + (k_{\text{onb}} + r_{\text{onb}} \cdot i_S) \cdot (T_j - T_a) \tag{23}$$

where k_{ona}, r_{ona}, k_{onb} and r_{onb} are fitting coefficients. Based on the above equations, the key parameters for different semiconductors in this paper can be extracted and illustrated in Table 3.

Table 3. Key fitting coefficients parameters of different semiconductors.

Parameter	k_{ga}	k_{gb}	k_{gTa}	k_{gTb}	k_{cea},k_{ona}	r_{cea},r_{ona}	k_{ceb},k_{onb}	r_{ceb},r_{onb}
Si IGBT	2.05	−6.96	14.2	−0.012	−0.001	$1 \cdot 10^{-4}$	0.972	0.0215
SiC MOSFET	0.42	−0.35	2.96	−0.086	$4.1 \cdot 10^{-4}$	$1.12 \cdot 10^{-6}$	0.073	$5.1 \cdot 10^{-4}$
Si MOSFET	3.72	−19.2	13	−0.57	0.048	$2.87 \cdot 10^{-3}$	0.034	0.0012
eGaN HEMT	0.23	−776.9	0.51	−0.31	0.067	$7.2 \cdot 10^{-5}$	$8.4 \cdot 10^{-4}$	$4 \cdot 10^{-6}$

- Parasitic capacitance and inductance

It is a fact that nonlinear capacitances are the key to the dynamic characteristic of the device. Typically, the capacitance curves provided in the datasheet is in the form of C_{iss}, C_{oss} and reverse capacitance (C_{rss}) which can be mathematically converted to junction capacitances. Generally, these capacitances are voltage-dependent and can be extracted by fitting the curves as

$$C(v) = f(v), \tag{24}$$

where f is a general fitting function for extraction of capacitance. In this paper, various f are used for different devices to fit the corresponding curves as shown in Figure 7. Notice that the nonlinear capacitance curves vary from different devices and thus it is reasonable to change f accordingly.

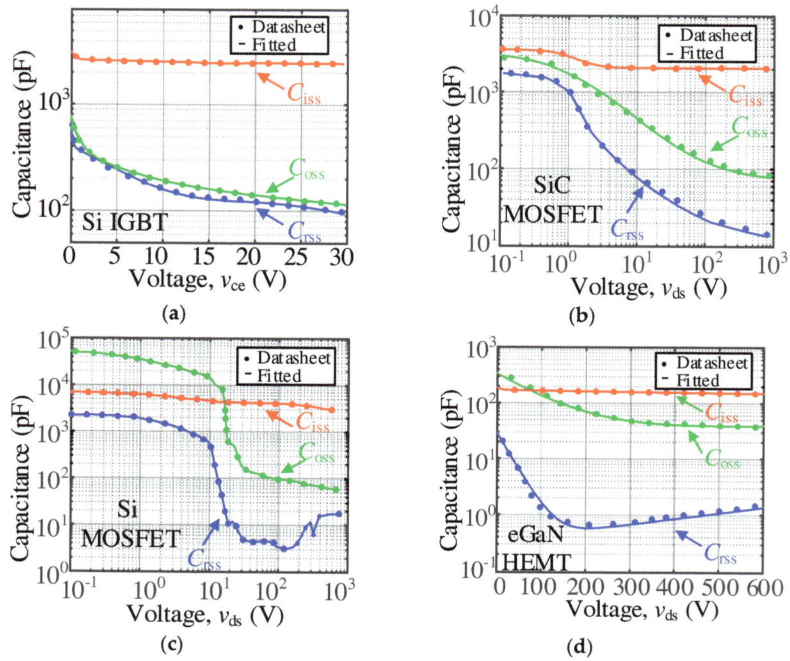

Figure 7. Parasitic capacitance extraction with the methods of (**a**) 5th order Gauss function; (**b**) Analytical equation [49]; (**c**) Interpolating look-up table (LUT); (**d**) 3rd order exponential function.

As for the parasitic inductance, only the internal inductance of the device is normally provided in the datasheet, while the stray parasitic inductance is highly related to the specific device package and PCB design. There are two widely used methods for inductance extraction, namely calculation method and experimental method. Based on the PCB and device package specification, the corresponding inductance can be computationally obtained with the help of calculation tools such as the Ansys Q3D Extractor software (v1.0, Canonsburg, PA, USA). According to (1), it also can be extracted from the slew rate of current along with v_L during S turn on transition or from the resonant frequency of the power loop in the experimental results. In this paper, the parasitic inductances are initially estimated based on the PCB trace length of the power loop and the gate loop [50] as well as the device package (e.g., 2–5 nH for TO-247 [51]) and further calibrated from the switching waveforms.

- Diode parameters

According to (4), R_F and v_{F0} are the key static parameters for D which can be extracted directly from the diode I-V curve in the datasheet and the corresponding values for various temperatures can be estimated by linear interpolation. Furthermore, the third quadrant operation of eGaN HEMT as D is of special concern. Since the corresponding voltage drop is dependent on the gate drive voltage of D, thus it should be fitted by (5) based on the output curves in the third quadrant from the datasheet. The diode I-V curve fitted results for different devices are compared with the datasheet and illustrated in Figure 8. Notice that the conduction performance of the body diode in SiC MOSFET is generally worse than the anti-parallel diode of IGBT and SiC SBD. As for the eGaN HEMT, the reverse voltage drop is highly dependent on the gate drive voltage and the typical value for turn off (i.e., −3 V) will result in considerable conduction loss.

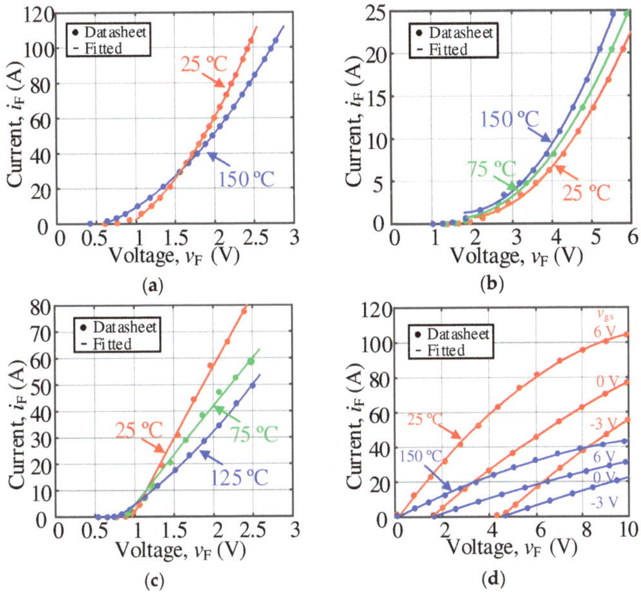

Figure 8. Diode static parameter extraction of (**a**) Si IGBT; (**b**) SiC MOSFET; (**c**) Si MOSFET; (**d**) eGaN HEMT.

In addition, C_F can be obtained using the same method as mentioned above for S from the capacitance curve in the datasheet. According to the previous switching transition analysis, it can be found that the reverse recovery behavior of diode plays a considerable role and the main parameters I_{rm} and Q_{rr} can be extracted from the diode curves as a function of T_j and di_F/dt [42],

$$I_{rm} = k_{rm0} + t_{rm0} \cdot di_F/dt + (k_{rm1} + t_{rm1} \cdot di_F/dt) \cdot (T_j - T_a) \tag{25}$$

$$Q_{rr} = k_{rr0} + t_{rr0} \cdot di_F/dt + (k_{rr1} + t_{rr1} \cdot di_F/dt) \cdot (T_j - T_a) \tag{26}$$

where k_{rm0}, t_{rm0}, k_{rm1}, t_{rm1}, k_{rr0}, t_{rr0}, k_{rr1} and t_{rr1} are all fitting coefficients. Besides, the reverse recovery time (t_{rr}) and τ_{re} can be further determined by

$$t_{rr} = 2\sqrt{Q_{rr}/(di_F/dt)} \tag{27}$$

$$\tau_{re} = 1/\ln 10 \cdot (t_{rr} - I_{rm}/(di_F/dt)) \tag{28}$$

4. Experimental Verification

The objective of the proposed method is to reproduce the switching performance of the SDC and generate the corresponding power loss look-up table (LUT) with reasonable accuracy and fast simulation speed. The models are implemented in PSCAD/EMTDC and validated by comparing with the experimental waveforms and power loss results in the DPT bench for different semiconductor combinations.

4.1. Setup Description

An automatic DPT bench was designed and built for device characterization and loss validation [17]. Since the gate-drive requirements and device package are different for all the devices, three daughter boards were designed accordingly with a power supply (up to 1 kV) and the inductive load (5 mH) as shown in Figure 9. Tektronix High voltage differential probe THDP0200 and current probe TCP0030A were used for Si IGBT/MOSFET voltage and current measurements, respectively, while passive voltage probe (TTP800) and 0.1 Ω current shunt resistor (SDN-414-01) were adopted for SiC MOSFET and eGaN HEMT measurements. The temperature was controlled by a heating block and monitored by a thermal imager (Fluke, TiS40). In the DPT, the desired test conditions (voltage, current, and temperature) were initially set in the personal computer (PC) and all the control signals were given by the microcontroller Arduino on the board. Afterwards, the DC capacitor bank was charged to the desired voltage by the power supply unit, and the device was heated to the desired temperature. When voltage and temperature conditions were ready, two gate pulses were given in sequence to turn on and turn off the device under test (DUT). The switching waveforms and data were obtained by oscilloscope and processed in the PC for transient information and power loss analysis. In order to mitigate the measurement error of power loss due to the asynchrony of voltage and current, it is necessary to calibrate the probes before conducting the DPTs. To further guarantee the accuracy, additional delay time adjustments for the transient waveforms are also needed for the turn-on and turn-off processes. Taking the current as the reference, the calibration time of probe and waveform are provided in Table 4. The calibrating fixture (067-1686-02) from Tektronix was used for calibration of the current and voltage probes. A 10 MHz sinusoidal signal was applied to both probes and the deskew time for voltage probe was adjusted until both measurements were synchronizing. Note that, these calibration times can be different for various probes and DUTs.

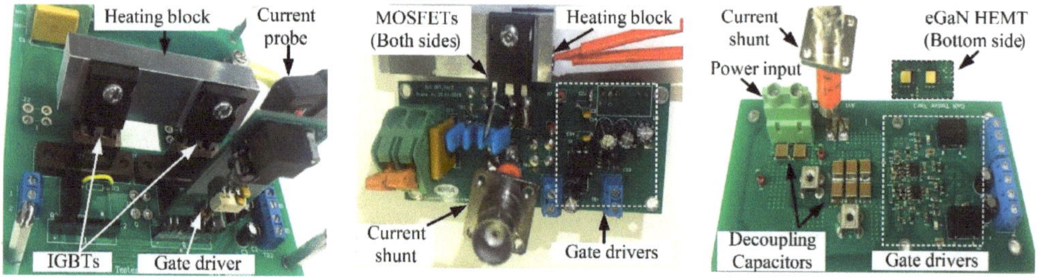

(a) (b) (c)

Figure 9. Device characterization DPT setups for (**a**) Si IGBT/MOSFET; (**b**) SiC MOSFET; (**c**) eGaN HEMT.

Table 4. Calibration time of probes and waveforms.

DUT Case	Si IGBT	SiC MOSFET	Si MOSFET	eGaN HEMT
Probes calibration (ns)	48	26	45	23
Turn-on calibration (ns)	35	13	26	14
Turn-off calibration (ns)	31	18	25	12

4.2. Switching Transient Verification

- Si IGBT

The daughterboard in Figure 9a was used for Si IGBT and MOSFET tests, and v_G was flipped between 15 V and 0 V to control the DUT's on and off, respectively. The simulated results of switching current and voltage waveforms for IKW40T120 are compared with the DPT measurements in Figure 10. The simulated results demonstrate good agreement with measurements for current and voltage switching waveforms under 25 °C and 150 °C. The switching details such as the tail current and the current spike resulting from the reverse recovery of D can be clearly observed. In addition, v_{ce} slightly drops to 500 V as current rising and reaches a peak of 700 V during turn-off transition due to parasitic inductance. Besides, as T_j increases from 25 °C to 150 °C, the reverse recovery behavior of D becomes more obvious resulting higher current peak (up to 60 A) and the rise of v_{ce} as well as the decline of i_c during turn off transition slows down which will increase the switching power loss.

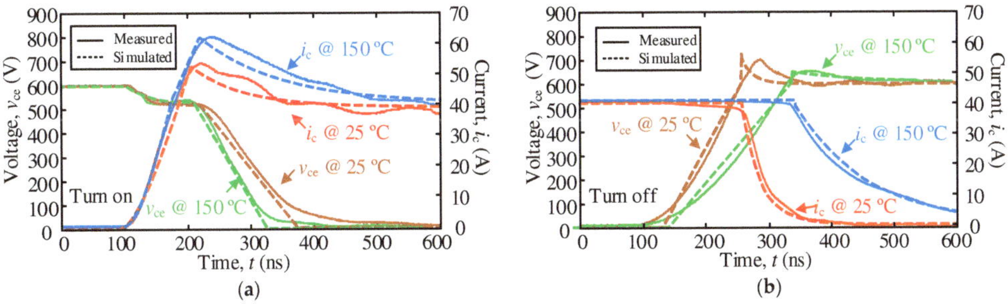

Figure 10. Switching waveforms of Si IGBTs (**a**) turn-on; (**b**) turn-off; @R_G = 15 Ω.

- SiC MOSFET

A more compact daughterboard as shown in Figure 9b was designed for testing SiC MOSFET. Additionally, the gate drive integrated circuit (IXDN609SI) was adopted as the gate driver onboard to provide 20 V/−5 V drive voltage for SiC MOSFET. Figure 11 shows the simulated switching waveforms of SiC MOSFET under the condition of 600 V and 20 A which match well with measured results. It can be observed that there is only a slight impact of T_j on the switching transients in terms of turn-on and turn-off time. Nevertheless, the current still can reach almost 40 A during the turn-on transition due to the reverse recovery behavior of the diode. Besides, it is found that a current ringing occurs during both turn-on and turn-off transitions because of the parasitic resonance. This ringing energy is generally consumed by the HF damping resistance in the circuit. Since either the voltage or current has typically dropped to a low level during the ringing period, thus this ringing loss is neglected in the model for simplicity.

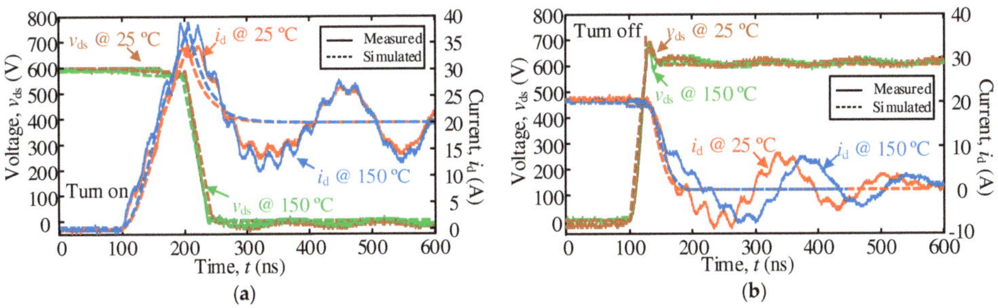

Figure 11. Switching waveforms of SiC MOSFETs (**a**) turn-on; (**b**) turn-off; @R_G = 5 Ω.

- Si MOSFET with SiC diode

The DPT results for Si MOSFET with SiC diode using the same daughterboard as for testing Si IGBT are shown in Figure 12. In general, the simulated results match well with the measured results for different R_G conditions. As R_G increases from 10 Ω to 33 Ω, a half less voltage drop of v_{ds} can be observed during turn-on transition due to the slower current rising speed. Likewise, only a slight increase of v_{ds} can be seen after v_{ds} climbs to V_{dd}. Moreover, it is noted that the current spike is significantly limited comparing with the previous testing using PIN diode due to the merit of zero reverse recovery for SiC SBD. Whereas, there is still a slight current bump causing by the resonance of parasitics as well as the capacitive displacement current as discussed previously.

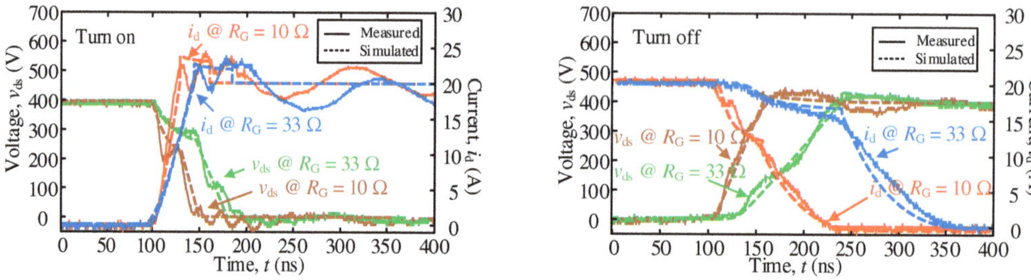

(a) (b)

Figure 12. Switching waveforms of Si MOSFET with SiC SBD (**a**) turn-on; (**b**) turn-off; @400 V, 20 A, 25 °C.

- eGaN HEMT

In order to test the eGaN HEMT which is a surface-mount device (SMD), a specific DPT daughterboard was used as shown in Figure 9c. The gate driver provided 6 V/0 V as gate drive voltage to control the lower side GaN switch S, while the upper side SiC SBD served as a freewheeling diode when S turned off. The simulated switching results are compared with the measurements for the two operating conditions as shown in Figure 13. It can be seen that the simulation results are consistent with the experimental results. In the turn-on waveforms, the current rising time is only tens of nanoseconds. After i_d reaches I_L, it behaves in the similar manner as previous test using Si MOSFET with SiC SBD. However, during the turn-off period, it can be clearly observed that, i_d declines significantly and drops to zero almost the same time as v_{ds} reaches steady state while for the other cases of devices, the fast decrease of current typically occurs after v_{ds} climbs to v_{dd}. This is mainly because the channel of eGaN shuts down very fast before v_{ds} increase significantly as discussed in Section 2.3. Hence, when the channel turns off completely, the apparent i_d is dominated by the capacitive displacement current which is highly related to the change of v_{ds}.

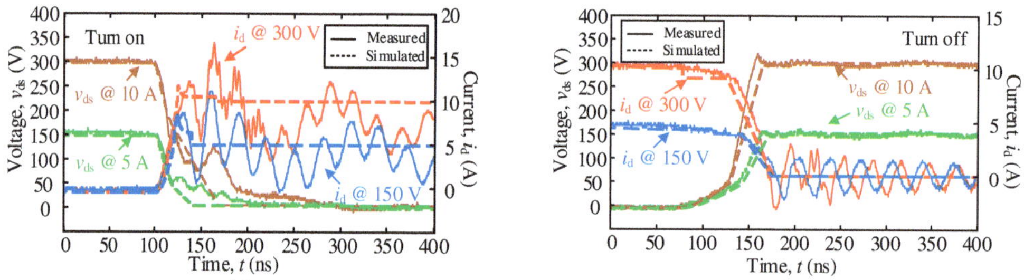

(a) (b)

Figure 13. Switching waveforms of eGaN HEMT with SiC SBD (**a**) turn-on; (**b**) turn-off; @25 °C, $R_G = 5\,\Omega$.

4.3. Power Loss Verification

With the aim of power loss verification for various devices, the switching losses were measured in the DPT and compared with the simulated results. During the switching transients, p_S, which is the product of voltage and current, can be obtained using the math function in oscilloscope and similarly the E_{onm} and E_{offm} can also be obtained by integrating p_S. As mentioned previously, the current and voltage probes are calibrated for each test and additional delay time is also added to the waveforms results to keep transient voltage and current synchronous. The captured waveforms and simulated waveforms under the same test conditions are demonstrated in Figure 14 taking Si IGBT as an example. By comparing the measured results with the simulated results, a good agreement can be clearly seen in terms of not only transient voltage and current waveforms but also computed p_S, E_{onm} and E_{offm}. Furthermore, the measured power loss results are compared with the simulated loss results for different devices under various operating conditions to validate the proposed method. The average error ($\bar{e}$) is calculated by averaging the absolute value of the error in each case.

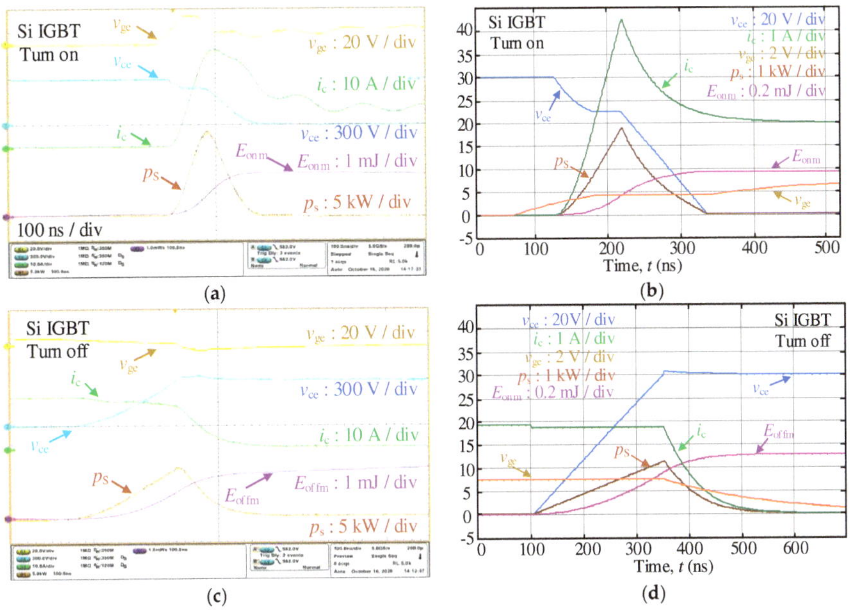

Figure 14. Si IGBT turn on waveforms (**a**) measured; (**b**) simulated; and turn off waveforms (**c**) measured; (**d**) simulated @600 V, 20 A, 25 °C, R_G15 Ω.

Figure 15 shows the power loss results of Si IGBT under different conditions of current, voltage, and temperature. Generally, the total power loss (E_{ts}) increases as the operating voltage and current increase, and E_{off} is less than E_{on} except for the high temperature condition. It can be seen that, the average errors of E_{ts} are within 7%, namely 5.1%, 5.5%, and 6.3% for different operating conditions of current, voltage, temperature, respectively.

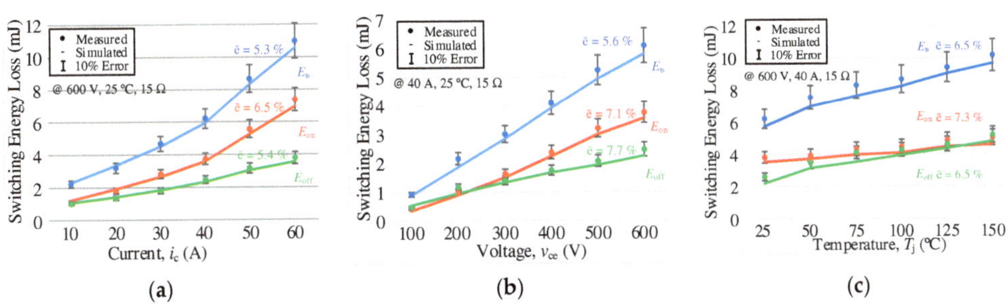

Figure 15. Si IGBT loss results under different values of (**a**) current; (**b**) voltage; (**c**) Temperature.

Likewise, the power loss comparison results for SiC MOSFET are illustrated in Figure 16. Note that the E_{ts} for SiC MOSFET is typically less than 1 mJ which is much less than the counterpart of Si IGBT for similar conditions. It also can be found in Figure 16c that, there is a negative correlation between E_{on} and T_j. Since E_{on} is the dominated loss, as T_j rises, E_{ts} reduces accordingly though E_{off} increases slightly. It is also noted that $\bar{e}$ of E_{off} for various conditions are more than 7% while $\bar{e}$ for E_{on} and E_{ts} are still within acceptable range. The reasons for the loss deviation can be the underestimation of parasitics and ringing loss as well as measurement error. Besides, a relatively small amount of loss deviation can still result in a high error percentage when the overall loss is relatively low.

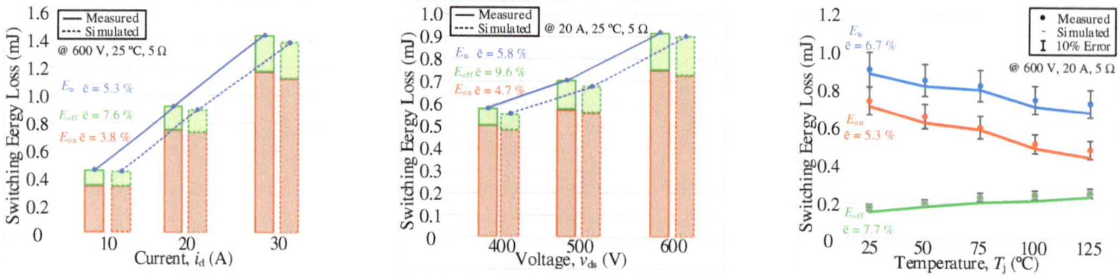

(a) (b) (c)

Figure 16. Power losses of SiC MOSFET under different values of (**a**) current; (**b**) voltage; (**c**) temperature.

Figure 17 shows E_{ts} results for the combination of Si MOSFET and SiC SBD as the SDC. It can be observed that E_{ts} increases as the operating voltage and current increase. In addition, a slight increase of E_{ts} can be found as the operating temperature rise from 25 °C to 150 °C, while E_{ts} increases significantly when 33 Ω R_G is used. Moreover, the power loss results for the case of eGaN HEMT with SiC SBD are illustrated in Figure 18. It should be mentioned that, in order to capture the switching waveforms and the power loss with reasonable accuracy, a 220 Ω gate resistance is used to relatively sacrifice the switching speed and avoid shoot through issue due to the very low v_{th} of GaN. Comparing with the simulated and measured results, a good agreement is achieved at various testing conditions and the error is within an acceptable range, although the $\bar{e}$ of E_{off} is slightly higher.

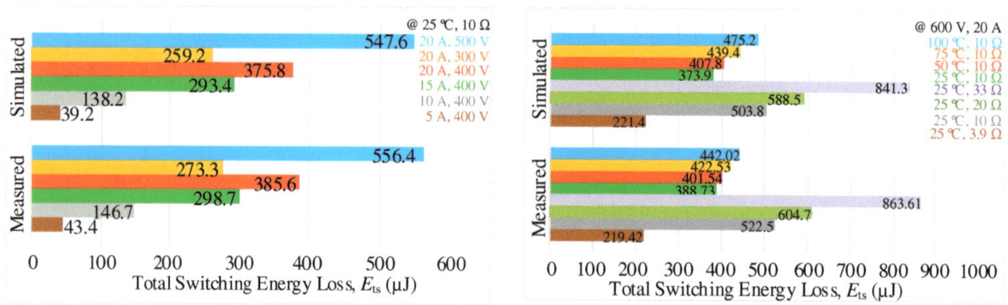

(a) (b)

Figure 17. Power losses of Si MOSFET under various values of (**a**) voltage and current, (**b**) temperature.

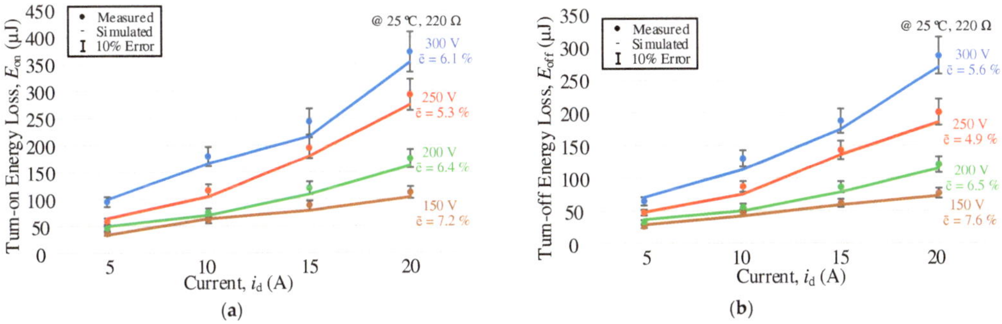

Figure 18. Power loss results of eGaN HEMT with SiC SBD (**a**) turn on, (**b**) turn off.

4.4. Discussion

- Accuracy

Comparing with the original two-state resistance switch model in PSCAD/EMTDC, the proposed model is capable to reproduce the switching transient waveforms considering various impacts of parasitics and interactive behavior of diode. In addition, the thermal effect is also considered to provide reasonably accurate results comparing with the measured results where the temperature is monitored by a thermal imager as shown in Figure 19a. Apart from the switching waveforms, multi-dimensional (i.e., voltage, current, temperature) power loss LUT as shown in Figure 19b also can be obtained simultaneously. The average error is within 10% comparing with measured results for various devices under various conditions. Comparing to the traditional physical model or analytical loss model, no significant advantage is found in the modelling accuracy using the proposed model due to the ignorance of parasitic resonance and some linear assumptions. Nevertheless, the complexity of the proposed model is reduced with no state equations and numerical calculations, and all the model parameters can be extracted from the datasheet.

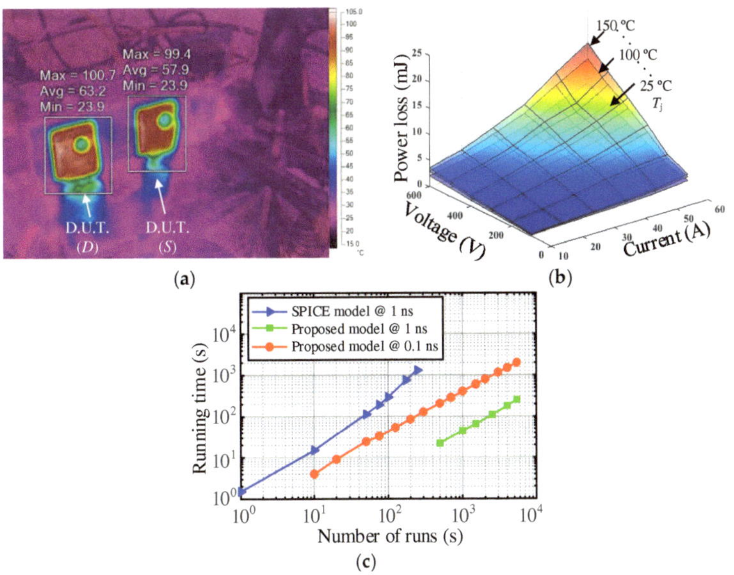

Figure 19. DPT results of (**a**) thermal image; (**b**) loss LUT for Si IGBT; (**c**) time cost comparison.

- Efficiency

The proposed model uses equivalent dependent voltage and current sources to represent the dynamic characteristics of devices based on the analytical equations for each sub-stage of the switching process. In addition, the gate loop and power loop are decoupled and the complicated numerical calculation as well as solving physical equations are not necessary which can boost the simulation efficiency and avoid convergence issue. In order to obtain an accurate power loss LUT with a wide range of operating conditions, there are numerous permutations to be taken into account and thousands of simulation runs are required instead of repetitive DPTs. For example, it requires around 5400 simulation runs to cover the operating range, namely voltage from 20 V to 600 V with 20 V step, current from 2 A to 60 A with a 2 A step, and temperature from 25 °C to 150 °C with a 25 °C step. Figure 19c demonstrates the time cost of using the SPICE model provided by the manufacturer and using the proposed model in PSCAD. In order to achieve reasonable accuracy, the simulation time step is typically one nanosecond or less. Notice that, it takes thousands of seconds to barely finish around 200 runs in SPICE, while more than 10 times less running time is needed to finish the same number of runs by the proposed model with even less time-step (i.e., 0.1 ns). Furthermore, to finish the whole 5400 simulation runs using the proposed model at 1 ns time-step, it takes less than 300 s which shows the merit of time-saving in generating the power loss LUT.

- Applicability

The proposed approach can be used to reproduce the switching waveforms and obtain the power loss LUT of SDC configured by various devices such as Si IGBT, Si/SiC MOSFET, eGaN HEMT, and SiC SBD. When it comes to other devices with a new structure such as Cascade GaN, the proposed model cannot be used directly and modifications are needed though the basic modelling method is still applicable. In addition, for each specific device, the curve fitting functions and algorithms for parameter extraction are needed to be adjusted for good fitting results. Apart from PSCAD/EMTDC, the proposed approach can also be applied to other simulators such as MATLAB/Simulink, PLECS, and Saber with respective modifications. Besides, the proposed model provides an insight into the device behavior and most of the elements have clear descriptions. As a result, it is more easily apprehensible than mathematical equations. Conversely, analytical loss models are normally limited to the specific device type or combination and it is difficult to extend the models to various PE applications for loss estimation. Besides, loss measurements are time-consuming, costly, and challenging especially for WBG devices due to the fast switching speed.

5. Conclusions

In this paper, a generalized behavioral modelling approach of the switch-diode cell for power loss prediction is proposed, implemented in PSCAD/EMTDC, and validated by experimental results in double-pulse tests. This proposed model consists of an active switch model and diode model and it can be used for different modern power semiconductors. The modelling approach along with power loss analysis is derived based on the comprehensive switching process analysis in a clamped inductive switching circuit. The static and dynamic characteristics of the switch-diode cell are modelled by dependent voltage and current sources with passive components. In addition, the proposed model is improved by considering the impacts of parasitic elements, interactive behavior of diode, and the temperature-dependent parameters. Besides, the extraction of the model parameters is introduced by curve fitting from the device datasheet. Moreover, the switching transient verification along with power verification is conducted for different devices under a wide range of operating conditions. A good agreement between the simulated results using the proposed model and experimental results can be achieved with less than 10% average error. Consequently, the proposed model provides a good balance in terms of accuracy, efficiency, and applicability.

Author Contributions: Conceptualization, Y.X. and C.N.M.H.; methodology, Y.X., C.N.M.H. and A.G.; software, C.N.M.H. and D.M.; validation, Y.X. and A.G.; formal analysis, Y.X.; investigation, Y.X.; resources, A.G., C.N.M.H. and D.M.; writing—original draft preparation, Y.X.; writing—review and editing, C.N.M.H. and A.G.; visualization, Y.X.; supervision, C.N.M.H.; project administration, C.N.M.H.; funding acquisition, D.M. All authors have read and agreed to the published version of the manuscript.

Funding: This research is funded by the NSERC Collaborative Research and Development Grants, Canada and in part by the Manitoba Hydro International, Canada.

Institutional Review Board Statement: Not applicable.

Informed Consent Statement: Not applicable.

Data Availability Statement: Data sharing not applicable.

Conflicts of Interest: The authors declare no conflict of interest.

References

1. Baliga, B.J. *Fundamentals of Power Semiconductor Devices*; Springer: New York, NY, USA, 2008.
2. Iwamuro, N.; Laska, T. IGBT History, State-of-the-Art, and Future Prospects. *IEEE Trans. Electron. Devices* **2017**, *64*, 741–752. [CrossRef]
3. Grant, D.A.; Gowar, J. *Power MOSFET Theory and Applications*; Wiley: New York, NY, USA, 1989.
4. Islam, M.; Mekhilef, S. Efficient Transformerless MOSFET Inverter for a Grid-Tied Photovoltaic System. *IEEE Trans. Power Electron.* **2016**, *31*, 6305–6316. [CrossRef]
5. Zhang, Y.; Yao, C.; Zhang, X.; Chen, H.; Li, H.; Wang, J. Power Loss Model for GaN-Based MHz Critical Conduction Mode Power Factor Correction Circuits. *IEEE Trans. Emerg. Sel. Top. Power Electron.* **2020**, *8*, 141–151. [CrossRef]
6. Yang, B.; Wang, J.; Xu, S.; Korec, J.; Shen, Z.J. Advanced Low-Voltage Power MOSFET Technology for Power Supply in Package Applications. *IEEE Trans. Power Electron.* **2013**, *28*, 4202–4215. [CrossRef]
7. Marroqui, D.; Garrigós, A.; Torres, C.; Orts, C.; Blanes, J.M.; Gutierrez, R. Interleaved, Switched Inductor and High-Gain Wide Bandgap Based Boost Converter Proposal. *Energies* **2021**, *14*, 800. [CrossRef]
8. Waradzyn, Z.; Stala, R.; Mondzik, A.; Skała, A.; Penczek, A. GaN-Based DC-DC Resonant Boost Converter with Very High Eciency and Voltage Gain Control. *Energies* **2020**, *13*, 6403. [CrossRef]
9. Millán, J.; Godignon, P.; Perpiñà, X.; Pérez-Tomás, A.; Rebollo, J. A Survey of Wide Bandgap Power Semiconductor Devices. *IEEE Trans. Power Electron.* **2014**, *29*, 2155–2163. [CrossRef]
10. Cha, K.H.; Ju, C.T.; Kim, R.Y. Analysis and Evaluation of WBG Power Device in High Frequency Induction Heating Application. *Energies* **2020**, *13*, 5351. [CrossRef]
11. Li, X.B.; Ma, H.B.; Yi, J.H.; Lu, S.; Xu, J.P. A Comparative Study of GaN HEMT and Si MOSFET-Based Active Clamp Forward Converters. *Energies* **2020**, *13*, 4160. [CrossRef]
12. Cooper, J.A.; Melloch, M.R.; Singh, R.; Agarwal, A.; Palmour, J.W. Status and prospects for SiC power MOSFETs. *IEEE Trans. Electron. Devices* **2002**, *49*, 658–664. [CrossRef]
13. Lidow, A.; Rooij, M.d.; Strydom, J.; Reusch, D.; Glaser, J. *GaN Transistors for Efficient Power Conversion*, 3rd ed.; Wiley: Hoboken, NJ, USA, 2020; pp. 1–23.
14. Jones, E.A.; Wang, F.F.; Costinett, D. Review of Commercial GaN Power Devices and GaN-Based Converter Design Challenges. *IEEE Trans. Emerg. Sel. Top. Power Electron.* **2016**, *4*, 707–719. [CrossRef]
15. Bryant, A.T.; Mawby, P.A.; Palmer, P.R.; Santi, E.; Hudgins, J.L. Exploration of power device reliability using compact device models and fast electrothermal simulation. *IEEE Trans. Ind. Appl.* **2008**, *44*, 894–903. [CrossRef]
16. Gole, A.M.; Keri, A.; Kwankpa, C.; Gunther, E.W.; Dommel, H.W.; Hassan, I.; Marti, J.R.; Martinez, J.A.; Fehrle, K.G.; Tang, L.; et al. Guidelines for modelling power electronics in electric power engineering applications. *IEEE Trans. Power Del.* **1997**, *12*, 505–514. [CrossRef]
17. Ghosh, A.; Ho, C.; Prendergast, J.; Xu, Y. Conceptual Design and Demonstration of an Automatic System for Extracting Switching Loss and Creating Data Library of Power Semiconductors. *IEEE Open J. Power Electron.* **2020**, *1*, 431–444. [CrossRef]
18. Mantooth, H.A.; Peng, K.; Santi, E.; Hudgins, J.L. Modelling of Wide Bandgap Power Semiconductor Devices—Part, I. *IEEE Trans. Electron. Devices* **2015**, *62*, 423–433. [CrossRef]
19. Kraus, R.; Castellazzi, A. A Physics-Based Compact Model of SiC Power MOSFETs. *IEEE Trans. Power Electron.* **2016**, *31*, 5863–5870. [CrossRef]
20. McNutt, T.R.; Hefner, A.R.; Mantooth, H.A.; Berning, D.; Ryu, S. Silicon Carbide Power MOSFET Model and Parameter Extraction Sequence. *IEEE Trans. Power Electron.* **2007**, *22*, 353–363. [CrossRef]
21. Mudholkar, M.; Ahmed, S.; Ericson, M.N.; Frank, S.S.; Britton, C.L.; Mantooth, H.A. Datasheet Driven Silicon Carbide Power MOSFET Model. *IEEE Trans. Power Electron.* **2014**, *29*, 2220–2228. [CrossRef]

22. Li, H.; Zhao, X.; Su, W.; Sun, K.; You, X.; Zheng, T.Q. Nonsegmented PSpice Circuit Model of GaN HEMT with Simulation Convergence Consideration. *IEEE Trans. Ind. Electron.* **2017**, *64*, 8992–9000. [CrossRef]
23. Li, H.; Zhao, X.; Sun, K.; Zhao, Z.; Cao, G.; Zheng, T.Q. A Non-Segmented PSpice Model of SiC mosfet With Temperature-Dependent Parameters. *IEEE Trans. Power Electron.* **2019**, *34*, 4603–4612. [CrossRef]
24. Turzynski, M.; Kulesza, W.J. A Simplified Behavioral MOSFET Model Based on Parameters Extraction for Circuit Simulations. *IEEE Trans. Power Electron.* **2016**, *31*, 3096–3105. [CrossRef]
25. Wen, Y.; Yang, Y.; Gao, Y. Model parameter calibration method of SiC power MOSFETs behavioural model. *IET Power Electron.* **2020**, *13*, 426–435. [CrossRef]
26. Ren, Y.; Xu, M.; Zhou, J.; Lee, F.C. Analytical loss model of power MOSFET. *IEEE Trans. Power Electron.* **2006**, *21*, 310–319.
27. Rąbkowski, J.; Skoneczny, H.; Kopacz, R.; Trochimiuk, P.; Wrona, G. A Simple Method to Validate Power Loss in Medium Voltage SiC MOSFETs and Schottky Diodes Operating in a Three-Phase Inverter. *Energies* **2020**, *13*, 4773. [CrossRef]
28. Mukunoki, Y.; Konno, K.; Tsubasa, M.; Horiguchi, T.; Nishizawa, A.; Kuzumoto, M.; Hagiwara, M.; Akagi, H. An Improved Compact Model for a Silicon-Carbide MOSFET, and its Application to Accurate Circuit Simulation. *IEEE Trans. Power Electron.* **2018**, *33*, 9834–9842. [CrossRef]
29. Jin, M.; Gao, Q.; Wang, Y.; Xu, D. A Temperature-Dependent SiC MOSFET Modelling Method Based on MATLAB/Simulink. *IEEE Access* **2017**, *6*, 4497–4505. [CrossRef]
30. Riccio, M.; D'Alessandro, V.; Romano, G.; Maresca, L.; Breglio, G.; Irace, A. A Temperature-Dependent SPICE Model of SiC Power MOSFETs for Within and Out-of-SOA Simulations. *IEEE Trans. Power Electron.* **2018**, *33*, 8020–8029. [CrossRef]
31. Shi, B.; Zhao, Z.; Zhu, Y. Piecewise Analytical Transient Model for Power Switching Device Commutation Unit. *IEEE Trans. Power Electron.* **2019**, *34*, 5720–5736. [CrossRef]
32. Yuan, D.; Zhang, Y.; Wang, X. An Improved Analytical Model for Crosstalk of SiC MOSFET in a Bridge-Arm Configuration. *Energies* **2021**, *14*, 683. [CrossRef]
33. Wang, X.; Zhao, Z.; Li, K.; Zhu, Y.; Chen, K. Analytical Methodology for Loss Calculation of SiC MOSFETs. *IEEE Trans. Emerg. Sel. Top. Power Electron.* **2019**, *7*, 71–83. [CrossRef]
34. Xie, R.; Wang, H.; Tang, G.; Yang, X.; Chen, K.J. An Analytical Model for False Turn-On Evaluation of High-Voltage Enhancement-Mode GaN Transistor in Bridge-Leg Configuration. *IEEE Trans. Power Electron.* **2017**, *32*, 6416–6433. [CrossRef]
35. Agrawal, B.; Freindl, M.; Bilgin, B.; Emadi, A. Estimating switching losses for SiC MOSFETs with non-flat miller plateau region. In Proceedings of the 2017 IEEE Applied Power Electronics Conference and Exposition (APEC), Tampa, FL, USA, 26–30 March 2017; pp. 2664–2670.
36. Wang, K.; Yang, X.; Li, H.; Ma, H.; Zeng, X.; Chen, W. An Analytical Switching Process Model of Low-Voltage eGaN HEMTs for Loss Calculation. *IEEE Trans. Power Electron.* **2016**, *31*, 635–647. [CrossRef]
37. Chen, J.; Luo, Q.; Huang, J.; He, Q.; Du, X. A Complete Switching Analytical Model of Low-Voltage eGaN HEMTs and Its Application in Loss Analysis. *IEEE Trans. Ind. Electron.* **2020**, *67*, 1615–1625. [CrossRef]
38. Gareau, J.; Hou, R.; Emadi, A. Review of Loss Distribution, Analysis, and Measurement Techniques for GaN HEMTs. *IEEE Trans. Power Electron.* **2020**, *35*, 7405–7418. [CrossRef]
39. Hou, R.; Lu, J.; Chen, D. Parasitic capacitance Eqoss loss mechanism, calculation, and measurement in hard-switching for GaN HEMTs. In Proceedings of the 2018 IEEE Applied Power Electronics Conference and Exposition (APEC), San Antonio, TX, USA, 4–8 March 2018; pp. 919–924.
40. Ho, C.; Breuninger, S.; Pettersson, G.; Escobar, L. Serpa, and A. Coccia. Practical design and implementation procedure of an interleaved boost converter using SiC diodes for PV applications. *IEEE Trans. Power Electron.* **2012**, *27*, 2834–2845. [CrossRef]
41. Ho, C.; Breuninger, H.; Pettersson, S.; Escobar, G.; Canales, F. A comparative performance study of an interleaved boost converter using commercial Si and SiC diodes for PV applications. *IEEE Trans. Power Electron.* **2013**, *28*, 289–299. [CrossRef]
42. Xu, Y.; Ho, C.N.M.; Ghosh, A.; Muthumuni, D. An Electrical Transient Model of IGBT-Diode Switching Cell for Power Semiconductor Loss Estimation in Electromagnetic Transient Simulation. *IEEE Trans. Power Electron.* **2020**, *35*, 2979–2989. [CrossRef]
43. Xu, Y.; Ho, C.N.M.; Ghosh, A.; Muthumuni, D. Design, Implementation and Validation of Electro-Thermal Simulation for SiC MOSFETs in Power Electronic Systems. *IEEE Trans. Ind. Appl.* **2021**. [CrossRef]
44. Reiner, R.; Waltereit, P.; Weiss, B.; Quay, R.; Ambacher, O. Investigation of GaN-HEMTs in Reverse Conduction. In Proceedings of the PCIM Asia, International Exhibition and Conference for Power Electronics, Intelligent Motion, Renewable Energy and Energy Management, Nuremberg, Germany, 16–18 March 2017; pp. 1–8.
45. Lauritzen, P.O.; Ma, C.L. A Simple Diode Model with Reverse Recovery. *IEEE Trans. Power Electron.* **1991**, *6*, 188–191. [CrossRef]
46. Jarndal, A.; Essaadali, R.; Kouki, A.B. A Reliable Model Parameter Extraction Method Applied to AlGaN/GaN HEMTs. *IEEE Trans. Comput. Aided Design Integr. Circuits Syst.* **2016**, *35*, 211–219. [CrossRef]
47. Jouha, W.; El Oualkadi, A.; Dherbécourt, P.; Joubert, E. Silicon Carbide Power MOSFET Model: An Accurate Parameter Extraction Method Based on Levenberg-Marquardt Algorithm. *IEEE Trans. Power Electron.* **2018**, *33*, 9130–9133. [CrossRef]
48. Bryant, A.T.; Kang, X.; Santi, E.; Palmer, P.R.; Hudgins, J.L. Two-step parameter extraction procedure with formal optimization for physics-based circuit simulator IGBT and p-i-n diode models. *IEEE Trans. Power Electron.* **2006**, *21*, 295–309. [CrossRef]

49. Ahmed, M.R.; Todd, R.; Forsyth, A.J. Predicting SiC MOSFET Behavior Under Hard-Switching, Soft-Switching, and False Turn-On Conditions. *IEEE Trans. Ind. Electron.* **2017**, *64*, 9001–9011. [CrossRef]
50. Howard, J.; Martin, G. *High-Speed Digital Design: A Handbook of Black Magic*; Pearson Education: London, UK, 1999.
51. Stueckler, F.; Vecino, E. *CoolMOS C7 650V Switch in a Kelvin Source Configuration*; Infineon Technology. Application Note AN 2013-05; Infineon Technologies: Neubiberg, Germany, 2013.

energies

Article

EMC Component Modeling and System-Level Simulations of Power Converters: AC Motor Drives

Bernhard Wunsch [1,*], Stanislav Skibin [1], Ville Forsström [2] and Ivica Stevanovic [3]

[1] ABB Research Switzerland, 5405 Baden, Switzerland; stanislav.skibin@ch.abb.com
[2] ABB Oy Drives, 00381 Helsinki, Finland; ville.forsstrom@fi.abb.com
[3] Federal Office of Communications OFCOM, 2501 Biel/Bienne, Switzerland; ivica.stevanovic@bakom.admin.ch
* Correspondence: bernhard.wunsch@ch.abb.com

Abstract: EMC simulations are an indispensable tool to analyze EMC noise propagation in power converters and to assess the best filtering options. In this paper, we first show how to set up EMC simulations of power converters and then we demonstrate their use on the example of an industrial AC motor drive. Broadband models of key power converter components are reviewed and combined into a circuit model of the complete power converter setup enabling detailed EMC analysis. The approach is demonstrated by analyzing the conducted noise emissions of a 75 kW power converter driving a 45 kW motor. Based on the simulations, the critical impedances, the dominant noise propagation, and the most efficient filter component and location within the system are identified. For the analyzed system, maxima of EMC noise are caused by resonances of the long motor cable and can be accurately predicted as functions of type, length, and layout of the motor cable. The common-mode noise at the LISN is shown to have a dominant contribution caused by magnetic coupling between the noisy motor side and the AC input side of the drive. All the predictions are validated by measurements and highlight the benefit of simulation-based EMC analysis and filter design.

Keywords: electromagnetic compatibility (EMC); electromagnetic interference (EMI) filter; SPICE simulations; AC drives; conducted noise emissions; EMC filter; transmission line; power cable; chokes

Citation: Wunsch, B.; Skibin, S.; Forsström, V.; Stevanovic, I. EMC Component Modeling and System-Level Simulations of Power Converters: AC Motor Drives. *Energies* **2021**, *14*, 1568. https://doi.org/10.3390/en14061568

Academic Editor: Anna Richelli

Received: 12 February 2021
Accepted: 9 March 2021
Published: 12 March 2021

1. Introduction

Switched-mode power converters transform the electrical power supplied to their inputs according to the requirements of the load connected at their output. While these converters provide high efficiency and control, their internal switching events also introduce high-frequency noise that pollutes the environment with conducted and radiated noise emissions [1,2]. In order to avoid any harm of such noise emissions to the device itself or to other equipment in the same environment, electromagnetic compatibility (EMC) norms limit the maximum allowed noise emissions of power converters and the minimum necessary immunity against external EM disturbances. In this work, we consider EMC norms on conducted noise emissions in the frequency band "B" ranging from 150 kHz to 30 MHz [3–5].

The objective of the EMC analysis of a converter is to design EMC filters that ensure compliance with these norms and to optimize these filters by minimizing cost and/or volume [6]. A powerful approach to EMC analysis is the usage of EMC simulations, where the entire power converter system including the external impedances is described in a form of an electrical circuit that captures the high-frequency noise sources as well as the noise propagation [7,8]. In principle, such simulations allow a detailed study of the system's electromagnetic behavior, including the spectrum and location of noise sources depending on the converter topology, the applied control scheme, and the semiconductor technology [9,10], or of the noise propagation depending on the internal power components or the externally connected load and grid impedances [11]. Simulations can also help in the assessment and optimization of interconnects or the locations of grounding connections [12]. Importantly, such simulations can be done at an early design stage, before the converter is built, which enables short development time and minimizes risks of late redesigns. With all these benefits, a simulation-based EMC design seems to clearly outperform a conventional measurement-based approach [13].

However, all these benefits crucially depend on the quality of the simulation model. All the relevant components need to be identified and represented by corresponding circuit models valid in the frequency range of interest, where at high frequencies, the components' behavior is far from ideal. Grounding connections and interconnects are no longer ideal shorts but introduce important coupling mechanisms [12]. Even the discrete components can couple to each other [1]. Ultimately, when the entire system is mapped to an accurate circuit model, the used circuit solver has to run numerically stable simulations within a reasonable time [13]. The active components such as semiconductor switches or the modeled nonlinear magnetic effects might require time-domain simulations spanning over a full fundamental period (20 ms for 50 Hz grid frequency), with a necessary resolution to resolve noise up to 30 MHz. These challenges explain why EMC analysis is still an art, despite all the advances in EMC technologies and in commercial software tools.

In this paper, we introduce and demonstrate a methodology to create, set up, and run an EMC simulation of power electronic converters. A state-of-the-art review of all EMC-critical components is given and a set of carefully developed component models is chosen [14–20] to build up an entire system-level model of industrial power converters. The used models not only accurately represent the EMC behavior of the power converter components, but also ensure stable and time-efficient simulations of the entire circuit. Prior work on EMC analysis of AC-drives has been reported in [21–27] and EMC analysis of DC-fed motors can be found in [9,10,13,28–30]. However, many of these publications are dealing with academic or low-power applications.

In this paper, the adopted methodology is applied to EMC analysis of conducted noise emissions in an industrial 75 kW AC motor drive (a variable-speed drive that controls the speed and torque of an induction motor connected at its output via a long, shielded motor cable). The system is represented by broadband stable circuits and is analyzed to identify the main noise propagation paths and the best noise filtering options. All effects predicted and explained by simulations are validated by measurements.

2. State-of-the-Art Review of EMC Modeling of Power Electronic Components

Efficient and accurate circuit models of power electronic components are necessary for EMC simulations of power electronic systems. In this section, we review the state-of-the-art in high-frequency modeling of all EMC-relevant power electronic components, such as chokes with non-linear magnetic cores, long cables, and motors, including resistors and capacitors, printed circuit boards (PCB) and housing, semiconductors, and line impedance stabilized networks (LISN).

Component modeling can be generally classified into two broad categories: physical or behavioral modeling [14,31]. Physical modeling is based on fundamental laws of physics, such as Maxwell equations. It requires the knowledge of geometry and electromagnetic characteristics of the modeled component and it can make use of analytical or numerical techniques to solve the equations and create an equivalent circuit model. The behavioral modeling, on the other hand, does not need this information and uses measured impedances of a component in various connection schemes to create an equivalent circuit that will reproduce that same behavior. In this paper, we follow this classification and designate the equivalent circuit models obtained using these two modeling procedures as physical and behavioral models, respectively.

2.1. Choke Modeling

Electromagnetic interference (EMI) filters in power electronic systems are based on passive elements with inductors (chokes) used to reduce both common-mode (CM) and differential-mode (DM) EMI levels. The frequency characteristics of chokes may be much different from ideal when operating at high frequencies, due to parasitic capacitance of windings, leakage inductance, and frequency-dependent permeability of the magnetic cores, among other high-frequency effects. Accurate high-frequency models of single- and three-phase chokes are, therefore, an important building block in EMC simulations used for the filter design of power systems.

2.1.1. Physical Models

Physical models are based on analytical or numerical approaches, or a combination of the two, where electromagnetic equations are constructed and solved using the geometry and electromagnetic characteristics of the choke's windings and the core, which are either available from datasheets or need to be separately measured.

Analytical equations exist to estimate the magnetic fields inside the magnetic core and in the air, as well as the electric fields between two turns and between turns and core. These calculations can then be used to create an equivalent circuit of the choke [32–35], including leakage inductances [36–39]and parasitic (turn-to-turn and turn-to-core) capacitances [40–46]. These models can be effectively used for the purpose of optimizing the common-mode choke geometry and core materials [47,48].

Numerical approaches use various numerical techniques to solve Maxwell's equations for the entire choke or for some of its parts. Such techniques include the Finite-Element Method (FEM) [49–51] or 3-D Partial Element Equivalent Circuit (PEEC) method combined with the boundary integral method (BIM) [52,53]. Numerical models are more suitable for choke designs and their frequency-domain analysis.

2.1.2. Behavioral Models

Behavioral choke models can be further broadly grouped into two subcategories [31,54]. The first subcategory is behavioral models with circuit topologies that reflect the physical effects inside the choke (e.g., magnetic characteristics of the core). The second subcategory is black-box behavioral models having circuit topologies without any relation to the choke's internal physical effects, but correctly reproducing the measured behavior at its terminals.

Behavioral modeling based on black-box measurements can be carried out on a choke in different connection schemes using a spectrum analyzer [55–58], impedance analyzer [14,59], or vector network analyzer (VNA) [60–62]. The parameters of the corresponding equivalent circuits (with either fixed or variable topologies) are then extracted from measurements by (iterative) analytical formulas [14] and/or optimization algorithms for fitting the measured data, such as rational function approximations [31,63], vector fitting [62,64,65], and evolutionary algorithms [14].

In behavioral models with circuit topologies reflecting the physical behavior of the core, the frequency-dependent core permeability (caused by eddy currents inside the core) is usually modeled using Cauer of Foster ladder networks, whose parameters are fitted to the measured complex permeability using different methods [54,66–73].

Recently, the Cauer RL ladder network from [66] has been extended by replacing linear inductors with Preisach [74] inductors. This model accurately captures high-frequency effects, saturation, and hysteresis in single- and three-phase chokes [15]. A description of the measurement setup used to characterize the magnetic properties and the parameter extraction procedure for the equivalent circuit has been provided. This model has been further extended in [16] to also account for dimensional resonance in ferrites. We use this methodology to create equivalent circuit models of the chokes for system-level simulations presented in this paper.

2.2. Multiconductor Cable Modeling

Power cables, found in power distribution systems and motor drives, are distributed elements whose lengths may by far exceed the operating wavelength. They can be modeled as multi-conductor transmission lines, where many frequency-dependent characteristics including per unit length (p.u.l.) parameters, skin and proximity effect, dielectric losses, and transmission line propagation, reflections, and delay need to be appropriately accounted for [75,76].

2.2.1. Determination of Frequency-Dependent per Unit Length Parameters

Typically, the first step in modeling cables is to determine their distribution or p.u.l. parameters. When cable material parameters and cross-sectional geometry are known, the p.u.l. parameters can be obtained using analytical formulas or numerical electromagnetic simulations.

Analytical formulas exist for a limited set of canonical cross-section geometries [77–80] or can be used in combination with cable cross-section partitioning for more general cross-section geometries [81,82]. Different numerical electromagnetic methods can also be used to extract the cable p.u.l. parameters, such as finite element method [80,83–89] or method of moments [90–92]. In cables designed primarily to carry high-frequency signals, the p.u.l. parameters are kept within strict tolerance levels during manufacture, which is not necessarily the case for power cables whose primary function is the distribution of power signals [79].

When the material and geometrical characteristics of the cable are not known, the p.u.l. parameters can also be extracted from measurements using an impedance meter [93,94], impedance analyzer [29,80,95,96], vector network analyzer (VNA) [18,97–100], and combining the VNA measurements with time-domain reflectometry (TDR) [101] or impedance meter measurements [18] to accurately capture both DC, low, and high-frequency behavior.

2.2.2. Cable Models Based on Lumped Segmentation of Transmission Line

Equivalent circuits of multiconductor cables to be used in SPICE (Simulation Program with Integrated Circuit Emphasis) circuit simulations can be created using conventional lumped segmentation of transmission lines. Each segment is represented by an elementary cell containing the extracted p.u.l. RLGC parameters to model the transmission line that may have an additional RL ladder network to model for frequency-dependent skin and proximity effects, and an additional RC ladder network to model for frequency-dependent dielectric losses [29,80,82,93–96,101–104].

Approaches based on conventional lumped segmentation of transmission lines provide an adequate solution to the problem of mixed frequency/time simulations for short cables or approximate predictions of overvoltage transients. However, these methods lead to large circuit matrices, rendering the simulations inefficient and/or unstable especially for system-level EMC simulations including long cables [105,106].

2.2.3. Cable Models Based on Matrix Transfer Functions

An equivalent circuit of transmission lines can also be obtained based on matrix transfer functions that interrelate terminal voltage and current vectors. They can be further emulated in SPICE using modal decoupling and voltage/current controlled generators [105] or represented in a form of matrix rational function in terms of poles and residues either analytically [107], or using various numerical algorithms such as a vector-fitting algorithm [108–111], matrix-rational approximation method [106], or method of characteristics [112]. Additionally, more advanced methods, such as extraction of the propagation delay in the method of characteristics [113], delayed vector fitting (DVF) [114], or multivariate orthonormal vector fitting (MOVF) [115,116] can also be used. These matrix rational function techniques are suited to generate state-space models and synthesize equivalent circuits, which can be embedded into conventional SPICE solvers by using lumped RLC elements [63,108,117], Laplace elements [17,105], and/or controlled sources [76,105,118].

The behavioral modeling procedure for long multiconductor power cables presented in [17] starts with measurements of the DC resistances of the cable and of its frequency-dependent admittance matrix up to the highest frequency of interest using a VNA. Rational function interpolation and passivity enforcement are used next to create an analytic representation of the tabulated measured data [108–110]. Finally, a SPICE equivalent circuit of the measured cable is produced by translating the rational function representation into the circuit domain [76,105,117]. This approach provides stable, causal, and passive equivalent circuits of multiconductor cables that can be used in both time- and frequency-domain SPICE simulations. An extension of this method is introduced in [18] to obtain length-scalable multi-conductor cable models by using modal decomposition [119,120], extracting the p.u.l. parameters from the measured admittance matrix, and by taking advantage of the rotational symmetries of the cable, which in turn result in reduced equivalent circuits that speed up the SPICE simulations of long multiconductor power cables (three- and four-conductor shielded cables of 150 m length were modeled). This method is used for creating models of multi-conductor cables in this paper.

2.3. Motor Modeling

High-frequency modeling of induction machines plays an important role in investigating motor drive electromagnetic interference issues such as stator winding reflected-wave overvoltage and bearing discharging current [121].

Physical modeling of motors utilizes electromagnetic field analysis based on 2-D or 3-D FEM to create the equivalent circuits [122–126]. This methodology requires complete geometry and material information and it enables a detailed EMC analysis and design of the motor itself.

Behavioral modeling needs real motor measurements to construct the equivalent circuit that reproduces the motor impedances. These impedances are usually measured using an impedance analyzer on a motor in several connection schemes (typically in common and differential modes). The circuit parameters can then be extracted from the measurements using either pre-defined equivalent circuit topologies that emulate the physical behavior of the motors [29,93,95,102,121,127–134] or more general circuit topologies based on rational function approximations [135,136] and vector fitting algorithm [17,19,137,138]. A combination of FEM and behavioral modeling may also be used to improve and fine-tune the model parameters [139].

It is useful to create equivalent circuit models of various motors, which can then be used to assess the conducted noise in the full converter system for different loads.

The behavioral motor model used in this paper is created by a parallel connection of a high-frequency motor model based on measurements of a standstill motor [19] and a low-frequency model capturing the operation condition of the motor [23]. A high pass filter placed in front of the high-frequency model ensures that depending on the frequency of excitation the correct motor model is used.

2.4. Interconnects, PCB, and Housing

From the EMC theory point of view, all low frequency (LF) coupling mechanisms (galvanic, capacitive, and inductive coupling) are present in a power converter system and must be adequately taken into account [1,2].

In the EMC analysis of real applications, the influence of PCB- and interconnects design is sometimes overlooked. It is however very important to include the above-described parasitic effects in the system level circuits [12,13,53,140], as will be demonstrated later in Section 3.3. Although the extraction of equivalent circuits from 3-D models of interconnects is quite a complex and time-consuming task, careful preparation of the CAD data can significantly accelerate the modeling so that it can be integrated into the EMC analysis process. For that purpose, the following features are expected from commercial design tools:

- import of the manufacturer CAD data with the automated placement of component models, ports, and terminals;
- extraction of models with a large number of ports (>100);
- extraction of passive, broadband models which are accurate in a preselected frequency range;
- Creating a standard, SPICE-compatible output model, suitable for time- and frequency-domain system level simulations.

Presently, there are several commercial tools, allowing a full data process chain from CAD import to the extraction of the wideband equivalent circuit. Tools like ANSYS Q3D Extractor [141] or Dassault CST studio suite (3D/PCB) [142] implement the full design chain described above, are easy to use and validated in a variety of applications by a large user community.

2.5. HF Capacitor and Resistor Models

Capacitors, together with inductors, are essential components of any filter design. Contrary to the complexity of the inductor modeling techniques (see Section 2.1), the impedance curves of nearly all types of capacitors can be accurately represented by a simple series RLC equivalent circuit. The RLC values can be obtained from measurements either directly using hardware (Impedance- or Vector Network Analyzer) or software fitting tools (e.g., Matlab Zfit [143]).

Analysis of impedance curves of different capacitors can be very helpful in the capacitor selection procedure, especially if certain resonances have to be damped by the appropriate selection of the filter capacitors.

2.6. Power Semiconductor Models

In most PE applications, the power semiconductors are operated in switching mode and their switching waveforms are the main EMC noise source. The spectrum of this noise source is determined by the applied control scheme that determines the switching frequency of the semiconductors, and by the switch-on/switch-off waveforms of the switching devices, which depend on the type of semiconductor (IGBT, MOSFET, BJT, WBG semiconductors [9,10]). The EMI noise spectrum increases with increasing switching frequency and with reduced turn-on and turn-off time (for frequencies above the inverse switching time). Due to the dominant role of semiconductors in EMC noise creation, accurate semiconductor models are needed in EMC simulations, that capture switching waveforms of controlled commutations and reverse recovery of free-wheeling diodes for spontaneous commutation.

A detailed analysis of semiconductor modeling techniques is beyond the scope of this paper, however, in the following we give a short overview of power semiconductor models used in system-level EMC simulations. Modeling of power diodes [144–147], MOSFETs [148,149], and BJTs [150] is well established, therefore, numerous models are readily available either in libraries of SPICE simulation tools (e.g., SIMetrix [151], LTSpice [152]) or on the web pages of most manufacturers. This does not seem to be the case for IGBT models. A detailed overview of the modeling techniques and different complexity levels of IGBT models can be found in [153,154]. Although several physical and behavioral models of the coupled MOSFET-BJT structure of an IGBT device are presented in the literature [145,146,155–167], the parameter extraction procedures ask for simultaneous fitting to both static and dynamic device measurement data, which makes the modeling of IGBTs a rather complex task even for major manufacturers. Quite a few of them provide models of some IGBT devices (e.g., [165,166]) but the accuracy and stability of these models differ and the selection of an appropriate model for system level EMC simulations is still challenging.

2.7. LISN and Noise Detector Modelling

The line impedance stabilized network (LISN) is a low pass filter device, connected between the mains and device under test (DUT). Application of a LISN in a test setup serves two purposes: it filters out the HF electrical noise from the grid and it provides a test port for EMC conducted emission measurements. A standard LISN must provide 50 Ω impedance between each line and the common ground in the frequency range 150 kHz–30 MHz.

A typical circuit diagram of one phase leg of a CISPR-16-1-2 [5] compatible LISN is shown in Figure 1. Depending on the power level and manufacturer, the values of filter components and the number of low-pass filter stages can vary but the circuit diagram from Figure 1 is a good starting point for creating a LISN equivalent circuit in system level EMC simulations.

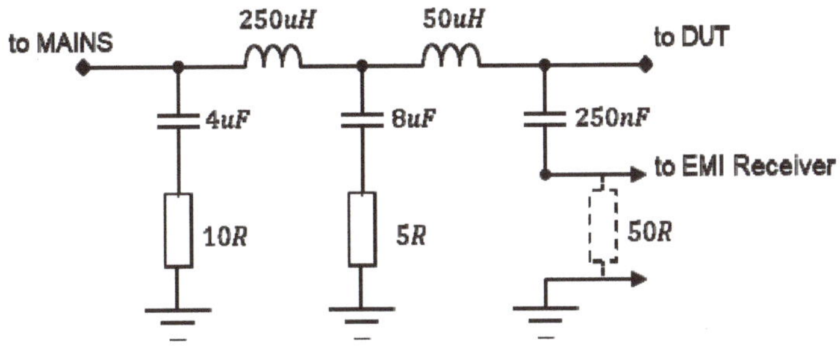

Figure 1. A CISPR-16-1-2 compatible one phase leg LISN circuit diagram [5].

EMC standards are referring to the EMC noise using two types of noise detectors: average (AV) and quasi-peak (QP). These noise detectors are implemented in a special type of spectrum analyzers called EMI receivers. Since system level EMC simulations are usually evaluated/compared with measured EMI noise, there is a need for having a software implementation of QP and AV detectors based on simulation data. The easiest way to obtain a QP or AV spectrum is the use of postprocessing routines. A digital time-domain data processing methodology with the emulation of AV, QP, etc. detectors can be found in [168] and has been used to post-process the simulations shown here.

3. System Level EMC Simulation and Analysis: Voltage-Source Inverter Driving a Motor

In this section, we analyze measured and simulated EMC noise spectra of an industrial power converter. The power converter is a 75 kW, 400 V three-phase variable speed drive with a two-level inverter topology and an input diode rectifier. The converter is connected via a long (10–200 m) shielded power cable to a 45 kW induction motor. The measurements of conducted EMC noise are performed in an EMC lab according to the norms specified in [3–5]. The simulation model of the EMC setup is created by combining the component models reviewed in Section 2 into a full system model in SPICE software [151]. We show the accuracy of the model in predicting the measured conducted EMC noise. Then we analyze and simplify the system model in order to identify the main noise propagation paths and the best noise filtering options. We found a dominant impact of the motor cable on the measured EMC noise as well as of the couplings caused by the interconnects.

3.1. Build-Up and Validation of System Level EMC Simulation

The schematic of the EMC setup is shown in Figure 2. The required component models have been discussed in the previous section. The common mode choke is realized by placing three nanocrystalline magnetic cores around the three-phase input and it is modeled using the physical equivalent circuit presented in [15], with parameters extracted from impedance measurements. The two identical DC reactors are modeled with a behavioral model created by applying vector fitting to the measured impedance [108].

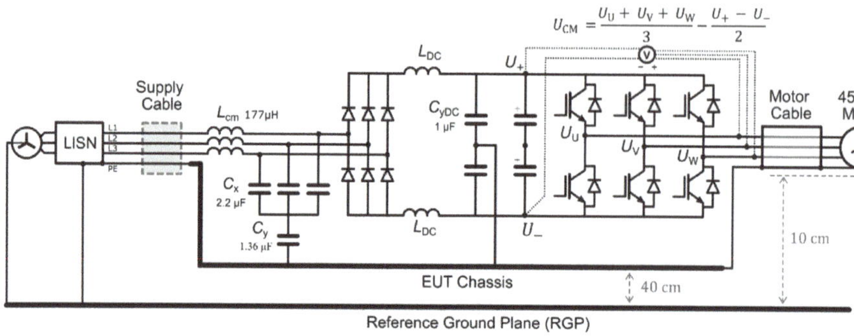

Figure 2. Simplified schematic representation of the EMC setup. Low-frequency values of filter components are specified, but corresponding EMC models are more complex as discussed in Section 2. The corresponding CM-circuit is derived in Section 3.2.

The model of the shielded motor cable is based on VNA measurements and created using vector fitting [17]. Models of the same cable with different lengths are obtained by first extracting the p.u.l. parameters of the cable which are then used to create the equivalent circuit of the same cable for any length [18].

The voltages across the IGBTs in the inverter power module with an average switching frequency of 4 kHz are the main noise source of the drive. The IGBTs were modeled using physical behavioral models obtained from the manufacturer [165,166]. The actual equipment under test (EUT) uses a complex vector control algorithm to determine the switching pattern. Given that the average switching frequency of 4 kHz is well below the EMC band of conducted emissions starting at 150 kHz, the exact switching pattern has only a minor impact on EMC noise in stationary operation. Therefore, a simplified scalar control scheme based on pulse width modulation is used in the simulations to approximate the real behavior.

Interconnects between the components are not ideal shorts and introduce relevant parasitic couplings that need to be included as was discussed in Section 2.4. Parasitic effects of wires, busbars, and the metal chassis are included in the full three-phase simulation by extracting an RLGC equivalent SPICE circuit of interconnects and housing using a quasi-static field solver based on the method of moments (MoM [141]) as described in Section 2.4. The simplified 3D model of the structure is shown in Figure 3. The extracted black-box model has 69 ports and includes parasitic capacitances, coupled inductances, and resistances. Instead of connecting the components of the schematic shown in Figure 2 by ideal shorts, they are then connected at the corresponding ports of the extracted SPICE circuit. The compact broadband model of interconnects and housing is not shown explicitly in the schematic of Figure 2 for better readability but is an important part of the full three-phase circuit model.

Figure 4a shows the measured EMC noise at the LISN for all three phases, labeled L1, L2, and L3 in the setup shown in Figure 2. The EMC noise shows repetitive resonances as functions of frequency that exist due to wave resonances in the motor cable that occur at certain ratios of wavelength to cable length, as we will discuss in more detail in Section 3.3. The schematic shown in Figure 2 does not include any auxiliary power circuit that powers fans or control and gate circuits. These auxiliary circuits also include switch mode power supplies with high switching frequency, however, they process much lower powers. Figure 4b shows that for the analyzed converter, the contribution of these auxiliary circuits to conducted noise emissions is small. The red curve in Figure 4b is the phase noise for normal operation, as also shown in Figure 4a. For comparison, the black curve in Figure 4b shows the EMC noise when only the auxiliary power is turned on, while the inverter is not switching, and the motor is at a standstill. The noise stemming from the auxiliary supply alone has a minor contribution to the total EMC noise and only becomes relevant for frequencies above 4 MHz, where for the setup shown, the noise is well below the norm. Therefore, the auxiliary power circuit is not included in the simulation, and simulation results above 4 MHz are not accurate.

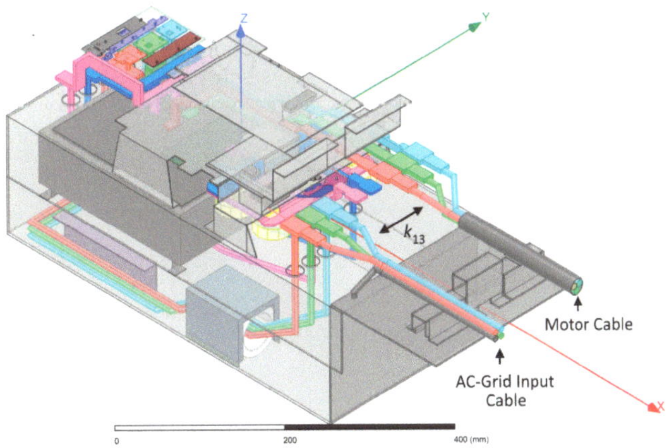

Figure 3. 3D model of the chassis and busbar geometry in the prototype for parasitic RLGC extraction (MoM) [141].

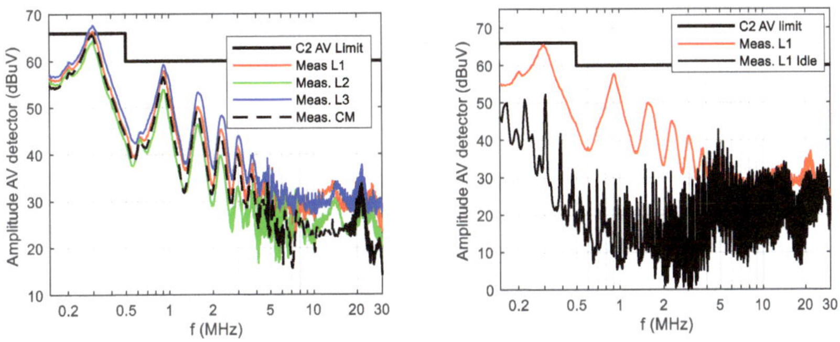

(**a**) (**b**)

Figure 4. Measured average detected grid noise of all three phases L1, L2, and L3 with a 100 m motor cable (16 mm^2) during (**a**) normal operation with motor rotation (**b**) comparison between the noise of normal operation and the motor at standstill with no switching at inverter (idle).

3.2. Common Mode Equivalent Circuit and Impact of Interconnects on EMI Noise

For EMC analysis it is useful to separate the phase noise in its common mode (CM) and differential mode (DM) components. CM current flows between the power conversion system and ground, while current flowing between the power lines is denoted as DM noise. As indicated in Figure 2 with the line voltages at the LISN denoted as $U_{Li}, i = 1,2,3$, the CM voltage U_{CM} and differential mode voltages U_{DMi} at the LISN can be defined as:

$$U_{CM} = \frac{U_{L1} + U_{L2} + U_{L3}}{3}; U_{DMi} = U_{Li} - U_{CM}; i = 1,2,3. \tag{1}$$

In measurements, the transformation from phase noise to common mode and differential mode noise is done using a CM/DM noise separator [169,170]. The measurement results shown in Figure 5a reveal that the EMC noise is dominated by the common mode noise, which will be the focus of the EMC analysis in this paper. Figure 5b shows the CM and DM noise obtained by postprocessing of time-domain simulations of the full setup. Simulated and measured EMC curves are in excellent agreement up to 4 MHz. For higher frequencies, the EMC noise is very low, and the auxiliary power supply noise becomes significant, which explains the deviation between simulations and measurements. As explained in Section 2.6, the dominant EMC noise sources of the converter shown in Figure 2 are the switching waveforms of the IGBTs in the inverter. The propagation of differential mode noise from the inverter towards the input (AC grid side) is well suppressed by the electrolytic DC capacitors, however, the DC capacitor cannot filter CM noise. The main reasons why measurements still show significant DM noise at the AC grid side are the noise mixing between CM and DM within the diode rectifier [20] and parasitic magnetic couplings from the noisy load side to the input side [171], which will be described later in Section 3.3.

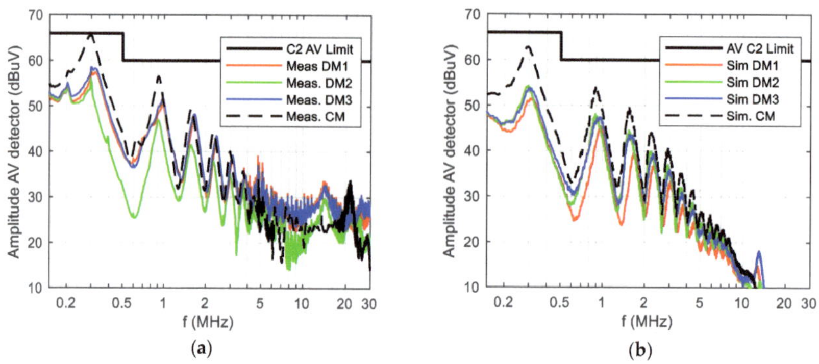

Figure 5. Average detector noise level of differential mode and common mode noise emissions as defined in Equation (1) with a 100 m motor cable (16 mm^2) (**a**) Measurement with CM/DM LISN separator. (**b**) Simulation results.

The propagation of the CM noise can be described by the equivalent circuit shown in Figure 6. Most of the impedances shown in the CM circuit of Figure 6 are the CM impedances of the components shown in the three-phase circuit of Figure 2. The frequency-dependent CM impedances are shown in Figure 7a and they are accurately reproduced by the circuit models. The impact of the diode rectifier either as a CM noise source or CM impedance is neglected in the CM circuit of Figure 6, while the HF noise introduced by the switching of the DC voltage in the inverter is represented as a fixed voltage noise source U_{cm}. As indicated in Figure 2, the noise source U_{cm} is the voltage difference between the DC bus midpoint and the average of the three output voltages, and it is determined only once, using a time-domain simulation of the full three-phase circuit (which uses accurate semiconductor models). The frequency spectrum of U_{cm} is shown in Figure 7b.

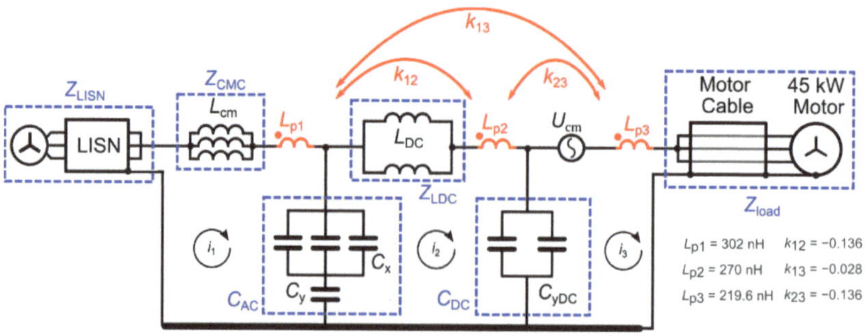

Figure 6. The equivalent CM circuit of the converter including magnetic parasitic couplings. The 3-phase circuit is shown in Figure 2.

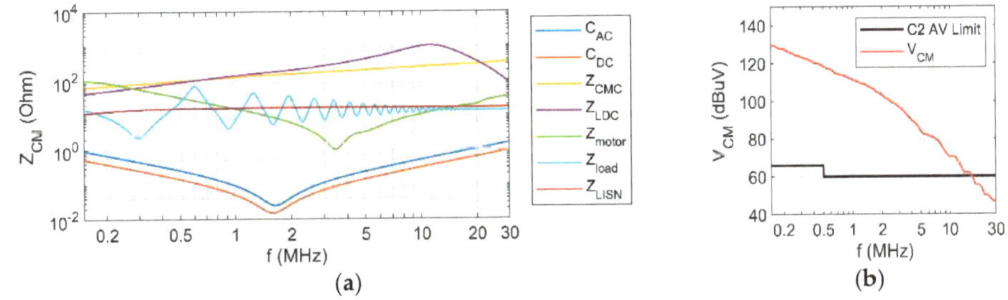

Figure 7. (**a**) The circuit impedances of the equivalent CM circuit shown in Figure 6 with a 100 m motor cable (16 mm^2). (**b**) The spectrum of the CM noise source (using AVG detector). The corresponding EMC noise is shown in Figure 5.

The parasitic CM loop inductances L_{p1}, L_{p2}, L_{p3} and their mutual couplings k_{12}, k_{13}, k_{23} capture the most relevant coupling mechanism of the interconnects and housing shown in Figure 3 and they can be directly extracted by defining the corresponding CM current paths in the quasi-static field solver [141] using matrix reduction which corresponds to a parallel connection of the AC-phases and DC buses. The loop inductances L_{p1} and L_{p3} and their coupling coefficient k_{13} quantify the important magnetic coupling between the noisy output of the converter on the load side and the quiet input of the converter on the AC-grid side. As indicated in Figure 3, this coupling is introduced by the pigtails of the input and output cables that are close to each other, run in parallel, and do not have any shielding. Why are the other types of parasitics introduced by the interconnects less important? Parasitic capacitances between different interconnects and between interconnects and chassis turn out to have little impact on CM noise propagation, due to the dominant shunt capacitances C_{AC} on the grid side, C_{DC} on the DC link, and the effective cable capacitance between the conductors of the motor cable and their grounded shield. Furthermore, resistances of the interconnects are small compared to the other impedances within each loop namely Z_{LISN} + Z_{CMC}, Z_{LDC}, or Z_{load}. In contrast, the magnetic coupling between the different current loops, which is modeled by the parasitic loop inductances L_{p1}, L_{p2}, L_{p3} with their mutual coupling factors k_{12}, k_{13}, k_{23} is highly relevant, because the induced voltages bypass the capacitive filters C_{AC} and C_{DC}. It turns out that the most important coupling is the one between the load loop with its high noise current between i_3 and the grid loop with its low noise current i_1.

The CM circuit shown in Figure 6 consists of three loops. The first loop contains the CM impedances on the grid side: LISN (Z_{LISN}), common mode choke (Z_{CMC}), and the AC side capacitance C_{AC} which consists of a series connection of X and Y capacitors. The loop current i_1 is subject to the EMC norms of the LISN side. The second loop consists of C_{AC}, the DC-reactors (Z_{LDC}), and the DC-Y-capacitors C_{DC}. As already stated above, the impact of the diode rectifier on the CM noise in the EMC range is neglected here. The third loop contains C_{DC}, the CM voltage noise source U_{cm}, and the load impedance Z_{load} given by the CM impedance of the cable terminated with the motor. The loop current i_3 is by far the largest CM current in the system and its propagation to the LISN is filtered by the CLC filter formed by C_{DC}, Z_{LDC}, C_{AC}. However, the magnetic couplings, most importantly k_{13}, bypass this filter.

The simple structure of the CM circuit helps to understand the noise propagation, furthermore, it is a linear circuit (if common mode choke and DC reactors are approximated by linear broadband models) which allows one to apply frequency-domain analysis, which is much faster than time-domain analysis.

The propagation of CM noise is assessed using the equivalent circuit shown in Figure 6 and the results are shown in Figure 8. As a validation of the simulation models, we show that the measured noise (curve (a) in Figure 8) is accurately predicted by time-domain simulations (curve (b) in Figure 8) using the three-phase circuit of Figure 2. Equivalent results are obtained by frequency-domain simulations (curve (c) in Figure 8) using the common mode circuit of Figure 6. By varying the three mutual coupling factors in the common mode circuit of Figure 6, the relative importance of the different couplings is analyzed. In the simulations, the coupling factors are simple parameters that can easily be changed. In the physical setup, the inductive couplings between the current loops depend on the layout of the interconnects and can be reduced by minimizing the area enclosed by the current loops, by increasing the distance between the loops, or by introducing electromagnetic shields. It is found that the coupling factor k_{13} between loops 3 and 1 has the biggest impact. In fact, as shown by curve (d) in Figure 8, if the coupling between the input and output loops is set to zero k_{13} = 0 (all other parameters stay the same), the grid side noise is reduced significantly below the norm limit. The load side loop (between inverter and load) has the largest EMI noise current, which explains the relevance of even a small magnetic coupling between output and input.

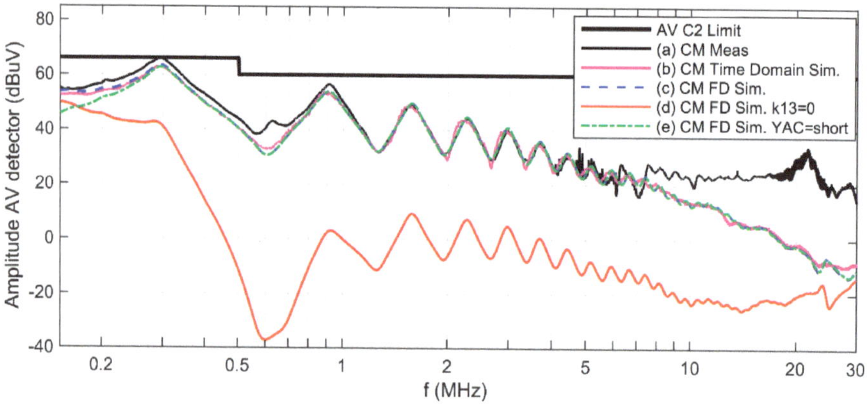

Figure 8. Simulated and measured conducted emissions of CM noise. Simulation (b) used the full three-phase circuit shown in Figure 2, simulations (c)–(e) used the common mode circuit shown in Figure 6 with frequency-domain (FD) solver. The setup uses a 100 m motor cable (16 mm^2).

In contrast, an increase of the AC-side capacitance does not reduce the CM noise. As shown by curve (e) in Figure 8, the noise level remains almost unchanged (except for a small reduction at 150 kHz) when the filter capacitance C_{AC} on the grid side is shorted completely but the magnetic couplings are present. This finding confirms that the grid side CM noise is mainly caused by the induced voltage source in the grid side current loop. In the presence of the induced voltage, the current on the grid side is most effectively suppressed by a large impedance within the grid side loop, which is achieved by placing a large CM choke impedance $Z_{CMC} \gg Z_{LISN} + Z_{CAC}$ see Figure 7a. Unfortunately, among the EMC filter components, the CM choke is the largest and most expensive component. Other CM filter options will be discussed in Section 4.

3.3. Impact of Cable Characteristics on EMI Noise

In this section, we show that the prominent maxima in the EMC noise measured at the LISN are caused by cable resonances and we discuss how these maxima depend on the length and the type of the motor cable. As shown in Figure 7, the load impedance Z_{load} is much larger than the impedance of the C_Y caps at the DC link, called C_{DC} in Figure 7. Therefore, the current i_3 in the output loop is largest at the minima of the load impedance Z_{load} (or maxima of its admittance Y_{load}). Furthermore, by comparing the impedances Z_{motor} of the motor alone, with the load impedance Z_{load} given by the CM impedance of the cable terminated with the motor, we notice that resonances of the load stem from the cable resonances. The measured noise at the LISN is determined by the current i_1 in the input loop. According to the circuit shown in Figure 6, the noise can propagate from the noisy output to the input either by conduction from output to DC link to input, or by the inductive coupling between the loop inductances. As we will see in the next section, the inductive coupling dominates the noise propagation. For both coupling paths, the EMC noise measured at the LISN shows the same maxima as the output noise (although with lower amplitude) and these maxima are essentially determined by the motor cable. Accurate cable models are therefore essential for EMC simulations of the current setup.

In order to further analyze the impact of the motor cable on EMC noise, measurements and simulations with two different cable types and for different cable lengths were done. The results are shown in Figure 9. Figure 9a compares the load admittance of the two cables, both 30 m long, connected to the same 45 kW motor. Both motor cables are three-phase, shielded power cables, but one uses conductors with a cross section of 16 mm^2 and the other with 70 mm^2 [172] The load admittances of the two cables shown in Figure 9a have the same qualitative behavior as the load admittances with the periodic cable resonances as functions of frequency. However, due to the different cable cross section, the p.u.l. parameters are different, which causes a different characteristic cable impedance and propagation constant. The measured and simulated EMC noise when these cables are used in the EMC setup of Figure 2 is shown in Figure 9c. Note that the maxima of the load admittance coincide with the maxima of the conducted noise emissions.

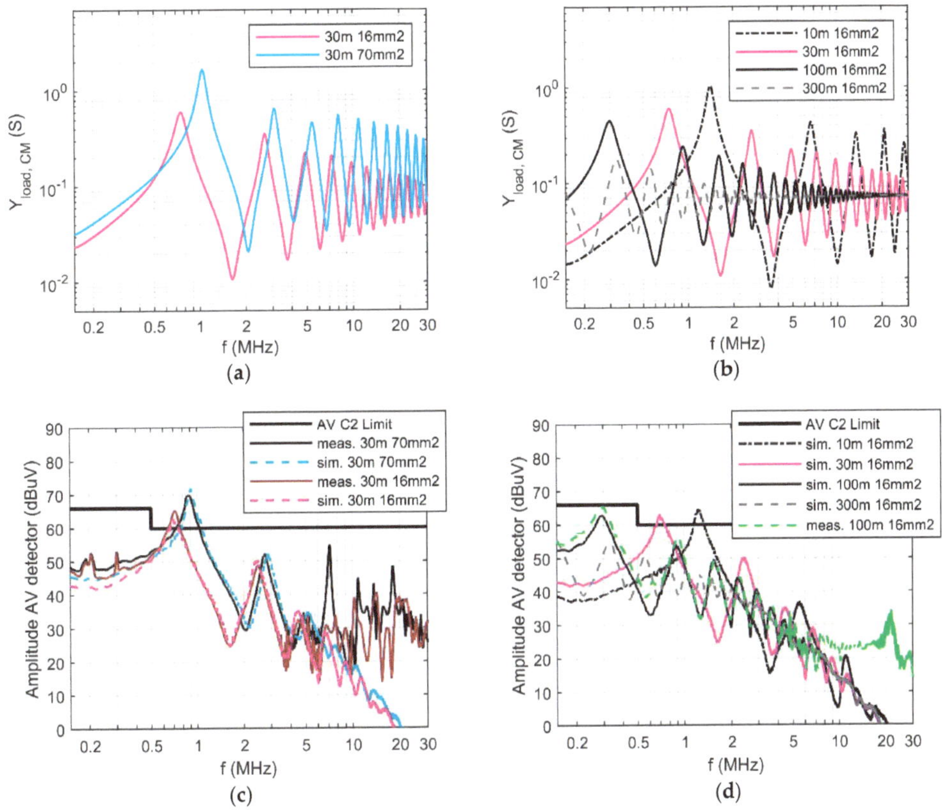

Figure 9. CM admittances Y_{load} of different motor cables terminated with 45 kW motor are shown in subfigures (**a**,**b**). The CM noise emissions of the full system when using these cables are shown in subfigures (**c**,**d**). The impact of cable type (for the same cable length of 30 m) on load admittance and noise emissions is shown in (**a**,**c**) and the impact of cable length (for same cable type with a 16 mm^2 cross section of each conductor) is shown in (**b**,**d**).

Figure 9c shows a good agreement between simulations and measurements up to about 5 MHz, however for higher frequencies the measured noise shows additional peaks that are not reproduced by the simulations. These peaks cannot be explained by the auxiliary supply alone but are most probably caused by coupling between the motor cable and the conductive ground plane present in the EMC setup [171]. Since the reference ground plane of the setup was not included in the simulation model, the effect is not reproduced in the simulation. In [171], the cable to ground coupling was analyzed and it was shown that changes in the cable layout relative to the ground plane can significantly affect the noise levels. However, the cable layout is currently not strictly specified in the EMC measurement standards, despite its large impact on the measured EMC noise.

Next, the impact of the cable length on the load admittance and EMC noise at the LISN is analyzed. Figure 9b compares the load admittances of motor cables of the same cable type but of different lengths, all connected to the same 45 kW motor. With increasing length, the frequency spacing between successive maxima as well as the amplitude of the oscillations of the admittance is reduced. Both observations are well explained by the transmission line behavior of the cable. In the limit of an infinitely long cable, the load impedance would be given by the characteristic cable impedance that shows no oscillations.

Figure 9d shows the EMC noise measured at the LISN, for the various cables shown in Figure 9b. Again, the resonances of the EMC noise stem from the cable resonances. The location of the first maximum in EMC noise (or load admittance) shown in Figure 9d moves to lower frequencies as the cable length increases, while the amplitude stays roughly the same (or might even be higher for shorter cables). Interestingly, for the data shown in Figure 9d, the EMC norm is violated for the shorter cables (10 m and 30 m) while the CM noise is below the limits prescribed by the norm for the longer cables (100 m and 300 m). The large EMC noise for short cables contradicts the common belief that the EMC emissions increase with increasing cable length, and this finding is discussed and explained in Section 4. Here we note that the length-scalable cable models are an effective and accurate tool to predict EMC noise emissions as shown by the good agreement between simulations and measurements in Figure 9.

4. Discussion

Before drawing general conclusions of this work and providing an outlook, in this section, we discuss two main findings of the EMC system simulations of the motor drive presented in Section 3.

The first finding was presented in Section 3.2 and concerns the impact of magnetic coupling on the filter performance of magnetic and capacitive filters. We showed that for the analyzed system, EMC noise is dominated by CM noise and we derived the equivalent circuit for CM noise propagation shown in Figure 6. Two main noise propagation paths were identified. The first noise path is the conventional galvanic noise propagation from the noise source at the inverter to the input side. This noise propagation is filtered by the CLCL-type filter formed by $C_{DC}, L_{DC}, C_{AC}, L_{CM}$. The CM choke L_{CM} is larger and more expensive than the capacitors C_{DC}, C_{AC} and its size could be reduced by increasing the capacitance C_{AC} to ground. However, there is a second noise propagation path introduced by the coupled loop inductances L_{p1}, L_{p2}, L_{p3} in Figure 6, which represents the magnetic couplings between the interconnects. The coupling gives rise to an induced voltage U_{ind} in the grid side current loop, which drives the grid current i_1. The induced voltage $U_{ind} = j\omega M i_3$, where $M = k_{13}\sqrt{L_{p1}L_{p3}}$ denotes the magnetic coupling between output and input loops and the dominant contribution was identified to be the cable connectors at the input and output of the convert as shown in Figure 3. The second noise propagation path dominates the emissions in the EMC frequency range. In fact, as shown in Figure 8, even when replacing the capacitive filter C_{AC} of the CM circuit of Figure 6 by an ideal short, the emissions are not reduced. The emissions caused by the induced voltage are most effectively suppressed by a large impedance within the grid side loop, which is achieved by placing a large CM choke. What about other filter options for CM noise? In the presented setup, most of the EMC noise on the grid side is caused by the induced voltage between the motor side and the grid input side, and therefore an increase of the DC link impedance Z_{LDC} value will have little effect on the grid noise. One could try to filter the noise on the load side, by placing a CM choke around the motor cable. However, the high CM current levels on the load side result in large losses in the core material and can cause saturation. Finally, the coupling factor k_{13} could be lowered by introducing a shielding or by the improved layout of the busbar geometry. If these parasitic mitigation methods are not possible, the grid side common mode impedance must be kept high enough to limit the coupled noise below standard requirements.

The second finding of the EMC analysis that we discuss here concerns the dependence of noise emissions on the type and length of the motor cable as analyzed in Section 3.3. The maxima of EMC noise are caused by resonances of the long motor cable. The interesting finding was that in the analyzed system, the conducted EMC emissions are larger for cables with a length of 30 m than for cables of 100 m (or even 300 m), which contradicts the common belief that emissions always increase with cable length. The common belief is based on the facts, that long cables have a larger capacitance to ground, leading to higher load admittance and, furthermore, the cable resonance occurs at a lower frequency where the noise source has a larger amplitude, see Figure 7b. The explanation for why the common belief is wrong here is based on a combination of EMC effects. First, while the noise source decreases in frequency as shown in Figure 7b, its propagation to the AC input is dominated by magnetic coupling (for the analyzed converter) leading to an induced voltage on the grid side $U_{ind} = j\omega M i_3$. This coupling is represented by the coupled loop inductances in circuit Figure 6. The magnetic coupling is more effective with increasing frequency, which improves noise propagation and compensates for the decrease of the noise source with increasing frequency. Furthermore, the cable resonances are better damped for longer cable lengths. Another aspect is that EMC norms are often stricter at higher frequencies (above 500 kHz in Figure 9d) and the frequency of the cable resonance increases with reducing cable length. Similarly, for sufficiently long cables, the first resonance moves below this frequency and for even longer cables (e.g., for the 300 m cable shown in Figure 9), the first resonance of the cable admittance is below 150 kHz, which is the lower frequency bound of the frequency band "B" relevant for conducted emissions. However, in addition to the impact of the cable length on the EMC spectrum above 150 kHz, longer cables also increase capacitive currents in the low-frequency range (DC up to low harmonics of switching frequencies) due to increased parasitic cable capacitance. The increased CM currents must be considered in the design of the inductive CM filter components to avoid saturation, which relativizes the surprising finding that longer cables can lead to smaller EMC problems.

5. Conclusions

EMC simulations are shown to be an excellent tool to analyze EMC noise propagation and to assess the best filtering options. However, the setup of EMC simulations is a difficult task and requires experience. First, the relevant components need to be identified and modeled in the required detail. The relevant models are not limited to the power components of the converter, but also include the couplings introduced by interconnects, PCB and housing, as well as the external impedance of the converter at the input and output sides, such as the power cable, motor, including the LISN and noise detector. The models can be represented as SPICE circuits to allow including them into a circuit model for the EMC analysis of the full power converter. Often the circuit representation of the full EMC setup is rather complex, which can lead to long simulation times and sometimes even to numerical instability. Therefore, a simplification of the circuit is very beneficial, which also helps with better understanding the noise propagation. For example, the complex semiconductor models can be replaced by a fixed voltage source, representing the switching waveform, and/or by separately analyzing common and differential mode noise, which can often be analyzed by simpler circuits. The simplified circuit is often a linear circuit and can then be analyzed using much faster frequency-domain simulations. Conducted noise emissions of a 75 kW power converter driving a 45 kW motor were analyzed. Based on the simulations, the critical impedances within the system were identified, as well as the dominant noise propagation and the most efficient filter component and location. The conducted noise was found to be dominated by common mode noise in the studied system and a simplified CM circuit was derived. The layout of the interconnects and housing was shown to introduce parasitic magnetic couplings between output and input grounding loops formed by the EMC shunt capacitors and multiple common mode chokes are required to suppress the EMC noise caused by induced voltages. Sharp maxima in the conducted noise emissions were shown to be caused by cable resonances that can be accurately predicted as functions of type, length, and layout of the motor cable. An interesting finding was that in the analyzed system, the conducted EMC emissions were larger for cables with a length of 30 m than for cables of 100 m length or more, which contradicts the common belief that emissions always increase with cable length. All effects predicted and explained by the simulations were validated by measurements, and the close agreement between simulations and measurements underlines the benefit of simulation-based EMC analysis and filter design.

There are various possible continuations of this work. Having identified the detrimental effect of magnetic couplings between input and output, possibilities of an improved layout of the busbar geometry need to be assessed as well as other shielding options. Furthermore, in order to reduce the need for expensive common mode chokes, the benefit of an active EMC filter should be assessed [11]. Finally, while EMC simulations are a powerful tool to analyze conducted noise emissions and to design the required filter, further efforts are needed to build up similar simulation tools for radiated noise emissions.

Author Contributions: Conceptualization, B.W., S.S., V.F. and I.S.; methodology, B.W., S.S., V.F. and I.S.; software, B.W., S.S., V.F. and I.S.; validation, B.W., S.S., V.F.; formal analysis, B.W., S.S., V.F.; investigation, B.W., S.S., V.F. and I.S.; resources, B.W., S.S., V.F. and I.S.; data curation, B.W., S.S., V.F. and I.S.; writing—original draft preparation, B.W., S.S., V.F. and I.S.; writing—review and editing, B.W., V.F. and I.S.; visualization, B.W., V.F.; project administration, B.W., S.S. All authors have read and agreed to the published version of the manuscript.

Funding: This research received no external funding.

Institutional Review Board Statement: Not applicable.

Informed Consent Statement: Not applicable.

Conflicts of Interest: The authors declare no conflict of interest.

References

1. Paul, C. *Introduction to Electromagnetic Compatibility*; Wiley: Hoboken, NJ, USA, 2006.
2. Ott, H. *Electromagnetic Compatibility Engineering*; Wiley: Hoboken, NJ, USA, 2011.
3. EN/IEC 61800-3, Ed. 3.0, Adjustable Speed Electrical Power Drive Systems–Part 3: EMC Requirements and Specific Test Methods, 2017. IEC 61800-3:2017 | IEC Webstore | Electromagnetic Compatibility, EMC, Smart City, Pump, Motor, Water Management. Available online: https://webstore.iec.ch/publication/31003 (accessed on 11 March 2021).
4. CISPR 11-Industrial Scientific and Medical Equipment-Radio-Frequency Disturbance Characteristics-Limits and Methods of Measurement. CISPR 11:2015 | IEC Webstore | Electromagnetic Compatibility, EMC, Smart City. Available online: https://webstore.iec.ch/publication/22643 (accessed on 11 March 2021).
5. CISPR 16-1-2 Specification for Radio Disturbance and Immunity Measuring Apparatus and Methods-Part 1-2: Radio Disturbance and Immunity Measuring Apparatus-Measuring Apparatus, 2014. CISPR 16-1-2:2014+AMD1:2017 CSV | IEC Webstore | Electromagnetic Compatibility, EMC, Smart City. Available online: https://webstore.iec.ch/publication/61924 (accessed on 11 March 2021).
6. Heldwein, M.L.; Kolar, J.W. Impact of EMC Filters on the Power Density of Modern Three-Phase PWM Converters. *IEEE Trans. Power Electron.* **2009**, *24*, 1577–1588. [CrossRef]
7. Mainali, K.; Oruganti, R. Conducted EMI mitigation techniques for switch-mode power converters: A survey. *IEEE Trans. Power Electron.* **2010**, *25*, 2344–2356. [CrossRef]
8. Kharanaq, F.A.; Emadi, A.; Bilgin, B. Modeling of Conducted Emissions for EMI Analysis of Power Converters: State-of-the-Art Review. *IEEE Access* **2020**, *8*, 189313–189325. [CrossRef]

9. Gong, X.; Ferreira, J.A. Comparison and Reduction of Conducted EMI in SiC JFET and Si IGBT-Based Motor Drives. *IEEE Trans. Power Electron.* **2014**, *29*, 1757–1767. [CrossRef]
10. Han, D.; Li, S.; Wu, Y.; Choi, W.; Sarlioglu, B. Comparative Analysis on Conducted CM EMI Emission of Motor Drives: WBG Versus Si Devices. *IEEE Trans. Ind. Electron.* **2017**, *64*, 8353–8363. [CrossRef]
11. Wang, S.; Maillet, Y.Y.; Wang, F.; Boroyevich, D.; Burgos, R. Investigation of hybrid EMI filters for common-mode EMI suppression in a motor drive system. *IEEE Trans. Power Electron.* **2010**, *25*, 1034–1045. [CrossRef]
12. Wang, S.; Wang, F.; Maillet, Y.Y.; Lai, R.; Luo, F.; Boroyevich, D. Parasitic effects of grounding paths on common-mode EMI filter's performance in power electronics systems. *IEEE Trans. Ind. Electron.* **2010**, *57*, 3050–3059. [CrossRef]
13. Lai, J.S.; Huang, X.; Pepa, E.; Chen, S.; Nehl, T.W. Inverter EMI modeling and simulation methodologies. *IEEE Trans. Ind. Electron.* **2006**, *53*, 736–744. [CrossRef]
14. Stevanovic, I.; Skibin, S.; Masti, M.; Laitinen, M. Behavioral Modeling of Chokes for EMI Simulations in Power Electronics. *IEEE Trans. Power Electron.* **2013**, *28*, 695–705. [CrossRef]
15. Wunsch, B.; Skibin, S.; Forsstrom, V.; Christen, T. Broadband Modeling of Magnetic Components with Saturation and Hysteresis for Circuit Simulations of Power Converters. *IEEE Trans. Magn.* **2018**, *54*, 1–5. [CrossRef]
16. Wunsch, B.; Christen, T.; Skibin, S.; Forsstrom, V. Broadband Circuit Model of a Ferrite Core, Including Dimensional Resonance, Saturation, and Hysteresis. *IEEE Trans. Magn.* **2019**, *55*, 1–5. [CrossRef]
17. Stevanovic, I.; Wunsch, B.; Madonna, G.L.; Skibin, S. High-Frequency Behavioral Multiconductor Cable Modeling for EMI Simulations in Power Electronics. *IEEE Trans. Ind. Inf.* **2014**, *10*, 1392–1400. [CrossRef]
18. Wunsch, B.; Stevanovic, I.; Skibin, S. Length-Scalable Multiconductor Cable Modeling for EMI Simulations in Power Electronics. *IEEE Trans. Power Electron.* **2017**, *32*, 1908–1916. [CrossRef]
19. Stevanovic, I.; Wunsch, B.; Skibin, S. Behavioral High-Frequency Modeling of Electrical Motors. In Proceedings of the 2013 Twenty-Eighth Annual IEEE Applied Power Electronics Conference and Exposition (APEC), Long Beach, CA, USA, 17–21 March 2013; pp. 2547–2550.
20. Wunsch, B.; Drofenik, U.; Skibin, S.; Forsstrom, V. Impact of Diode-Rectifier on EMC-Noise Propagation and Filter Design in AC-Fed Motor Drives. In Proceedings of the IEEE International Symposium on Electromagnetic Compatibility, Washington, DC, USA, 7–11 August 2017; pp. 237–242. [CrossRef]
21. Skibinski, G.L.; Kerkman, R.J.; Schlegel, D. EMI emissions of modern PWM AC drives. *IEEE Ind. Appl. Mag.* **1999**, *5*, 47–80. [CrossRef]
22. Tallam, R.M.; Skibinski, G.L.; Shudarek, T.A.; Lukaszewski, R.A. Integrated differential-mode and common-mode filter to mitigate the effects of long motor leads on AC drives. *IEEE Trans. Ind. Appl.* **2011**, *47*, 2075–2083. [CrossRef]
23. Ran, L.; Gokani, S.; Clare, J.; Bradley, K.J.; Christopoulos, C. Conducted electromagnetic emissions in induction motor drive systems. I. Time domain analysis and identification of dominant modes. *IEEE Trans. Power Electron.* **1998**, *13*, 757–767. [CrossRef]
24. Ran, L.; Gokani, S.; Clare, J.; Bradley, K.J.; Christopoulos, C. Conducted electromagnetic emissions in induction motor drive systems. II. Frequency domain models. *IEEE Trans. Power Electron.* **1998**, *13*, 768–776. [CrossRef]
25. Ogasawara, S.; Akagi, H. Analysis and reduction of EMI conducted by a PWM inverter-fed AC motor drive system having long power cables. In Proceedings of the 2000 IEEE 31st Annual Power Electronics Specialists Conference. Conference Proceedings (Cat. No.00CH37018), Galway, Ireland, 18–23 June 2000; Volume 2, pp. 928–933.
26. Costa, F.; Vollaire, C.; Meuret, R. Modeling of conducted common mode perturbations in variable-speed drive systems. *IEEE Trans. Electromagn. Compat.* **2005**, *47*, 1012–1021. [CrossRef]
27. Gubia, E.; Sanchis, P.; Ursúa, A.; López, J.; Marroyo, L. Frequency domain model of conducted EMI in electrical drives. *IEEE Power Electron. Lett.* **2005**, *3*, 45–49. [CrossRef]
28. Moreau, M.; Idir, N.; Le Moigne, P. Modeling of conducted EMI in adjustable speed drives. *IEEE Trans. Electromagn. Compat.* **2009**, *51*, 665–672. [CrossRef]
29. Revol, B.; Roudet, J.; Schanen, J.L.; Loizelet, P. EMI study of three-phase inverter-fed motor drives. *IEEE Trans. Ind. Appl.* **2011**, *47*, 223–231. [CrossRef]
30. Bishnoi, H.; Mattavelli, P.; Boroyevich, D. Un-terminated common-mode EMI model of DC-fed motor drives. In Proceedings of the 2012 15th International Power Electronics and Motion Control Conference (EPE/PEMC), Novi Sad, Serbia, 4–6 September 2012; pp. DS2a.15-1–DS2a.15-8. [CrossRef]
31. Tan, W.; Cuellar, C.; Margueron, X.; Idir, N. A High Frequency Equivalent Circuit and Parameter Extraction Procedure for Common Mode Choke in the EMI Filter. *IEEE Trans. Power Electron.* **2013**, *28*, 1157–1166. [CrossRef]
32. Hague, B. *The Principles of Electromagnetism Applied to Electrical Machines*; Dover: New York, NY, USA, 1962.
33. Grover, F.W. *Inductance Calculations: Working Formulas and Tables*; Dover: New York, NY, USA, 2004.
34. Kazimierczuk, M.K. *High-Frequency Magnetic Components*; Wiley: Chichester, UK, 2009.
35. Massarini, A.; Kazimierczuk, M.K.; Grandi, G. Lumped parameter models for single- and multiple-layer inductors. In Proceedings of the 27th Annual IEEE Power Electronics Specialists Conference, Baveno, Italy, 23–27 June 1996; Volume 1, pp. 295–301.
36. Wilcox, D.J.; Hurley, W.G.; Conlon, M. Calculation of self and mutual impedances between sections of transformer windings. *IEE Proc. C Gener. Transm. Distrib. UK* **1989**, *136*, 308. [CrossRef]
37. Wilcox, D.J.; Conlon, M.; Hurley, W.G. Calculation of self and mutual impedances for coils on ferromagnetic cores. *IEE Proc. A Phys. Sci. Meas. Instrum. Manag. Educ. Rev. UK* **1988**, *135*, 470. [CrossRef]

38. Hurley, W.G.; Wilcox, D.J. Calculation of leakage inductance in transformer windings. *IEEE Trans. Power Electron.* **1994**, *9*, 121–126. [CrossRef]
39. Ouyang, Z.; Zhang, J.; Hurley, W.G. Calculation of Leakage Inductance for High-Frequency Transformers. *IEEE Trans. Power Electron.* **2015**, *30*, 5769–5775. [CrossRef]
40. Massarini, A.; Kazimierczuk, M.K. Self-capacitance of inductors. *IEEE Trans. Power Electron.* **1997**, *12*, 671–676. [CrossRef]
41. Grandi, G.; Kazimierczuk, M.K.; Massarini, A.; Reggiani, U. Stray capacitances of single-layer solenoid air-core inductors. *IEEE Trans. Ind. Applicat.* **1999**, *35*, 1162–1168. [CrossRef]
42. Pasko, S.W.; Kazimierczuk, M.K.; Grzesik, B. Self-Capacitance of Coupled Toroidal Inductors for EMI Filters. *IEEE Trans. Electromagn. Compat.* **2015**, *57*, 216–223. [CrossRef]
43. Ayachit, A.; Kazimierczuk, M.K. Self-Capacitance of Single-Layer Inductors with Separation Between Conductor Turns. *IEEE Trans. Electromagn. Compat.* **2017**, *59*, 1642–1645. [CrossRef]
44. Biela, J.; Kolar, J.W. Using Transformer Parasitics for Resonant Converters—A Review of the Calculation of the Stray Capacitance of Transformers. *IEEE Trans. Ind. Applicat.* **2008**, *44*, 223–233. [CrossRef]
45. Wu, B.; Zhang, X.; Liu, X.; He, C. An Analytical Model for Predicting the Self-Capacitance of Multi-Layer Circular-Section Induction Coils. *IEEE Trans. Magn.* **2018**, *54*, 1–7. [CrossRef]
46. de Freitas Gutierres, L.F.; Cardoso, G. Analytical Technique for Evaluating Stray Capacitances in Multiconductor Systems: Single-Layer Air-Core Inductors. *IEEE Trans. Power Electron.* **2018**, *33*, 6147–6158. [CrossRef]
47. Heldwein, M.L.; Dalessandro, L.; Kolar, J.W. The Three-Phase Common-Mode Inductor: Modeling and Design Issues. *IEEE Trans. Ind. Electron.* **2011**, *58*, 3264–3274. [CrossRef]
48. Kovacic, M.; Hanic, Z.; Stipetic, S.; Krishnamurthy, S.; Zarko, D. Analytical Wideband Model of a Common-Mode Choke. *IEEE Trans. Power Electron.* **2012**, *27*, 3173–3185. [CrossRef]
49. Henglin, C.; Zhaoming, Q.; Shaodong, Y.; Wolf, C. Finite-Element Modeling of Saturation Effect Excited by Differential-Mode Current in a Common-Mode Choke. *IEEE Trans. Power Electron.* **2009**, *24*, 873–877. [CrossRef]
50. Kovacic, M.; Stipetic, S.; Hanic, Z.; Zarko, D. Small-Signal Calculation of Common-Mode Choke Characteristics Using Finite-Element Method. *IEEE Trans. Electromagn. Compat.* **2015**, *57*, 93–101. [CrossRef]
51. Traub, F.; Wunsch, B.; Skibin, S. A high frequency model of toroidal chokes for EMC filtering. In Proceedings of the 2015 IEEE International Symposium on Electromagnetic Compatibility (EMC), Dresden, Germany, 16–22 August 2015; pp. 902–907.
52. Kovacevic, I.F.; Friedli, T.; Musing, A.M.; Kolar, J.W. Full PEEC Modeling of EMI Filter Inductors in the Frequency Domain. *IEEE Trans. Magn.* **2013**, *49*, 5248–5256. [CrossRef]
53. Kovacevic, I.F.; Friedli, T.; Musing, A.M.; Kolar, J.W. 3-D Electromagnetic Modeling of Parasitics and Mutual Coupling in EMI Filters. *IEEE Trans. Power Electron.* **2014**, *29*, 135–149. [CrossRef]
54. Liu, X.; Grassi, F.; Spadacini, G.; Pignari, S.A.; Trotti, F.; Mora, N.; Hirschi, W. Behavioral Modeling of Complex Magnetic Permeability With High-Order Debye Model and Equivalent Circuits. *IEEE Trans. Electromagn. Compat.* **2020**, 1–9. [CrossRef]
55. Dominguez-Palacios, C.; Bernal, J.; Prats, M.M. Characterization of Common Mode Chokes at High Frequencies with Simple Measurements. *IEEE Trans. Power Electron.* **2018**, *33*, 3975–3987. [CrossRef]
56. Dominguez-Palacios, C.; Mendez, J.B.; Prats, M.A.M. Characterization of Three-Phase Common-Mode Chokes at High Frequencies. *IEEE Trans. Power Electron.* **2018**, *33*, 6471–6475. [CrossRef]
57. Dominguez-Palacios, C.; Gonzalez-Vizuete, P.; Martin-Prats, M.A.; Mendez, J.B. Smart Shielding Techniques for Common Mode Chokes in EMI Filters. *IEEE Trans. Electromagn. Compat.* **2019**, *61*, 1329–1336. [CrossRef]
58. González-Vizuete, P.; Domínguez-Palacios, C.; Bernal-Méndez, J.; Martín-Prats, M.A. Simple Setup for Measuring the Response to Differential Mode Noise of Common Mode Chokes. *Electronics* **2020**, *9*, 381. [CrossRef]
59. Kotny, J.-L.; Margueron, X.; Idir, N. High-Frequency Model of the Coupled Inductors Used in EMI Filters. *IEEE Trans. Power Electron.* **2012**, *27*, 2805–2812. [CrossRef]
60. Li, Z.; Pommerenke, D.; Shimoshio, Y. Common-mode and differential-mode analysis of common-mode chokes. In Proceedings of the 2003 IEEE Symposium on Electromagnetic Compatibility, Boston, MA, USA, 18–22 August 2003; Volume 1, pp. 384–387.
61. Sánchez, A.M.; Pérez, A.; Regué, J.R.; Ribó, M.; Rodríguez-Cepeda, P.; Pajares, F.J. A Modal Model of Common-Mode Chokes for Conducted Interference Prediction. *IEEE Trans. Electromagn. Compat.* **2010**, *52*, 749–752. [CrossRef]
62. Nguyen, T.N.; Blanchette, H.F.; Wang, R. General Impedance Representation of Passive Devices Based on Measurement. *IEEE Trans. Power Electron.* **2018**, *33*, 6699–6709. [CrossRef]
63. Antonini, G. Spice equivalent circuits of frequency domain responses. *IEEE Trans. Electromagn. Compat.* **2003**, *45*, 502–512. [CrossRef]
64. Gustavsen, B.; Semlyen, A. Enforcing passivity for admittance matrices approximated by rational functions. *IEEE Trans. Power Syst.* **2001**, *16*, 97–104. [CrossRef]
65. Gustavsen, B. Wide band modeling of power transformers. *IEEE Trans. Power Del.* **2004**, *9*, 414–422. [CrossRef]
66. Sullivan, C.R.; Muetze, A. Simulation Model of Common-Mode Chokes for High-Power Applications. *IEEE Trans. Ind. Appl.* **2010**, *46*, 884–891. [CrossRef]
67. Nomura, K.; Kojima, T.; Hattori, Y. Straightforward Modeling of Complex Permeability for Common Mode Chokes. *IEEJ J. IA* **2018**, *7*, 462–472. [CrossRef]

68. Cuellar, C.; Idir, N.; Benabou, A. High-Frequency Behavioral Ring Core Inductor Model. *IEEE Trans. Power Electron.* **2016**, *31*, 3763–3772. [CrossRef]
69. Illia, M.; Koleff, L.; Griepentrog, G. Non-ideal model of the common mode choke for EMI filters. In Proceedings of the 2017 IEEE Applied Power Electronics Conference and Exposition (APEC), Tampa, FL, USA, 26–30 March 2017; pp. 938–944.
70. Blanken, P.G.; Van Vlerken, J.J.L.M. Modeling of electromagnetic systems. *IEEE Trans. Magn.* **1991**, *27*, 4509–4515. [CrossRef]
71. Takahashi, S.; Ogasawara, S. A Novel Simulation Model for Common-Mode Inductors Based on Permeance-Capacitance Analogy. In Proceedings of the 2020 IEEE Energy Conversion Congress and Exposition (ECCE), Detroit, MI, USA, 11–15 October 2020; pp. 5862–5869.
72. Nomura, K.; Kikuchi, N.; Watanabe, Y.; Inoue, S.; Hattori, Y. Novel SPICE model for common mode choke including complex permeability. In Proceedings of the 2016 IEEE Applied Power Electronics Conference and Exposition (APEC), Long Beach, CA, USA, 20–24 March 2016; pp. 3146–3152.
73. Sixdenier, F.; Yade, O.; Martin, C.; Bréard, A.; Vollaire, C. How to include frequency dependent complex permeability into SPICE models to improve EMI filters design? *AIP Adv.* **2018**, *8*, 056604. [CrossRef]
74. Szabo, Z. Preisach functions leading to closed form permeability. *Physica B* **2006**, *372*, 61–67. [CrossRef]
75. Paul, C.R. *Analysis of Multiconductor Transmission Lines*; Wiley-IEEE Press: Hoboken, NJ, USA, 2007.
76. Achar, R.; Nakhla, M.S. Simulation of High-Speed Interconnects. *Proc. IEEE* **2001**, *89*, 693–728. [CrossRef]
77. Ametani, A. A General Formulation of Impedance and Admittance of Cables. *IEEE Trans. Power Apparatus Syst.* **1980**, *PAS-99*, 902–910. [CrossRef]
78. Ametani, A.; Fuse, K. Approximate method for calculating the impedances of multiconductors with cross section of arbitrary shapes. *Elect. Eng. Jpn.* **1992**, *112*, 117–123. [CrossRef]
79. Dickinson, J.; Nicholson, P.J. Calculating the High Frequency Transmission Line Parameters of Power Cables. In Proceedings of the ISPLC, Minneapolis, MN, USA, 2–4 April 1997; pp. 127–133.
80. Idir, N.; Weens, Y.; Franchaud, J.-J. Skin Effect and Dielectric Loss Models of Power Cables. *IEEE Trans. Dielect. Electr. Insul.* **2009**, *16*, 147–154. [CrossRef]
81. Rivas, R.A.; Marti, J.R. Calculation of Frequency-Dependent Parameters of Power Cables: Matrix Partitioning Techniques. *IEEE Trans. Power Deliv.* **2002**, *17*, 1085–1092. [CrossRef]
82. Hoshmeh, A.; Schmidt, U. A Full Frequency-Dependent Cable Model for the Calculation of Fast Transients. *Energies* **2017**, *10*, 1158. [CrossRef]
83. Weens, Y.; Idir, N.; Bausiere, R.; Franchaud, J.J. Modeling and Simulation of Unshielded and Shielded Energy Cables in Frequency and Time Domains. *IEEE Trans. Magn.* **2006**, *42*, 1876–1882. [CrossRef]
84. Weens, Y.; Idir, N.; Franchaud, J.J.; Bausiére, R. HF Model of an Unshielded 2-Wire Energy Cable Submitted to Power Electronic Converter Constraints. *Int. J. Model. Simul.* **2008**, *28*, 292–298. [CrossRef]
85. Gustavsen, B.; Bruaset, A.; Bremnes, J.J.; Hassel, A. A Finite-Element Approach for Calculating Electrical Parameters of Umbilical Cables. *IEEE Trans. Power Delivery* **2009**, *24*, 2375–2384. [CrossRef]
86. Habib, S.; Kordi, B. Calculation of Multiconductor Underground Cables High-Frequency Per-Unit-Length Parameters Using Electromagnetic Modal Analysis. *IEEE Trans. Power Deliv.* **2013**, *28*, 276–284. [CrossRef]
87. Hafner, A.A.; Caballero, P.T.; Monteiro, J.H.A.; Costa, E.C.M.; Kurokawa, S.; Luz, M.V.F.; Carpes, W.P. Modeling of Power Cables with Arbitrary Cross Section: From Parameter Calculation to Electromagnetic Transients Simulation. *J. Control Autom. Electr. Syst.* **2017**, *28*, 405–417. [CrossRef]
88. Huangfu, Y.; Di Rienzo, L.; Wang, S. Frequency-Dependent Multi-Conductor Transmission Line Model for Shielded Power Cables Considering Geometrical Dissymmetry. *IEEE Trans. Magn.* **2018**, *54*, 1–4. [CrossRef]
89. Greedy, S.; Smartt, C.; Basford, M.J.; Thomas, D.W.P. Open Source Cable Models for EMI Simulations. *IEEE Electromagn. Compat. Mag.* **2018**, *7*, 69–81. [CrossRef]
90. Kruizinga, B.; Wouters, P.A.A.F.; Steennis, E.F. High Frequency Modeling of a Shielded Four-Core Low Voltage Underground Power Cable. *IEEE Trans. Dielect. Electr. Insul.* **2015**, *22*, 649–656. [CrossRef]
91. Patel, U.R.; Triverio, P. MoM-SO: A Complete Method for Computing the Impedance of Cable Systems Including Skin, Proximity, and Ground Return Effects. *IEEE Trans. Power Deliv.* **2015**, *30*, 2110–2118. [CrossRef]
92. Shafieipour, M.; Chen, Z.; Menshov, A.; De Silva, J.; Okhmatovski, V. Efficiently Computing the Electrical Parameters of Cables with Arbitrary Cross-Sections Using the Method-of-Moments. *Electr. Power Syst. Res.* **2018**, *162*, 37–49. [CrossRef]
93. Moreira, A.F.; Lipo, T.A.; Venkataramanan, G.; Bernet, S. High-Frequency Modeling for Cable and Induction Motor Overvoltage Studies in Long Cable Drives. *IEEE Trans. Ind. Appl.* **2002**, *38*, 1297–1306. [CrossRef]
94. Knockaert, J.; Peuteman, J.; Catrysse, J.; Belmans, R. A Vector Impedance Meter Method to Characterize Multiconductor Transmission-Line Parameters. *IEEE Trans. Electromagn. Compat.* **2010**, *52*, 1019–1025. [CrossRef]
95. Wang, L.; Ngai-Man Ho, C.; Canales, F.; Jatskevich, J. High-Frequency Modeling of the Long-Cable-Fed Induction Motor Drive System Using TLM Approach for Predicting Overvoltage Transients. *IEEE Trans. Power Electron.* **2010**, *25*, 2653–2664. [CrossRef]
96. Bade, T.G.; Roudet, J.; Guichon, J.-M.; Sartori, C.A.F.; Kuo-Peng, P.; Schanen, J.-L.; Derbey, A. Frequency-Domain Modeling of Unshielded Multiconductor Power Cables for Periodic Excitation with New Experimental Protocol for Wide Band Parameter Identification. *Electr. Eng.* **2019**, *101*, 333–343. [CrossRef]

97. Williams, D.F. Multiconductor Transmission Line Characterization. *IEEE Trans. Comp. Packag. Manuf. Technol. B* **1997**, *20*, 129–132. [CrossRef]
98. Williams, D.F.; Rogers, J.E.; Holloway, C.L. Multiconductor Transmission-Line Characterization: Representations, Approximations, and Accuracy. *IEEE Trans. Microw. Theory Tech.* **1999**, *47*, 403–409. [CrossRef]
99. Wang, L.; Avolio, G.; Deconinck, G.; Lil, E.V.; Lai, L.L. Estimation of Multi-Conductor Powerline Cable Parameters for the Modelling of Transfer Characteristics. *IET Sci. Meas. Technol.* **2014**, *8*, 39–45. [CrossRef]
100. Gustavsen, B. Eliminating Measurement Cable Effects From Transformer Admittance Measurements. *IEEE Trans. Power Deliv.* **2016**, *31*, 1609–1617. [CrossRef]
101. Kim, J.-H.; Oh, D.; Kim, W. Accurate Characterization of Broadband Multiconductor Transmission Lines for High-Speed Digital Systems. *IEEE Trans. Adv. Packag.* **2010**, *33*, 857–867. [CrossRef]
102. De Paula, H.; de Andrade, D.A.; Chaves, M.L.R.; Domingos, J.L.; de Freitas, M.A.A. Methodology for Cable Modeling and Simulation for High-Frequency Phenomena Studies in PWM Motor Drives. *IEEE Trans. Power Electron.* **2008**, *23*, 744–752. [CrossRef]
103. Marlier, C.; Videt, A.; Idir, N. NIF-Based Frequency-Domain Modeling Method of Three-Wire Shielded Energy Cables for EMC Simulation. *IEEE Trans. Electromagn. Compat.* **2015**, *57*, 145–155. [CrossRef]
104. Bogarra, S.; Riba, J.-R.; Sala-Caselles, V.; Garcia, A. Optimal Fitting of High-Frequency Cable Model Parameters by Applying Evolutionary Algorithms. *Int. J. Electr. Power Energy Syst.* **2017**, *87*, 16–26. [CrossRef]
105. Djordjevic, A.R. SPICE-Compatible Models for Multiconductor Transmission Lines in Laplace-Transform Domain. *IEEE Trans. Microw. Theory Tech.* **1997**, *45*, 569–579. [CrossRef]
106. Dounavis, A.; Achar, R.; Nakhla, M. A General Class of Passive Macromodels for Lossy Multiconductor Transmission Lines. *IEEE Trans. Microw. Theory Tech.* **2001**, *49*, 1686–1696. [CrossRef]
107. Tanji, Y.; Ushida, A. Closed-Form Expression of RLCG Transmission Line and Its Application. *Electron. Comm. Jpn. Pt. III* **2004**, *87*, 1–11. [CrossRef]
108. Gustavsen, B. Computer Code for Rational Approximation of Frequency Dependent Admittance Matrices. *IEEE Trans. Power Deliv.* **2002**, *17*, 1093–1098. [CrossRef]
109. Gustavsen, B. Computer Code for Passivity Enforcement of Rational Macromodels by Residue Perturbation. *IEEE Trans. Adv. Packag.* **2007**, *30*, 209–215. [CrossRef]
110. Gustavsen, B. Fast Passivity Enforcement for Pole-Residue Models by Perturbation of Residue Matrix Eigenvalues. *IEEE Trans. Power Deliv.* **2008**, *23*, 2278–2285. [CrossRef]
111. Krim, A.; Lakrim, A.; Tahri, D. Two VHDL-AMS-Based Models of Multi-Conductor Power Cables for EMI Simulations. *Electr. Eng.* **2020**. [CrossRef]
112. Grivet-Talocia, S.; Huang, H.-M.; Ruehli, A.E.; Canavero, F.; Elfadel, I.M.A. Transient Analysis of Lossy Transmission Lines: An Efficient Approach Based on the Method of Characteristics. *IEEE Trans. Adv. Packag.* **2004**, *27*, 45–56. [CrossRef]
113. Dounavis, A.; Pothiwala, V.A. Passive Closed-Form Transmission Line Macromodel Using Method of Characteristics. *IEEE Trans. Adv. Packag.* **2008**, *31*, 190–202. [CrossRef]
114. Chinea, A.; Triverio, P.; Grivet-Talocia, S. Delay-Based Macromodeling of Long Interconnects from Frequency-Domain Terminal Responses. *IEEE Trans. Adv. Packag.* **2010**, *33*, 246–256. [CrossRef]
115. Grivet-Talocia, S.; Acquadro, S.; Bandinu, M.; Canavero, F.G.; Kelander, I.; Rouvala, M. A Parameterization Scheme for Lossy Transmission Line Macromodels With Application to High Speed Interconnects in Mobile Devices. *IEEE Trans. Electromagn. Compat.* **2007**, *49*, 18–24. [CrossRef]
116. Ferranti, F.; Antonini, G.; Dhaene, T.; Knockaert, L. Parametric Macromodeling of Lossy and Dispersive Multiconductor Transmission Lines. *IEEE Trans. Adv. Packag.* **2010**, *33*, 481–491. [CrossRef]
117. Morched, A.S.; Ottevangers, J.H.; Marti, L. Multi-Port Frequency Dependent Network Equivalents for the EMTP. *IEEE Trans. Power Delivery* **1993**, *8*, 1402–1412. [CrossRef]
118. Neumayer, R.; Haslinger, F.; Stelzer, A.; Weigel, R. Synthesis of SPICE-Compatible Broadband Electrical Models from n-Port Scattering Parameter Data. In Proceedings of the IEEE International Symposium on Electromagnetic Compatibility, Minneapolis, MN, USA, 19–23 August 2002; Volume 1, pp. 469–474.
119. Paul, C.R. Decoupling the Multiconductor Transmission Line Equations. *IEEE Trans. Microwave Theory Tech.* **1996**, *44*, 1429–1440. [CrossRef]
120. AbuShaaban, M.; Scanlan, S.O. Modal Circuit Decomposition of Lossy Multiconductor Transmission Lines. *IEEE Trans. Microwave Theory Tech.* **1996**, *44*, 1046–1056. [CrossRef]
121. Toulabi, M.S.; Wang, L.; Bieber, L.; Filizadeh, S.; Jatskevich, J. A Universal High-Frequency Induction Machine Model and Characterization Method for Arbitrary Stator Winding Connections. *IEEE Trans. Energy Convers.* **2019**, *34*, 1164–1177. [CrossRef]
122. Suresh, G.; Toliyat, H.A.; Rendusara, D.A.; Enjeti, P.N. Predicting the Transient Effects of PWM Voltage Waveform on the Stator Windings of Random Wound Induction Motors. *IEEE Trans. Power Electron.* **1999**, *14*, 23–30. [CrossRef]
123. Mohammed, O.A.; Ganu, S.; Abed, N.; Liu, S.; Liu, Z. High Frequency PM Synchronous Motor Model Determined by FE Analysis. *IEEE Trans. Magn.* **2006**, *42*, 1291–1294. [CrossRef]
124. Maki, K.; Funato, H.; Shao, L. Motor Modeling for EMC Simulation by 3-D Electromagnetic Field Analysis. In Proceedings of the 2009 IEEE International Electric Machines and Drives Conference, Miami, FL, USA, 3–6 May 2009; pp. 103–108.

125. De Gersem, H.; Muetze, A. Finite-Element Supported Transmission-Line Models for Calculating High-Frequency Effects in Machine Windings. *IEEE Trans. Magn.* **2012**, *48*, 787–790. [CrossRef]
126. Rahimi, A.; Kanzi, K. Improved High-Frequency Modeling of PMSM Using 3-D Finite Element Analysis. In Proceedings of the 2019 International Power System Conference (PSC), Tehran, Iran, 9–11 December 2019; pp. 71–78.
127. Zhong, E.; Lipo, T.A. Improvements in EMC Performance of Inverter-Fed Motor Drives. *IEEE Trans. Ind. Appl.* **1995**, *31*, 1247–1256. [CrossRef]
128. Weber, S.-P.; Hoene, E.; Guttowski, S.; John, W.; Reichl, H. Modeling Induction Machines for EMC-Analysis. In Proceedings of the 2004 IEEE 35th Annual Power Electronics Specialists Conference, Aachen, Germany, 20–25 June 2004; pp. 94–98.
129. Schinkel, M.; Weber, S.; Guttowski, S.; John, W.; Reichl, H. Efficient HF Modeling and Model Parameterization of Induction Machines for Time and Frequency Domain Simulations. In Proceedings of the Twenty-First Annual IEEE Applied Power Electronics Conference and Exposition, Dallas, TX, USA, 19–23 March 2006; pp. 1181–1186.
130. Mirafzal, B.; Skibinski, G.L.; Tallam, R.M.; Schlegel, D.W.; Lukaszewski, R.A. Universal Induction Motor Model with Low-to-High Frequency-Response Characteristics. *IEEE Trans. Ind. Appl.* **2007**, *43*, 1233–1246. [CrossRef]
131. Mirafzal, B.; Skibinski, G.L.; Tallam, R.M. Determination of Parameters in the Universal Induction Motor Model. *IEEE Trans. Ind. Appl.* **2009**, *45*, 142–151. [CrossRef]
132. Sun, J.; Xing, L. Parameterization of Three-Phase Electric Machine Models for EMI Simulation. *IEEE Trans. Power Electron.* **2014**, *29*, 36–41. [CrossRef]
133. Ryu, Y.; Park, B.-R.; Han, K.J. Estimation of High-Frequency Parameters of AC Machine from Transmission Line Model. *IEEE Trans. Magn.* **2015**, *51*, 1–4. [CrossRef]
134. Vidmar, G.; Miljavec, D. A Universal High-Frequency Three-Phase Electric-Motor Model Suitable for the Delta- and Star-Winding Connections. *IEEE Trans. Power Electron.* **2015**, *30*, 4365–4376. [CrossRef]
135. Sudhoff, S.D.; Tichenor, J.L.; Drewniak, J.L. Wide-Bandwidth Multi-Resolutional Analysis of a Surface-Mounted PM Synchronous Machine. *IEEE Trans. Energy Convers.* **1999**, *14*, 1011–1018. [CrossRef]
136. Idir, N.; Weens, Y.; Moreau, M.; Franchaud, J.J. High-Frequency Behavior Models of AC Motors. *IEEE Trans. Magn.* **2009**, *45*, 133–138. [CrossRef]
137. Jia, K.; Bohlin, G.; Enohnyaket, M.; Thottappillil, R. Modeling an AC Motor with High Accuracy in a Wide Frequency Range. *IET Electr. Power Appl.* **2013**, *7*, 116–122. [CrossRef]
138. Zhao, D.; Shen, K.; Liu, W.; Lang, L.; Liang, P. A Measurement-Based Wide-Frequency Model for Aircraft Wound-Rotor Synchronous Machine. *IEEE Trans. Magn.* **2019**, *55*, 1–8. [CrossRef]
139. Rahimi, A.; Kanzi, K. High-frequency modelling of permanent magnet synchronous motor for conducted EMI studies. *IET Electr. Power Appl.* **2020**, *14*, 2027–2036. [CrossRef]
140. Wang, S.; Lee, F.C.; Odendaal, W.G. Characterization and Parasitic Extraction of EMI Filters Using Scattering Parameters. *IEEE Trans. Power Electron.* **2005**, *20*, 502–510. [CrossRef]
141. Ansys Q3D Extractor, Version 2020 R2. Available online: https://www.ansys.com/products/electronics/ansys-q3d-extractor (accessed on 21 January 2021).
142. CST Studio Suite. Available online: https://www.3ds.com/products-services/simulia/products/cst-studio-suite/ (accessed on 21 January 2021).
143. Dellis, J.-L. Zfit. Available online: https://www.mathworks.com/matlabcentral/fileexchange/19460-zfit (accessed on 21 January 2021).
144. Ma, C.L.; Lauritzen, P.O.; Sigg, J. Modeling of Power Diodes with the Lumped-Charge Modeling Technique. *IEEE Trans. Power Electron.* **1997**, *12*, 398–405. [CrossRef]
145. Palmer, P.R.; Santi, E.; Hudgins, J.L.; Kang, X.; Joyce, J.C.; Eng, P.Y. Circuit Simulator Models for the Diode and IGBT with Full Temperature Dependent Features. *IEEE Trans. Power Electron.* **2003**, *18*, 1220–1229. [CrossRef]
146. Bryant, A.T.; Kang, X.; Santi, E.; Palmer, P.R.; Hudgins, J.L. Two-Step Parameter Extraction Procedure with Formal Optimization for Physics-Based Circuit Simulator IGBT and p-i-n Diode Models. *IEEE Trans. Power Electron.* **2006**, *21*, 295–309. [CrossRef]
147. Prada, D.; Bellini, M.; Stevanovic, I.; Lemaitre, L.; Victory, J.; Vobecky, J.; Sacco, R.; Lauritzen, P.O. On the Performance of Multiobjective Evolutionary Algorithms in Automatic Parameter Extraction of Power Diodes. *IEEE Trans. Power Electron.* **2015**, *30*, 4986–4997. [CrossRef]
148. Budihardjo, I.; Lauritzen, P.G. The Lumped-Charge Power MOSFET Model, Including Parameter Extraction. *IEEE Trans. Power Electron.* **1995**, *10*, 379–387. [CrossRef]
149. McNutt, T.R.; Hefner, A.R.; Mantooth, H.A.; Berning, D.; Ryu, S.-H. Silicon Carbide Power MOSFET Model and Parameter Extraction Sequence. *IEEE Trans. Power Electron.* **2007**, *22*, 353–363. [CrossRef]
150. Gachovska, T.; Hudgings, J.L.; Bryant, A.; Santi, E.; Mantooth, H.A.; Agarwal, A.K. Modeling, Simulation, and Validation of a Power SiC BJT. *IEEE Trans. Power Electron.* **2012**, *27*, 4338–4346. [CrossRef]
151. SIMetrix/SIMPLIS SPICE Simulator Version 8.3. Available online: https://www.simetrix.co.uk/ (accessed on 21 January 2021).
152. LTSpice SPICE Simulator. Available online: https://www.analog.com/en/design-center/design-tools-and-calculators/ltspice-simulator.html# (accessed on 21 January 2021).
153. Sheng, K.; Williams, B.W.; Finney, S.J. A Review of IGBT Models. *IEEE Trans. Power Electron.* **2000**, *15*, 1250–1266. [CrossRef]

154. Cavaiuolo, D. Compact Electro-Thermal Modeling of IGBT for Application Circuit Effective Design. Ph.D. Thesis, University of Naples Federico II, Naples, Italy, 2014.
155. Lauritzen, P.O.; Andersen, G.K.; Helsper, M. A Basic IGBT Model with Easy Parameter Extraction. In Proceedings of the IEEE Annual Power Electronics Specialists Conference, Vancouver, BC, Canada, 17 June 2001; pp. 2160–2165.
156. Chibante, R.; Araujo, A.; Carvalho, A. Finite-Element Modeling and Optimization-Based Parameter Extraction Algorithm for NPT-IGBTS. *IEEE Trans. Power Electron.* **2009**, *24*, 1147–1427. [CrossRef]
157. Chimento, F.; Mora, N.; Bellini, M.; Stevanovic, I.; Tomarchio, S. A Simplified Spice Based IGBT Model for Power Electronics Modules Evaluation. In Proceedings of the IECON 2011—37th Annual Conference of the IEEE Industrial Electronics Society, Melbourne, VIC, Australia, 7–10 November 2011; pp. 1155–1160.
158. Duan, Y.; Xiao, F.; Luo, Y.; Iannuzzo, F. A Lumped-Charge Approach Based Physical SPICE-Model for High Power Soft-Punch Through IGBT. *IEEE J. Emerg. Sel. Top. Power Electron.* **2019**, *7*, 62–70. [CrossRef]
159. Duan, Y.; Iannuzzo, F.; Blaabjerg, F. A New Lumped-Charge Modeling Method for Power Semiconductor Devices. *IEEE Trans. Power Electron.* **2020**, *35*, 3989–3996. [CrossRef]
160. Han, L.; Liang, L.; Kang, Y.; Qui, Y. A Review of SiC IGBT: Models, Fabrications, Characteristics, and Applications. *IEEE Trans. Power Electron.* **2021**, *36*, 2080–2093. [CrossRef]
161. Xu, Y.; Ngai-Man Ho, C.; Ghosh, A.; Muthumuni, D. An Electrical Transient Model of IGBT-Diode Switching Cell for Power Semiconductor Loss Estimation in Electromagnetic Transient Simulation. *IEEE Trans. Power Electron.* **2020**, *35*, 2979–2989. [CrossRef]
162. Cao, H.; Ning, P.; Wen, X.; Yuan, T.; Li, H. An Electrothermal Model for IGBT Based on Finite Differential Method. *IEEE J. Emerg. Sel. Topics Power Electron.* **2020**, *8*, 673–684. [CrossRef]
163. Jing, L.; Du, M.; Wei, K.; Hurley, W.G. An Improved Behavior Model for IGBT Modules Driven by Datasheet and Measurement. *IEEE Trans. Electron Devices* **2020**, *67*, 230–236. [CrossRef]
164. Miyaoku, Y.; Tone, A.; Matsuura, K.; Miura-Mattausch, M.; Mattausch, H.J.; Ikoma, D. Compact Modeling of IGBT Charging/Discharging for Accurate Switching Prediction. *IEEE J. Electron Devices Soc.* **2020**, *8*, 1373–1380. [CrossRef]
165. Infineon IGBT Devices and Models. Available online: https://www.infineon.com/cms/en/product/power/igbt/igbt-discretes/#!simulation (accessed on 11 March 2021).
166. Kraus, R.; Turkes, P.; Sigg, J. Physics-based models of power semiconductor devices for the circuit simulator SPICE. In Proceedings of the 29th Annual IEEE Power Electronics Specialists Conference, Fukuoka, Japan, 22 May 1998; Volume 2, pp. 1726–1731. [CrossRef]
167. Shi, B.; Zhao, Z.; Zhu, Y. Piecewise Analytical Transient Model for Power Switching Device Commutation Unit. *IEEE Trans. Power Electron.* **2019**, *34*, 5720–5736. [CrossRef]
168. Krug, F.; Russer, P. The Time-Domain Electromagnetic Interference Measurement System. *IEEE Trans. Electromagn. Compat.* **2003**, *45*, 330–338. [CrossRef]
169. Heldwein, M.L.; Biela, J.; Ertl, H.; Nussbaumer, T.; Kolar, J.W. Novel Three-Phase CM/DM Conducted Emission Separator. *IEEE Trans. Ind. Electron.* **2009**, *56*, 3693–3703. [CrossRef]
170. Wang, S.; Lee, F.C.; Odendaal, W.G. Characterization, evaluation, and design of noise separator for conducted EMI noise diagnosis. *IEEE Trans. Power Electron.* **2005**, *20*, 974–982. [CrossRef]
171. Forsstrom, V.; Skibin, S.; Wunsch, B. Influence of Parasitic Coupling to Ground Plane on EMC Noise of Power Converters. In Proceedings of the International Symposium on Electromagnetic Compatibility-EMC EUROPE, Rome, Italy, 23–25 September 2020; pp. 1–5.
172. The Power Cables are the MCCMK 3x16/16 RM 1 kV and the MCMK 3x70/35 AN 1 kV Listed on Page 28 and 31 of the Catalogue. Available online: http://media.drakakeila.ee/2016/01/Draka_Cataloque_EN.pdf (accessed on 11 March 2021).

energies

Article

Optimization of a Gate Distribution Layout to Compensate the Current Imbalance Generated by the 3D Geometry of a Railway Inverter

Andressa Nakahata-Medrado [1,2], Jean-Luc Schanen [1,*], Jean-Michel Guichon [1], Pierre-Olivier Jeannin [1], Emmanuel Batista [2] and Guillaume Desportes [2]

[1] G2ELab, University Grenoble Alpes, 21 Av des Martyrs, 38000 Grenoble, France; andressa_nakahata@hotmail.com (A.N.-M.); jean-michel.guichon@g2elab.grenoble-inp.fr (J.-M.G.); pierre-olivier.jeannin@g2elab.grenoble-inp.fr (P.-O.J.)
[2] Alstom Transport, 21 rue Dr Guinier, 65600 Semeac, France; emmanuel.batista@alstomgroup.com (E.B.); guillaume.desportes@alstomgroup.com (G.D.)
* Correspondence: jean-luc.schanen@grenoble-inp.fr

Abstract: The impact of the stray inductances originated from interconnects in power electronics becomes crucial with the next generation of SiC devices. This paper shows that the existing layout of a railway inverter, operating with Si IGBTs already exhibits a dynamic current imbalance between paralleled modules. This will not allow using this geometry with SiC MOSFETs. A complete investigation of the electromagnetic origin of this issue has been performed. A generic circuit model has been proposed to establish a cabling rule to design a Gate Distribution Printed Circuit Board (PCB) in such a way that it compensates the power dissymmetry. An optimization strategy has been used to obtain a new geometry of this PCB, which has been validated with a time domain simulation.

Keywords: power electronics; PEEC method; optimization; paralleling devices; 3D layout; stray inductance; coupling

Citation: Nakahata-Medrado, A.; Schanen, J.-L.; Guichon, J.-M.; Jeannin, P.-O.; Batista, E.; Desportes, G. Optimization of a Gate Distribution Layout to Compensate the Current Imbalance Generated by the 3D Geometry of a Railway Inverter. *Energies* **2021**, *14*, 1891. https://doi.org/10.3390/en14071891

Academic Editor: Andrea Mariscotti

Received: 18 February 2021
Accepted: 25 March 2021
Published: 29 March 2021

1. Introduction

The European project "Shift2rail" [1] is investigating the advantages of the SiC wide bandgap devices in railway applications. The potential benefits of using SiC MOSFETs in future converters are explored, but also the mandatory impacts on the 3D layout design. With the high speed commutation of these new switches, there is a strong need of properly designing a geometrical arrangement of the various parts of a converter, in order to prevent all consequences of the stray elements originated from the cabling. Voltage overshoots, ringing, and power drive interactions are major concerns of power electronics designers in the "wide bandgap era" [2–4].

Among the issues arising from high switching speed in high current applications, the current division among paralleled power modules is a key topic, which has to be addressed carefully to avoid any derating of the devices [5].

Several parameters may generate a bad current division among power modules [6]:

- difference between device's parameters due to industrial processes or different operating temperatures,
- the power layout which provides various current paths among the paralleled devices and therefore different loop inductances, leading to a different current commutation speed (dI/dt),
- the gate circuit layout, including the common emitter and the mutual coupling between gate and power circuits: both effects modify the gate-to-emitter voltage, and consequently the dI/dt.

All these phenomena have been investigated for many years for Si IGBTs, and more recently for SiC MOSFETs. The impact of power layout dissymmetry has been reported in [7–10]. It was shown that an electrical symmetry (i.e., equivalent loop inductance seen from each devices) can be obtained even with non-symmetrical geometries, taking into account the effect of mutual inductances. However, this necessitates many degrees of freedom for the 3D placement of the power modules, which are not always available in a railway inverter. Indeed, compact solutions are requested, and the cold plate of the cooling system often imposes the modules arrangement.

Another very important origin of the bad dynamic current division is due to the gate driver. Many papers have addressed the issue of the common inductance between power and gate circuits on the emitter (or source) pin [11–14]. Of course, this phenomena has mainly been studied for the internal layout of the power module, and has more impact on the current division among the dies than the difference between several modules. However, these papers also described the impact of the mutual coupling between gate and power circuits, which is another origin of power-drive interaction. In other words, reducing the common emitter impedance is not sufficient to avoid the feedback effect from power to gate during transients, and the impact of the magnetic couplings has to be considered. This has been done in [15], where a metric has been defined in order for all the dies inside the power module to be submitted to the same feedback voltage from the power circuit. Therefore, using mutual couplings, and despite different values of common emitter inductances, all gate to emitter voltages during transients were identical, leading to the same dI/dt. Defining a gate circuit layout according to this metric was obtained, using an optimization process, for a fixed power layout.

Nevertheless, in a general situation, the current imbalance among power devices during transients combines both effects of power layout and power-drive interaction. To elucidate these joint effects, it is necessary to add the device's behavior. This was done in [16] to account for electro-thermal effects, the device's parameters being modified with temperature.

The present paper will propose a full investigation of the impact of all inductive stray effects generated by both power and gate layout. It will be shown that a dynamic current imbalance generated by the power layout can be compensated by properly designing a gate circuit layout. The electrical equations developed in this paper are more or less similar to the recent work reported in [17] and will be used in an optimization process to obtain an improved current division, despite the current imbalance caused by the power layout.

The approach to get the circuit equations must start with a clear link between the 3D geometry and the electrical representation. Partial Element Equivalent Circuit (PEEC) [18] is today the classical modeling approach to reach this goal. It allows for the representation of any conductor with a partial inductance, and for mutual couplings with the rest of the circuit. However, if the transfer from a simple interconnection linking two points to its electrical model is straightforward, multiple port connections, as busbars, must be clarified. This paper will present a generic electrical representation for this kind of multiport conductors, which is implemented in many PEEC software but not reported in the literature so far.

To support the proposed method, a 1700 V 1200 A railway inverter, *ONIX 671*, will be used as an example. Even if this inverter is not based on SiC devices, it exhibits some current imbalance, whose origin will be investigated thanks to simulation (Section 2). In Section 3, the generic circuit model derived from the 3D layout will be presented. Equations got from the model will be used to obtain a cabling rule to guarantee balanced currents during transients, despite a non-symmetric power layout. This cabling rule will be the objective function of the optimization process carried out in Section 4.

2. Inverter Description

The 1700 V 1200 A studied inverter is a 3 phases + 1 brake chopper topology. The electrical representation of the inverter is given in Figure 1, and the 3D layout in Figure 2. Several parts can be identified:

- An Electrolytic Capacitor Bank made of 54 devices (18 groups of three series capacitors), connected to the two HT+ and the two HT− terminals (DC bus);
- two identical subparts, each of them realizing two legs. One sub part is composed of six power modules (three paralleled devices for top and bottom switches), eight decoupling capacitors, and three busbars for linking the aforementioned elements: one power busbar (DC bus) and two phase busbars for connection to the output phases.
- Each subpart contains drivers for controlling the IGBTs. A specific Printed Cicuit Board (PCB) carries the driver signal to the three paralleled devices for the top and bottom switch of each phase. Two Gate Distribution PCBs are thus used for each subpart.
- A cold plate is supporting all 12 power modules (dry natural cooling system).

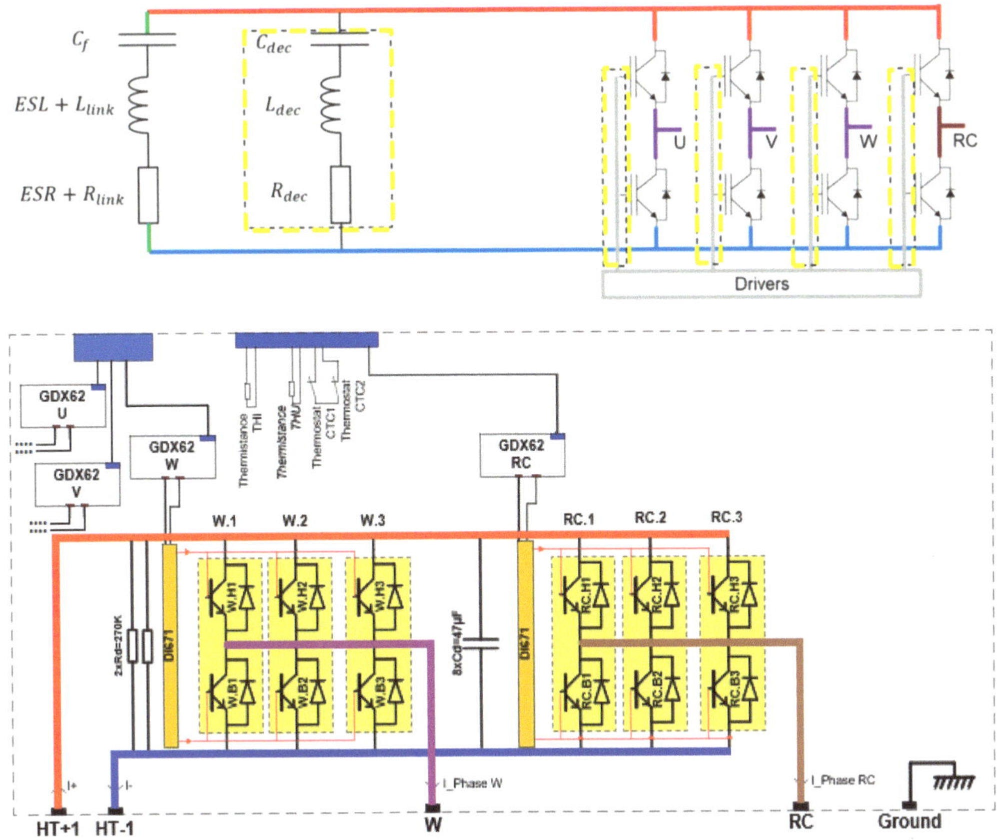

Figure 1. Railway inverter composition. The global function is a three-phase inverter with one additional non-reversible leg for braking. Each leg uses three paralleled IGBT modules. The full converter is built from two identical subparts composed of two legs and decoupling capacitors (bottom of the figure). A common DC capacitor tank is used to provide the energy storage at DC-bus level.

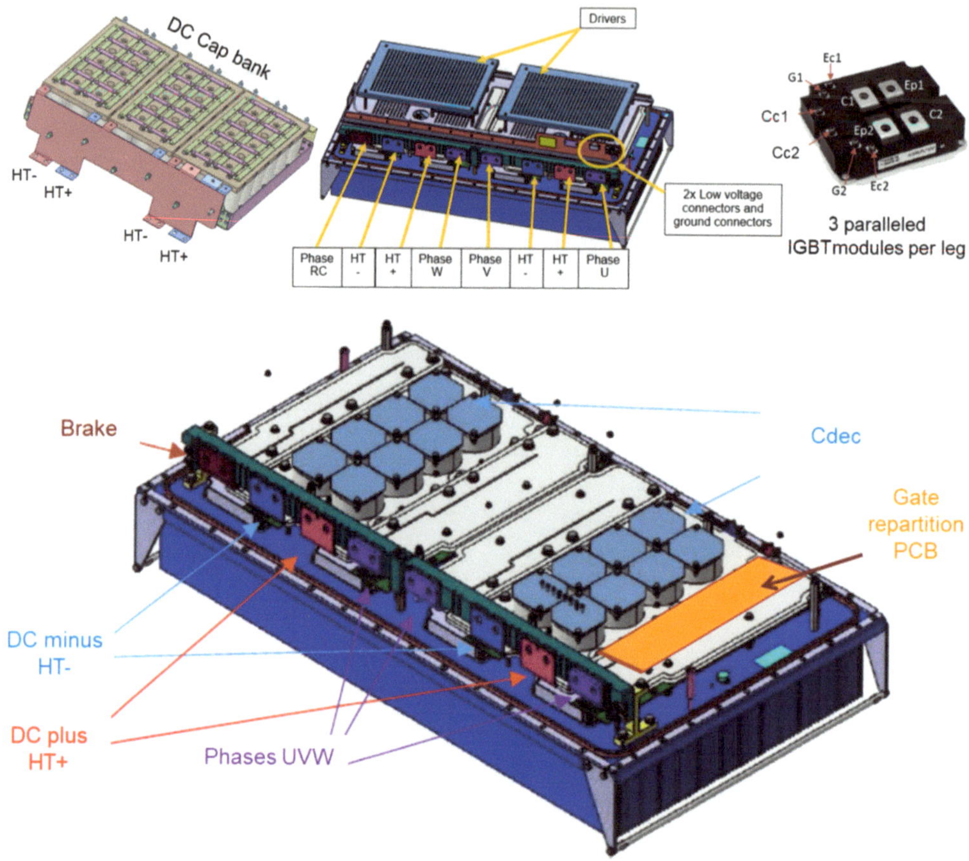

Figure 2. (**Top**: **left**), DC capacitor bank with duplicated DC bus connections, one for each subpart. (**Center**), overview of the global 3D layout. (**Right**), IGBT power Module. (**Bottom**): 3D layout of the power part.

Figure 2 shows the 3D representation of the inverter. The Electrolytic Capacitor Bank associates three groups of capacitors realized with simple copper bars and is linked to the two inverter subparts through a dedicated busbar. The HT+ and HT− connections are therefore duplicated. The IGBT module is also displayed in Figure 2. It offers separate Emitter and Collector pins for top and bottom switches (E1, C1 and E2, C2), the series connection being made by the phase busbar. A Kelvin connection is of course provided for each emitter of each IGBT (Ec1 and Ec2). Gate pins G1 and G2 as well as separate collectors Cc1 and Cc2 for control purposes are also indicated in the figure. On the bottom of Figure 2, one can see the inverter power layout, which is split into two parts, each of them associating two inverter legs. For each subpart, six power modules are fixed on the cold plate and interconnected with a power busbar in the center and two phase busbars on both sides, which also provide the connection between all top emitters and bottom collectors of the three paralleled IGBTs. The power busbar associates the power module and eight decoupling capacitors, and provides links to the DC capacitor tank. The gate drivers are on the top, and a specific PCB is used for interfacing them with the three paralleled IGBTs. These four Gate Distribution PCBs are localized directly over the phase busbars.

The cold plate is not much larger than the 12 power modules. Therefore, there are almost no degrees of freedom for their geometrical implementation. The DC power busbar, linking the decoupling capacitors Cdec to the power modules, has to provide both small loop inductance and as much symmetry as possible for ensuring an equal current division between power modules. As mentioned before, the power modules are not providing the series association of the two IGBTs of the leg, and the electrical link between the Emitter of switch 1 and the Collector of switch 2 is performed through a part of the phase busbar. The switching cell is thus composed of Decoupling Capacitors, Power Busbar, and a part of the Phase Busbar.

It is worth noting that all phase busbars exhibit a slit, visible on Figure 3. This one has been inserted for managing the static current division (DC resistance) between the three paralleled IGBTs. Indeed, the current path from the power modules to the phase output will have almost the same length, whereas without the slit the module close to the power connections would see a much shorter path than the most distant one.

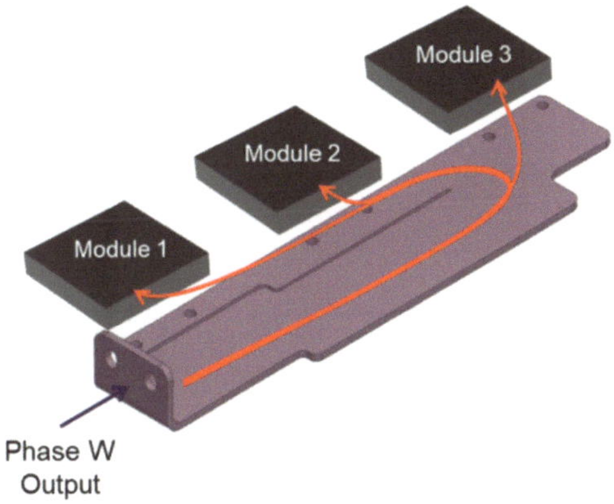

Figure 3. Zoom of the slit in the phase busbar to provide almost equal length for the current path between the phase output and the power modules.

In the experiment, the operations of this converter were not fully satisfying regarding the current division among the three paralleled modules. A huge experimental investigation was carried out to check if this current imbalance was due to the devices' properties, but even when carefully sorting IGBTs with the same characteristics, the issue remained.

The 3D layout is thus at the origin of the bad current distribution among the three switches. Moving to SiC MOSFET with this kind of layout is not possible because the influence of the cabling will be much more important with a higher commutation speed. Understanding the root cause of the issue is therefore mandatory. For this purpose, an investigation using a simulation has been carried out, using Ansoft Q3D [19] and the associated time domain simulator Simplorer.

For this purpose, it was decided to focus on the bottom switch of phase W, composed of 3 paralleled IGBTs. The power layout is composed of the DC busbars and the phase W busbar. IGBT power modules have been simplified and represented by simple lumped inductances, to avoid overloading the 3D simulation model. This simplification is not a real issue, since all modules are identical: this is not the origin of the current dissymmetry. The eight decoupling capacitors have been identified through impedance measurement (simple esl–esr model [20]) and their equivalent circuit is connected to the appropriate ports of the 3D PEEC model. The DC capacitor bank has also be simplified: from Figure 2, it can be seen that the parallel association of 18 groups of three series capacitors is realized with three blocks of 6 * 3 series components. An equivalent model of each block has been identified (again, equivalent series inductance and resistance), and only the DC capacitor busbar (Figure 2) has been modeled in Ansoft Q3D. The power busbar and phase W busbar are also included in the 3D description. It is worth noting that to save memory space, the DC capacitor busbar has been modeled in a separate file, since its mutual coupling with the rest of the power structure is negligible.

The gate driver is replaced by a perfect voltage source, common to each of the three paralleled IGBTs. The link between this voltage source and each of the three devices is made by the Gate Distribution PCB (Figure 4 represents the part of this PCB corresponding to the bottom switches). It is a pure passive PCB, connecting the drive signal to each IGBT, and providing external gate resistances (some additional feedback and protection circuits are not studied here). It was included in the full 3D model, together with power and phase W busbars.

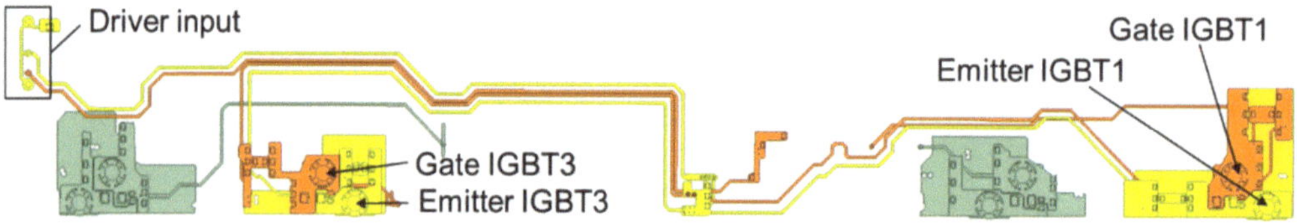

Figure 4. Gate Distribution PCB, collecting the driver signal and linking it to each of the three paralleled IGBTs.

An illustration of the electrical simulation with all aforementioned blocks is provided in Figure 5.

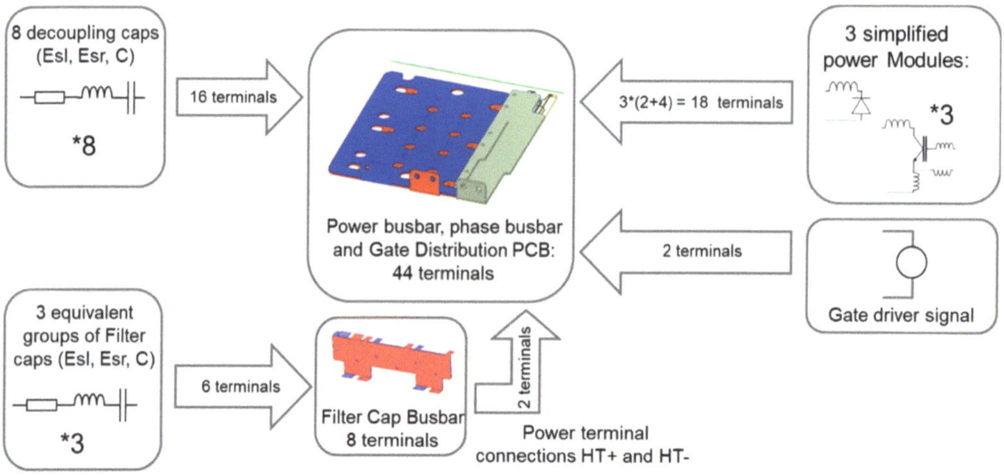

Figure 5. Time domain simulation scheme. Two Partial Element Equivalent Circuit (PEEC) models used from separate files: power and phase busbar plus Gate Distribution PCB and DC capacitor tank busbar. Equivalent esr–esl models of the decoupling capacitors and the three blocks of the DC Capacitor tank. "Home made" IGBT and diode models are used. Idealized drive signal is a simple voltage source.

Figure 6 shows the simulation results for the three IGBTs. It is obvious in this figure that IGBT 3 (far from the phase output and close to the driver input) is slower than the two other ones, resulting in higher switching losses at turn off. IGBT 2 (in the middle) is the quickest one, and consequently takes the major part of the diode recovery current.

To understand the root cause of the imbalance and determine if was due to the power layout or to the power-drive interaction, a second simulation was done, removing the Gate Distribution PCB from the 3D representation and directly controlling the IGBTs by the ideal driver voltage source. Therefore, only the power layout was impacting the current division. Simulation results are given in Figure 7. Again, current division is not perfect, but this time the slowest device is not IGBT3, as in Figure 6, but IGBT 1. Therefore, the current imbalance is necessarily affected by both effects, power and gate circuits.

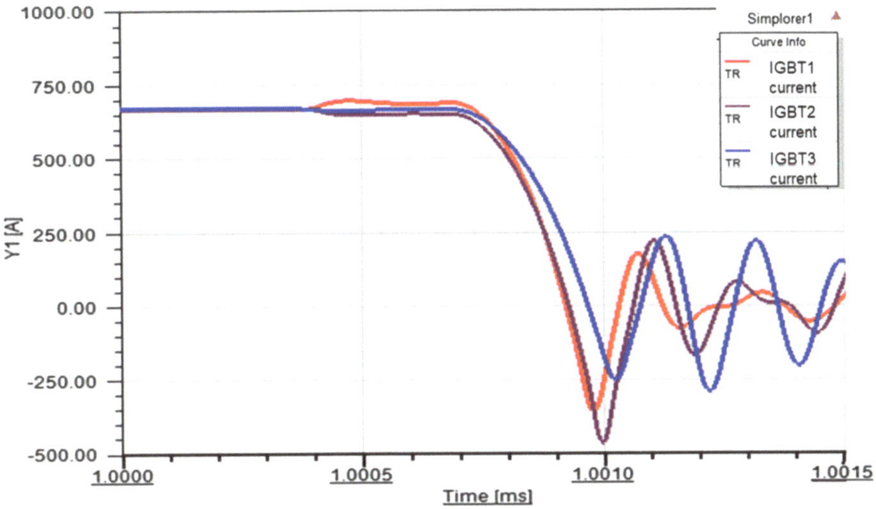

Figure 6. Simulated results of current division among the three bottom switches of phase W. Full 3D model.

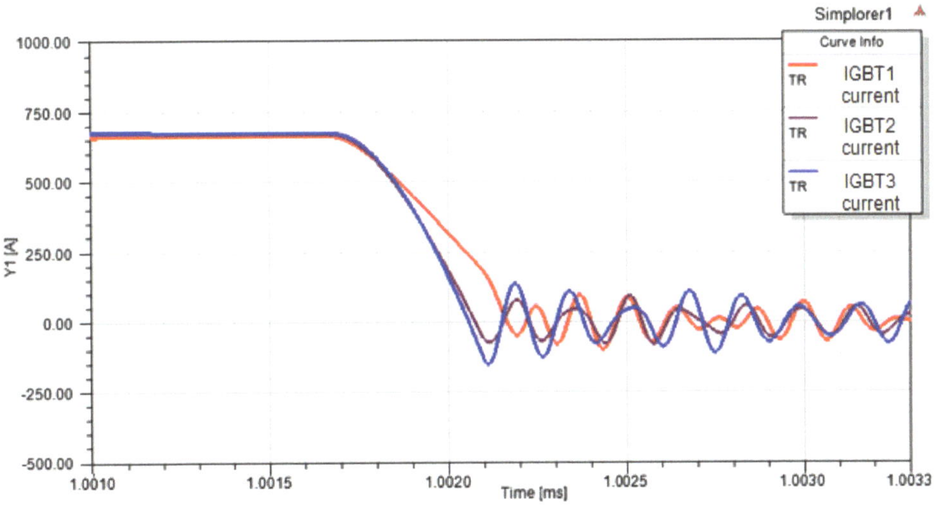

Figure 7. Simulated results of current division among the three bottom switches of phase W with power layout only, identical idealized gate signals.

A last simulation was also carried out to confirm the impact of mutual couplings in the power-drive interaction. The Gate Distribution PCB was described in a separate file in Q3D, therefore keeping the effects of its impedance, but not considering the mutual couplings with the power layout. The results of Figure 8 show that IGBT1 is still the slowest, as with the perfect gate circuit, which clearly shows that the most impacting effect is due to this coupling between Power and Gate Distribution PCB. This is not surprising considering the position of this PCB, directly on the top of the phase busbar, which is part of the switching cell, so carrying high dI/dt.

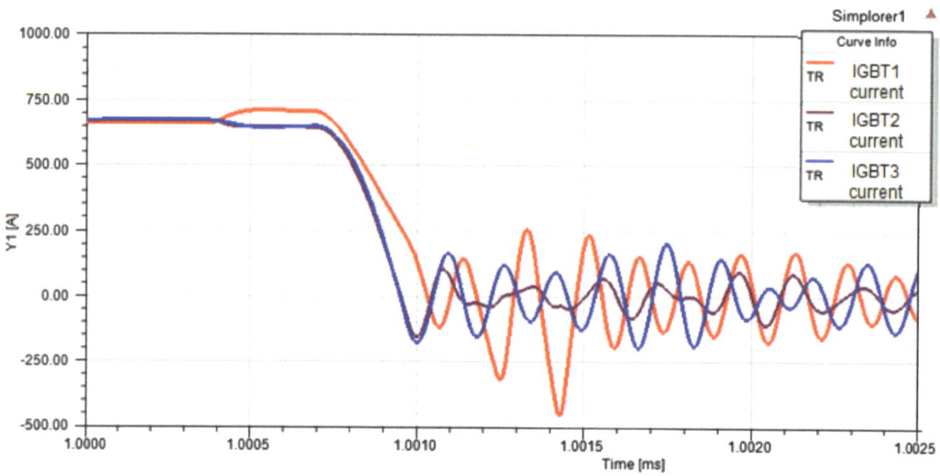

Figure 8. Simulated results of current divisions among the three bottom switches of phase W with power layout and Gate Distribution PCB layout—no coupling considered.

From these simulations, it can be concluded that:

- the power layout generates a current imbalance, IGBT 3 being slower than the two others
- the couplings between this PCB and the power part actually modify the current division, since taking them into account leads to a real change in the slowest device (IGBT 1 instead of IGBT 3). They must therefore be considered.

The power layout is hard to modify, due to a lack of space and the mandatory position of the power modules on the cold plate. It has been shown that it is generating a current imbalance, but that this one is reversed when considering the Gate Driver PCB. Therefore, there is a possibility to compensate the imbalance caused by the power layout, using the coupling with the gate circuit. However, to guarantee this compensation, it is first necessary to consider both effects in a circuit model, combining an electrical representation of the power layout, the gate layout, their couplings, and the IGBT model. This is the aim of the following section.

3. Electrical Model of Paralleled IGBTs

Obtaining the circuit model from a 3D layout is quite straightforward using the PEEC approach for a simple conductor linking two points: the conductor is replaced by a resistance, an inductance, and mutual coupling with the rest of the circuit. However, in the kind of 3D geometry used in Figure 2, some conductors—as busbars, for instance—are not linking only two points but several points. Consequently, the electrical representation becomes tricky. In some cases, the geometrical layout is organized in such a way that the user can "guess" a circuit representation almost corresponding to the current path and the layout (Figure 9 top). However, in the case of plates, for instance, the current path is unknown, and the equivalent circuit topology not straightforward. A possibility would be to connect all points with R-L coupled circuits, as in Figure 9 bottom, but this is clearly not the representation using the minimal number of elements. The first subsection will propose a generic approach to get an electrical circuit from the 3D layout, based on a "terminal behavior" vision. This generic representation is used in several existing PEEC softwares, as Ansoft Q3D [19] or Altair Flux-PEEC [21], but has not been explained in such detail so far.

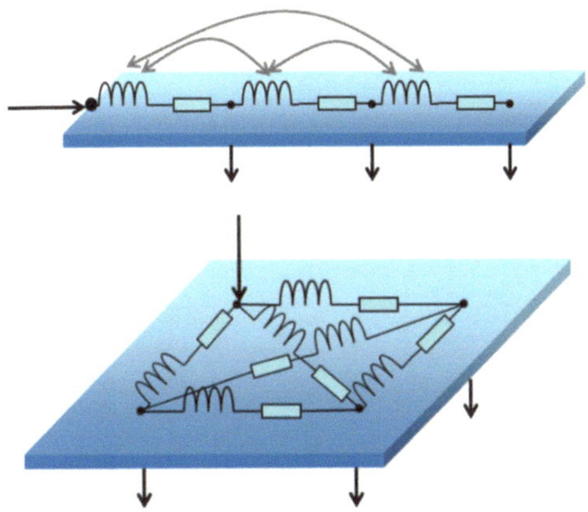

Figure 9. Electrical representation of conductors linking more than two points. (**Top**), the current path is quite obvious, and an electrical representation can be "guessed". (**Bottom**), the case of a plate with no straightforward representation, except for the "heavy" full connection of each point with R-L coupled elements (mutual coupling not displayed for clarity).

This circuit representation of both power and gate circuits will be combined with an IGBT conventional model in order to investigate the current division in the simplified case of two paralleled IGBTs. From this analysis, a rule on the impedance matrix will be derived to reach balanced currents in each IGBT.

3.1. Generic Circuit Representation

The results of the previous sections show the importance of considering all stray inductive elements involved in the dynamic behavior of the switching, including power, layout, and mutual couplings with the gate circuits. It is thus mandatory to take all of them into account. For simplification and illustration purposes, only two paralleled power modules will be considered in this section, even if, obviously, the generalization to three and more is quite straightforward with the proposed equations and matrix writing. We will focus on the IGBT 1 and the IGBT 3 of the bottom switch of phase W to illustrate the geometry.

The chosen electrical circuit representation of a conductor linking N_port > 2 will be composed of N_port-1 inductors and resistors. One point among the I/O ports will be considered as a reference and linked with the others with R-L circuits (and couplings). The reference point can be chosen arbitrarily. Figure 10 shows a simple illustration for the basic case of two plates associating three capacitors in parallel.

It is worth noting that this method is the one used in most PEEC-based softwares to generate Pspice-compatible netlists. One further remark is that in this representation, the full PEEC matrix exhibits both a real and an imaginary part. In other words, not only the inductors but also the resistors are coupled. Physically, this corresponds to a common resistive path within the same conductor and eddy currents induced in the rest of the geometry. These "resistive couplings" can be represented in circuit equations with current-controlled voltage sources (Figure 11), or by writing differential equations in a programming language (VHDL AMS or MAST for Saber software).

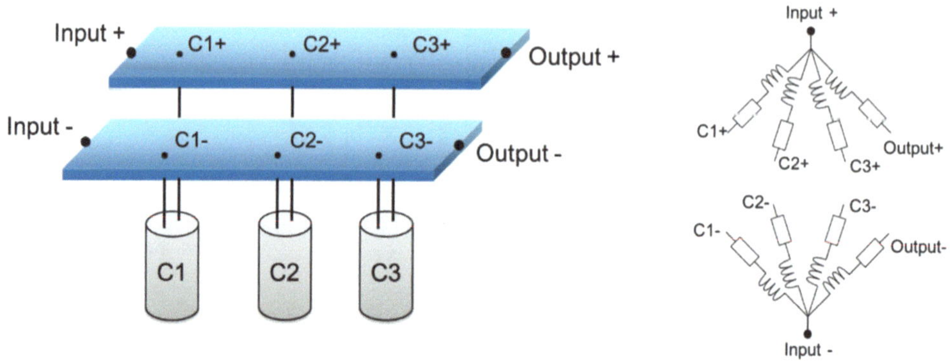

Figure 10. Generic representation, based in an input/output port choosing an arbitrary reference port. This representation is NOT linked with any geometrical current path.

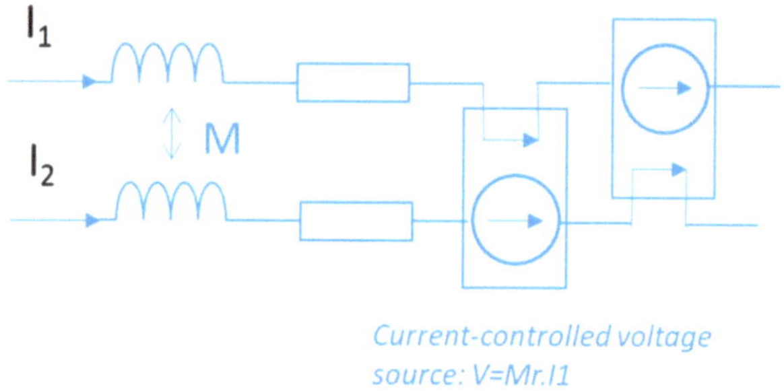

Figure 11. A possible way to reproduce "resistive couplings" in Pspice-compatible representations.

3.2. Application to Two Paralleled IGBTs

Let's consider two paralleled IGBTs connected between phase W and the HT− negative potential (IGBT 1 and IGBT 3 of the full geometry). The PEEC method was applied to model the power busbar as well as the phase busbar W. The power module was also described using a PEEC model.

For the power busbar, the phase W connecting terminal was chosen as reference point for linking it to both collectors C1 and C2, and the HT− potential for linking to both power emitters, E1 and E2. For the power module, collectors and gate pins are simple connections linking the die to the external pin. Only Emitters have to be linked to the Die, Control Emitter, and Power Emitter pins. In this case, the reference point was the emitter die. Therefore, the circuit of Figure 12 left can be obtained, which can be reduced to the circuit of Figure 12 right by associating all impedances in series.

An important note is that this arbitrary circuit representation does not correspond to the geometrical layout: for instance, a part of the circuit can be common to the power and gate parts in the module (the contribution of a portion of DBC or bondings). This is not seen as a common inductance in the equivalent circuit of Figure 12, but it is taken into account in the value of mutual impedance between stray elements. This is illustrated further in Figure 13 for more clarity. In this figure, a part of the emitter path is common to the power and gate circuits. A geometrical representation would be as in Figure 13 left, but the generic representation is always the same whatever the layout geometry.

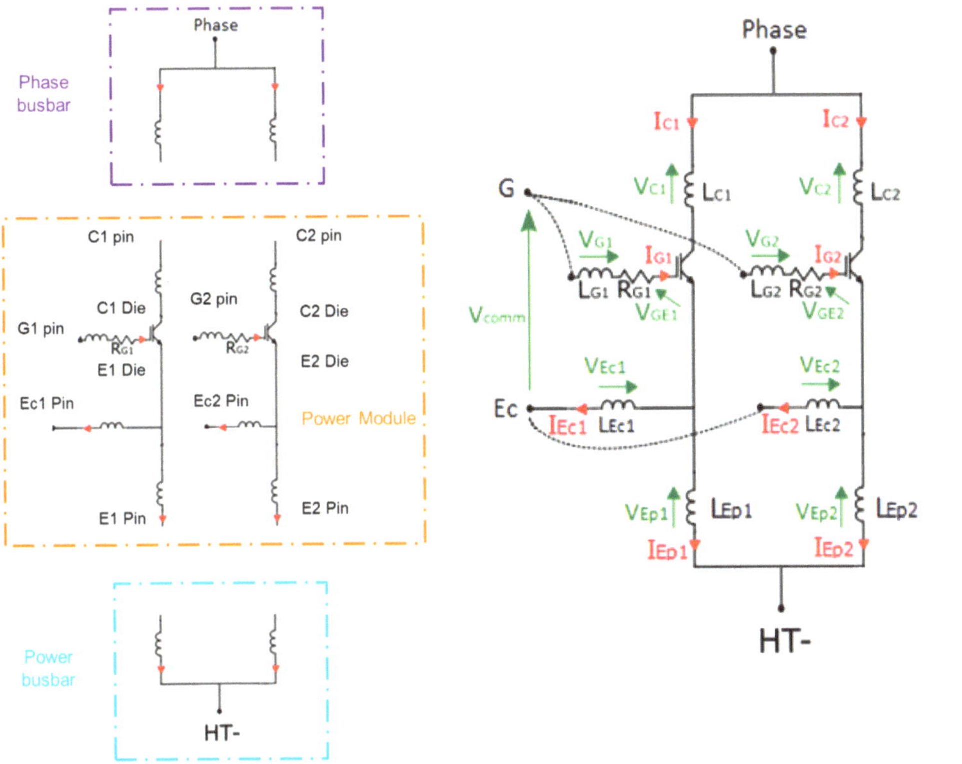

Figure 12. (**Left**): individual models of each part of the system. (**Right**): reduction of the model by associating series elements.

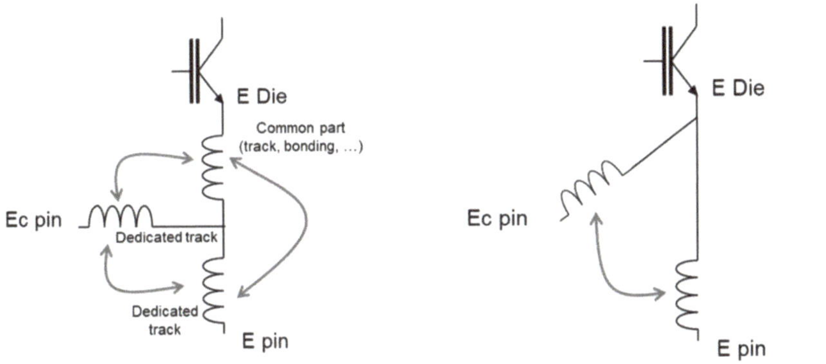

Figure 13. Example of a PEEC representation in the case of classical common emitter coupling. The (**left**) representation is based on the geometry and the proposed generic representation; (**right**) is based on terminals only and is independent from the geometry. Both account for exactly the same phenomena.

3.3. Circuit Equations

The method starts with considering the power circuit only, including a semiconductor model composed of a current source and the output capacitor, as illustrated in Figure 14.

$$I_{mos} = gm(V_{GE} - V_{th}) \quad (1)$$

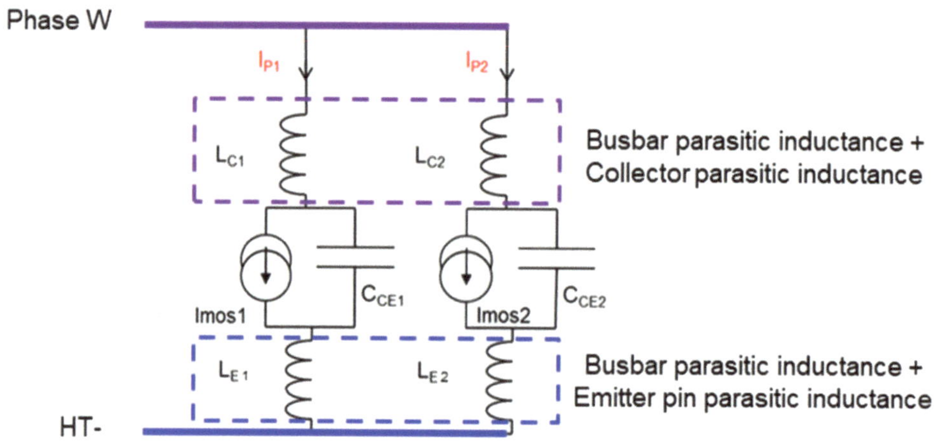

Figure 14. Representation of the power part only, including a simple IGBT model.

To simplify writing, only inductive elements are displayed and considered in the equations, but all resistive elements and resistive couplings could be easily taken into account.

The expressions of the two power currents, I_{P1} and I_{P2}, can be written as follows.

$$I_{P1} = \mathrm{gm}_1(V_{GE1} - V_{th1}) + C_{CE1}\left(-a\frac{d^2 I_{P1}}{dt^2} - b\frac{d^2 I_{P2}}{dt^2}\right) \quad (2)$$

$$I_{P2} = \mathrm{gm}_2(V_{GE2} - V_{th2}) + C_{CE2}\left(-c\frac{d^2 I_{P1}}{dt^2} - d\frac{d^2 I_{P2}}{dt^2}\right) \quad (3)$$

with

$a = L_{C1} + L_{Ep1} + 2M_{C1Ep1}$
$b = M_{C1C2} + M_{C1Ep2} + M_{Ep1C2} + M_{Ep1Ep2}$
$c = M_{C2C1} + M_{C2Ep1} + M_{Ep2C1} + M_{Ep2Ep1}$
$d = L_{C2} + L_{Ep2} + 2M_{C2Ep2}$

according to the notations of Figure 10 right, keeping the power elements only.

Considering identical devices, $C_{CE1} = C_{CE2}$, $\mathrm{gm}_1 = \mathrm{gm}_2$, and $V_{th1} = V_{th2}$. Furthermore, instead of solving the equation in the time domain, the frequency domain is preferred. Therefore, it leads to

$$I_{P1} = \frac{gm(V_{GE1} - V_{th}) - C_{CE} b\omega^2 I_{P2}}{1 + C_{CE} a\omega^2} \quad (4)$$

$$I_{P2} = \frac{gm(V_{GE2} - V_{th}) - C_{CE} c\omega^2 I_{P1}}{1 + C_{CE} d\omega^2} \quad (5)$$

Finally, combining (8) and (9), we obtain

$$I_{P1} = \frac{gm}{C_{CE}\omega^2(ad - bc)}[d(V_{GE1} - V_{th}) - b(V_{GE2} - V_{th})] \quad (6)$$

$$I_{P2} = \frac{gm(V_{GE2} - V_{th})}{C_{CE}\omega^2 d} - \frac{c}{d}\left[\frac{gm}{C_{CE}\omega^2(ad - bc)}[d(V_{GE1} - V_{th}) - b(V_{GE2} - V_{th})]\right] \quad (7)$$

Considering that the dynamic current share phenomena involves only high frequencies, $C_{DS}.a.\omega^2 >> 1$ and $C_{DS}.d.\omega^2 >> 1$. If we want to have $I_{P1} = I_{P2}$, it leads to

$$V_{GE1} = \left(\frac{a+b}{c+d}\right)V_{GE2} + V_{th}\left(1 - \left(\frac{a+b}{c+d}\right)\right) \quad (8)$$

This means that to obtain an equal dynamic share despite dissymmetric layouts, it is possible to act on the gate voltages in the following way:

$$V_{GE1} = \alpha V_{GE2} + V_{th}(1 - \alpha) \quad (9)$$

with $\alpha = \left(\frac{a+b}{c+d}\right)$

It is worth noting that if the power layout is symmetrical, $a = d$ and $b = c$; therefore, $\alpha = 1$ and thus $V_{GS1} = V_{GS2}$: it is not necessary to compensate the power currents from the gate side.

Once the power layout has been defined, the current dissymmetry is defined (i.e., a, b, c, d, and thus α), and it is possible to modify the gate circuit, in order to obtain different values of V_{GE1} and V_{GE2}.

These gate voltages can be expressed as a function of gate and power currents and the various stray impedances of the circuit from Figure 12 right:

$$\begin{bmatrix} V_{GE1} \\ V_{GE2} \end{bmatrix} = \begin{bmatrix} V_{driver} \\ V_{driver} \end{bmatrix} - j\omega[(Z_G|Z_{PG})]\begin{bmatrix} I_{G1} \\ I_{G2} \\ I_{P1} \\ I_{P2} \end{bmatrix} \quad (10)$$

Z_G being the stray impedance matrix on the gate distribution PCB, V_{driver} being the driver output voltage (common for all devices), and Z_{PG} being the coupling matrix between gate distribution PCB and power circuit. This coupling matrix contains all mutual impedances the between power and the gate circuit, as already explained to introduce Figure 12 right.

Neglecting gate currents in comparison with power currents leads to focus on Z_{PG} only. Therefore, the following expressions can be obtained for V_{GE1} and V_{GE2}:

$$V_{GE1} = V_{driver} - j\omega Z_{PG\ line1} I_{P1} \quad (11)$$

$$V_{GE2} = V_{driver} - j\omega Z_{PG\ line2} I_{P2} \quad (12)$$

If we want to obtain $I_{P1} = I_{P2}$, combining (11) and (12) with (9) leads to:

$$j\omega I_P\left[\left(\sum Z_{PG\ line1}\right) - \left(\alpha \sum Z_{PG\ line2}\right)\right] - (V_{th} - V_{driver})(1 - \alpha) = 0 \quad (13)$$

the summation term is just here to show that more than two paralleled devices can be considered.

Therefore, to reach equal currents, the Z_{PG} matrix has to be designed in such a way that (13) is fulfilled, in order to compensate the power dissymmetry, expressed by a non-unity value of α. In other words, all couplings between gate PCB and power layout should be designed in order to compensate the non-ideal behavior of the power part. This is a major improvement in comparison to previous works where only power dissymmetry [7–10] or power-drive interaction [11–15] were considered, but not the possibility of compensating power dissymmetry through the gate circuit layout, using the device's model, as in [16,17]. This target will be reached through optimization in the following section.

4. Optimization of the Gate Distribution PCB

Finding the layout of the Gate Distribution PCB which, coupling with the power circuit, fits (13) is not possible with a human brain, even with years of trial-and-error work! Therefore, an optimization algorithm has been set up. The principle is illustrated in Figure 15: a home-made optimization environment (RSM, developed by the Alstom group, Semeac France), controls the Ansoft Q3D PEEC software, in which a parametrized geometry of the Gate Distribution PCB and the power layout is described. By varying the geometrical parameters of the PCB, the objective function defined by (13) is minimized. The optimization used in the RSM environment is based on genetic algorithms NSGA2 [22] and will not be described in this paper. The parametrized geometry of the Gate Distribution PCB is given in Figure 16. For simplicity, a multilayer description has been chosen, in order to draw the two gate tracks and the two emitter tracks independently. Track crossing in each layer is also allowed, to simplify the description. After optimization, the number of layers can be easily reduced, and track crossing can be avoided using vias. The geometrical description of the gate connections involves four parameters per gate, whereas the source connections are simpler and defined using two parameters per IGBT. Parameter description and an example of corresponding geometrical constraints are given in Figure 16 and Table 1, respectively. Choosing these 12 parameters allows the layout building some "loops", which can provide positive or negative couplings with the power part, and thus allow sufficient degrees of freedom for the algorithm to reach the goal defined by the objective function. All connection points to the power modules are obviously fixed, since the power layout and the module location remain unchanged.

The optimization function is frequency-dependent: first of all, because equations were written in the frequency domain, and also because impedance evaluation in the PEEC software is carried out at a specific frequency. Since the goal is to balance currents during the switching phase, an equivalent frequency corresponding to the switching time, $0.35/t_{switching}$, has been used [23]. It has been evaluated to 50 MHz. The impact of this choice will be discussed later.

The main interest of the proposed method is that it does not imply any time simulation in the optimization loop, which is really time-consuming and often generates convergence issues: some configurations generated by the optimizer may indeed not be able to converge in the time domain, and the optimization process may thus fail.

After several hours of optimization, which is quite a reasonable amount of time, since the process is automatic, a result has been obtained. It is displayed in Figure 17. The geometry displayed on the top of the figure is compatible with the imposed constraints (fixed connection points and maximum size of PCB).

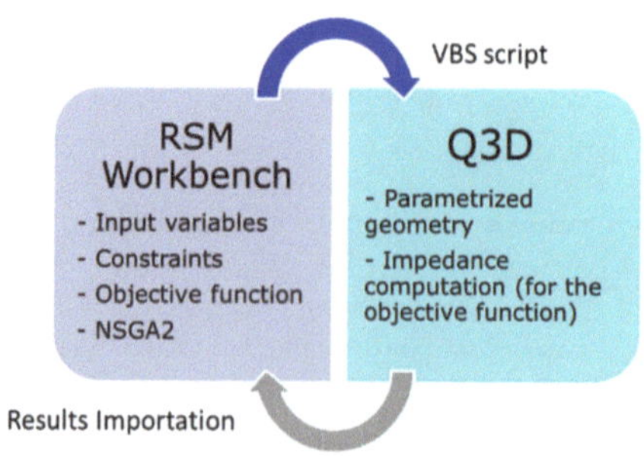

Figure 15. Optimization methodology.

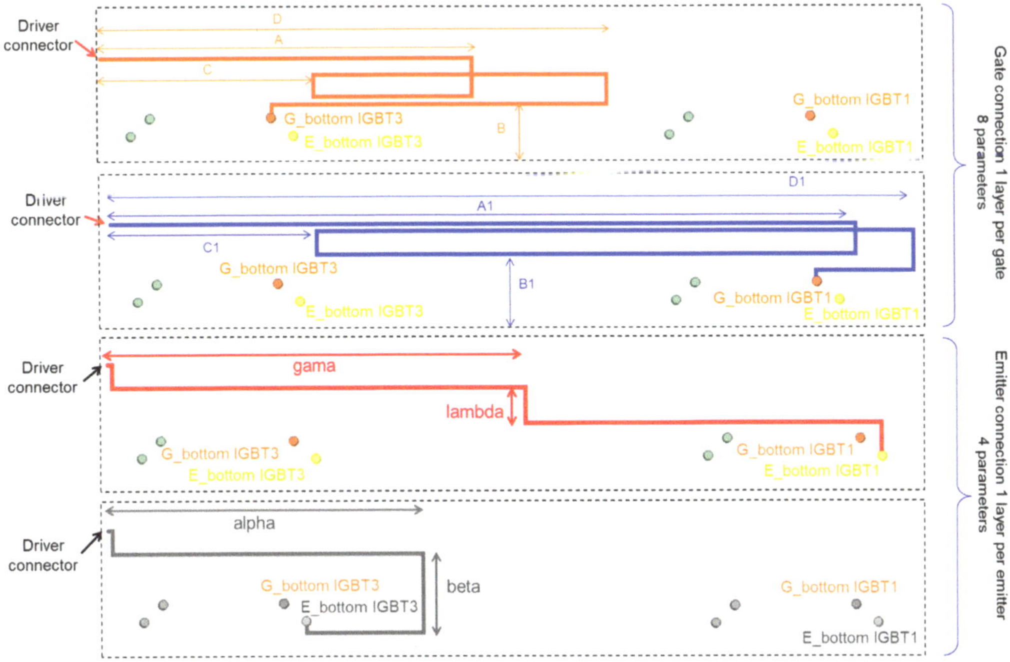

Figure 16. Gate Distribution PCB: parametrized geometry.

Table 1. Example of geometrical constraints associated to parametrized geometry: example for A, B, C, D.

Parameter Name	Min Value	Max Value
A	0	Amax = PCB max size X
B	0	Bmax = PCB max size Y
C	0	Cmax = PCB max size X
D	0	Dmax = PCB max size X

On the bottom of Figure 17, the evolution of the objective function is plotted as a function of the number of iterations of the algorithm. This representation helps in evaluating the convergence of the NSGA2 algorithm. It can be seen, for instance, that after iteration 400, the minimal value of the objective function has been reached and is not really modified between 400 and 1050, where it has been stopped. This minimal value converges up to 225,000, which is lower than the initial value, around 260,000. This objective function is the left part of (13). It has not reached zero due to all the constraints, but it has been reduced compared to the initial situation.

Since some assumptions have been made to build the objective function, and since it has been evaluated at a specific frequency (50 MHz), there is a strong need to validate the obtained result in the time domain. For this purpose, the PEEC model of the 3D layout (power and gate circuit in the same file for including couplings) has been combined with precise models of IGBTs in the Simplorer time domain simulator (Figure 18).

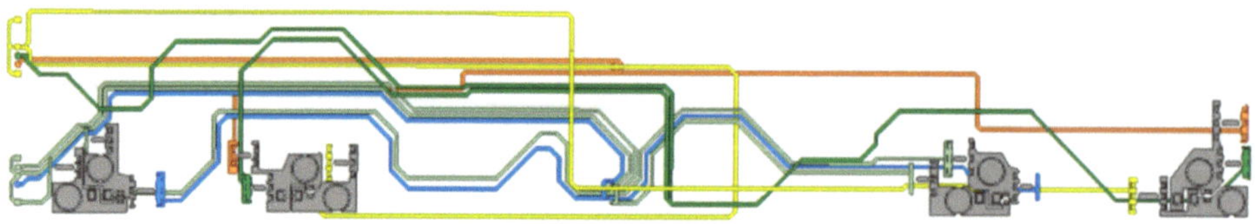

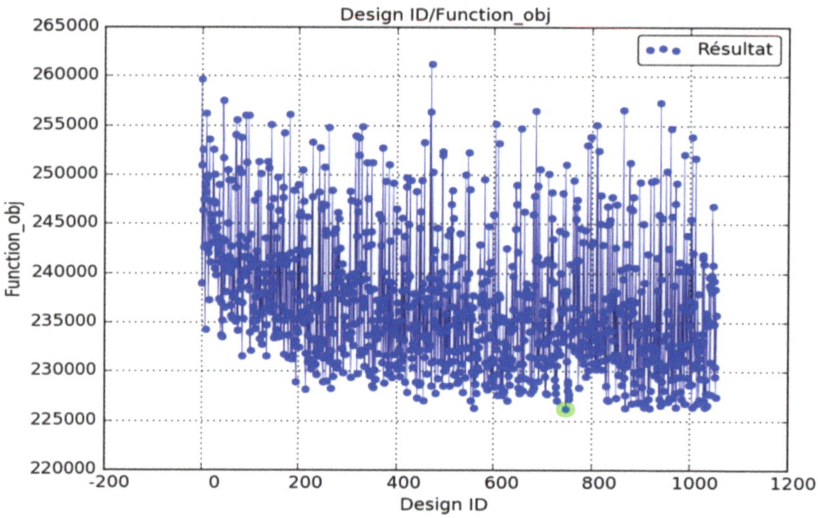

Figure 17. (**Top**): Optimized Gate Distribution PCB corresponding to the optimal point. (**Bottom**): evolution of the objective function vs iterations. The green point corresponds to the minimum of the objective function.

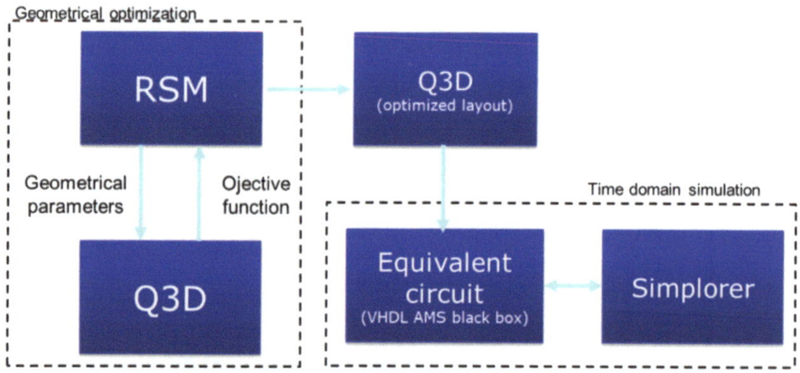

Figure 18. Post-processing of the optimization results: after geometric optimization, the full geometry is exported to Simplorer for a time domain simulation.

The initial geometry of the gate distribution PCB and the power layout has been simulated as well (light/dark blue curves in Figure 19). Results are compared to the optimized gate distribution PCB (magenta/red in Figure 19). The improvement is really obvious, which validates both the methodology and the optimized geometry.

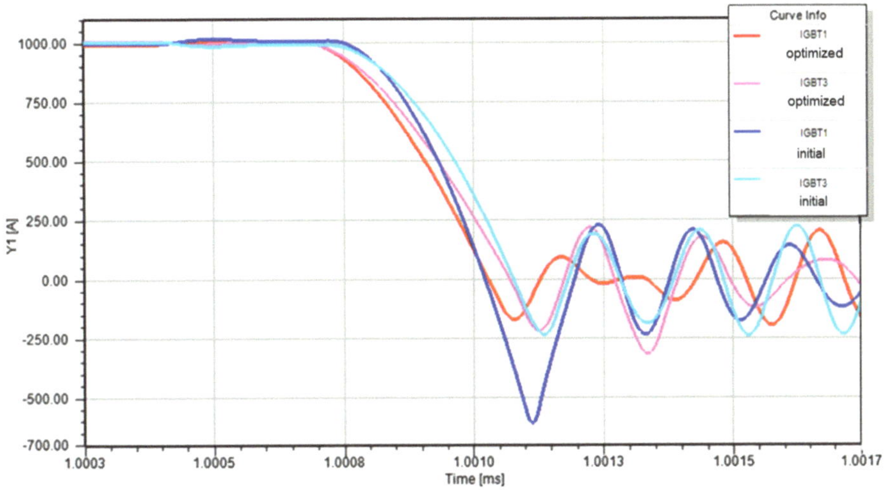

Figure 19. Results of the time domain simulation: in blue, the initial geometry of the gate distribution PCB; in red/pink, the optimized PCB.

5. Conclusions

This paper has presented a 3D layout of a power inverter used in a railway application. Due to the high-speed switching of candidate SiC MOSFETs, specific care has to be taken with regard to stray inductances and mutual couplings generated by the interconnections. In particular, it has been shown that the present 3D layout of the existing Si inverter was exhibiting uneven current division among the paralleled IGBTs. It was therefore mandatory to improve this layout before going to the next step with SiC devices. Using the PEEC approach, it has been demonstrated that the power layout was at the origin of a dynamic current imbalance, and that the power-gate interaction caused another imbalance. Modifying the power layout was not possible due to several other constraints. However, it was possible to design a Gate Distribution PCB in such a way that the couplings between this PCB and the power layout compensate the non-ideal behavior of the power part. For this purpose, a detailed analysis of the circuit equations has been carried out, using a generic representation independent from the geometry. This has led to an objective function to be minimized, which has been implemented in an optimization environment controlling a PEEC based software. The optimized layout has been validated with a time domain simulation.

Author Contributions: Conceptualization, J.-L.S., J.-M.G. and P.-O.J.; methodology, J.-M.G. and E.B.; software, J.-M.G. and E.B.; validation, P.-O.J. and A.N.-M.; formal analysis, J.-M.G. and A.N.-M.; investigation, A.N.-M.; resources, A.N.-M. and E.B.; data curation, A.N.-M. and E.B.; writing—original draft preparation, A.N.-M. and J.-L.S.; writing—review and editing, J.-L.S.; visualization, A.N.-M.; supervision, J.-L.S., J.-M.G. and P.-O.J.; project administration, G.D. and J.-L.S.; funding acquisition, G.D. All authors have read and agreed to the published version of the manuscript.

Funding: This research was funded by ALSTOM, within the European Project Shift2rail.

Institutional Review Board Statement: Not applicable.

Informed Consent Statement: Not applicable.

Data Availability Statement: Data sharing not applicable.

Conflicts of Interest: The authors declare no conflict of interest.

References

1. Shift2Rail Project. European Funding for Research & Innovation. Available online: https://shift2rail.org/ (accessed on 7 March 2021).
2. Rabkowski, J.; Peftitsis, D.; Nee, H. Silicon Carbide Power Transistors: A New Era in Power Electronics Is Initiated. *IEEE Ind. Electron. Mag.* **2012**, *6*, 17–26. [CrossRef]
3. Seal, S.; Mantooth, H.A. High Performance Silicon Carbide Power Packaging—Past Trends, Present Practices, and Future Directions. *Energies* **2017**, *10*, 341. [CrossRef]
4. Schanen, J.; Jeannin, P. Integration solutions for clean and safe switching of high speed devices. In Proceedings of the CIPS 2018 10th International Conference on Integrated Power Electronics Systems, Stuttgart, Germany, 20–22 March 2018; pp. 1–11.
5. Hatanaka, A.; Kageyama, H.; Masuda, T. A 160-kW high-efficiency photovoltaic inverter with paralleled SiC-MOSFET modules for large-scale solar power. In Proceedings of the 2015 IEEE International Telecommunications Energy Conference (INTELEC), Osaka, Japan, 18–22 October 2015; pp. 1–5. [CrossRef]
6. Schrader, R.; Cordt, D.; Münster, P.; Eckel, H. Influences of Gate-Circuit and Parasitic Inductances on Turn-Off Current Imbalances of Paralleled IGBTs Due to Differences in Their Switching Behaviour. In Proceedings of the 2018 20th European Conference on Power Electronics and Applications (EPE'18 ECCE Europe), Riga, Latvia, 17–21 September 2018; pp. P.1–P.10.
7. Jeannin, P.; Schanen, J.; Clavel, E. Original cabling conditions to insure balanced current during switching transitions between paralleled semiconductors. *IEEE Trans. Ind. Appl.* **2002**, *38*, 181–188. [CrossRef]
8. Zhang, Y.; Zang, Z.; Hu, B.; Alsaif, F.; Alkhalid, K.; Li, X.; Wang, J. Current Sharing Analysis of SiC Power Modules in Parallel Operation. In Proceedings of the2020 IEEE Energy Conversion Congress and Exposition (ECCE), Detroit, MI, USA, 15–20 October 2020; pp. 711–716. [CrossRef]
9. Wissen, M.; Domes, D.; Brekel, W.; Yilmaz, K. How asymmetric busbar design causes symmetric switching behavior of paralleled IGBT modules. In Proceedings of the CIPS 2018, 10th International Conference on Integrated Power Electronics Systems, Stuttgart, Germany, 20–22 March 2018; pp. 1–3.
10. Mainali, K.; Wang, R.; Sabate, J.; Klopman, S. Current Sharing and Overvoltage Issues of Paralleled SiC MOSFET Modules. In Proceedings of the 2019 IEEE Energy Conversion Congress and Exposition (ECCE), Baltimore, MD, USA, 29 September–3 October 2019; pp. 2413–2418. [CrossRef]
11. Li, H.; Zhou, W.; Wang, X.; Nielsen-Munk, S.; Li, D.; Wang, Y.; Dai, X. Influence of Paralleling Dies and Paralleling Half-Bridges on Transient Current Distribution in Multichip Power Modules. *IEEE Trans. Power Electron.* **2018**, *33*, 6483–6487. [CrossRef]
12. Bęczkowski, S.; Jørgensen, A.B.; Li, H.; Uhrenfeldt, C.; Dai, X.; Munk-Nielsen, S. Switching current imbalance mitigation in power modules with parallel connected SiC MOSFETs. In Proceedings of the 2017 19th European Conference on Power Electronics and Applications (EPE'17 ECCE Europe, Warsaw, Poland, 11–14 September 2017; pp. P.1–P.8. [CrossRef]
13. Li, H.; Munk-Nielsen, S.; Bęczkowski, S.; Wang, X. A Novel DBC Layout for Current Imbalance Mitigation in SiC MOSFET Multichip Power Modules. *IEEE Trans. Power Electron.* **2016**, *31*, 8042–8045. [CrossRef]
14. Li, H.; Munk-Nielsen, S.; Bęczkowski, S.; Wang, X.; Eni, E. Effects of auxiliary source connections in multichip power module. In Proceedings of the 2016 IEEE Applied Power Electronics Conference and Exposition (APEC), Long Beach, CA, USA, 21–24 March 2016; pp. 3101–3106. [CrossRef]
15. Martin, C.; Guichon, J.; Schanen, J.; Pasterczyk, R. Gate Circuit Layout Optimization of Power Module Regarding Transient Current Imbalance. *IEEE Trans. Power Electron.* **2006**, *21*, 1176–1184. [CrossRef]
16. Tang, Y.; Ma, H. Dynamic Electrothermal Model of Paralleled IGBT Modules with Unbalanced Stray Parameters. *IEEE Trans. Power Electron.* **2017**, *32*, 1385–1399. [CrossRef]
17. Zeng, Z.; Zhang, X.; Li, X. Layout-Dominated Dynamic Current Imbalance in Multichip Power Module: Mechanism Modeling and Comparative Evaluation. *IEEE Trans. Power Electron.* **2019**, *34*, 11199–11214. [CrossRef]
18. Ruehli, A.E. Equivalent Circuit Models for Three-Dimensional Multiconductor Systems. *IEEE Trans. Microw. Theory Tech.* **1974**, *22*, 216–221. [CrossRef]
19. Ansys Q3D. Available online: https://www.ansys.com/products/electronics/ansys-q3d-extractor (accessed on 7 March 2021).
20. Nunnally, W.C.; Kristiansen, M.; Hagler, M.O. Differential Measurement of Fast Energy Discharge Capacitor, Inductance, and Resistance. *IEEE Trans. Instrum. Meas.* **1975**, *24*, 112–114. [CrossRef]
21. Altair Flux PEEC. Available online: https://www.altair.com/resource/conductor-impedance-and-near-field-simulation-using-altair-flux (accessed on 7 March 2021).
22. Deb, K.; Pratap, A.; Agarwal, S.; Meyarivan, T. A fast and elitist multiobjective genetic algorithm: NSGA-II. *IEEE Trans. Evol. Comput.* **2002**, *6*, 182–197. [CrossRef]
23. Brooks, D. *Signal Integrity Issues and Printed Circuit Board Design*; Prentice Hall: Upper Saddle River, NJ, USA, 2004; ISBN 139780133359473.

energies

MDPI

Article

Multi-Electrode Architecture Modeling and Optimization for Homogeneous Electroporation of Large Volumes of Tissue

Borja López-Alonso *, Héctor Sarnago, José M. Burdío, Pablo Briz and Oscar Lucía

Department Electronic Engineering and Communications, I3A, Universidad de Zaragoza, C/María de Luna, 1-50018 Zaragoza, Spain; hsarnago@unizar.es (H.S.); burdio@unizar.es (J.M.B.); 723669@unizar.es (P.B.); olucia@unizar.es (O.L.)

* Correspondence: blopez@unizar.es

Abstract: Electroporation is a phenomenon that consists of increasing the permeability of the cell membrane by means of high-intensity electric field application. Nowadays, its clinical application to cancer treatment is one of the most relevant branches within the many areas of electroporation. In this area, it is essential to apply homogeneous treatments to achieve complete removal of tumors and avoid relapse. It is necessary to apply an optimized transmembrane potential at each point of the tissue by means of a homogenous electric field application and appropriated electric field orientation. Nevertheless, biological tissues are composed of wide variety, heterogeneous and anisotropic structures and, consequently, predicting the applied electric field distribution is complex. Consequently, by applying the parallel-needle electrodes and single-output generators, homogeneous and predictable treatments are difficult to obtain, often requiring several repositioning/application processes that may leave untreated areas. This paper proposes the use of multi-electrode structure to apply a wide range of electric field vectors to enhance the homogeneity of the treatment. To achieve this aim, a new multi-electrode parallel-plate configuration is proposed to improve the treatment in combination with a multiple-output generator. One method for optimizing the electric field pattern application is studied, and simulation and experimental results are presented and discussed, proving the feasibility of the proposed approach.

Keywords: electroporation; finite element methods; electromagnetic fields

Citation: López-Alonso, B.; Sarnago, H.; Burdío, J.M.; Briz, P.; Lucía, O. Multi-Electrode Architecture Modeling and Optimization for Homogeneous Electroporation of Large Volumes of Tissue. *Energies* **2021**, *14*, 1892. https://doi.org/10.3390/en14071892

Academic Editor: Ivica Stevanovic

Received: 18 February 2021
Accepted: 23 March 2021
Published: 29 March 2021

1. Introduction

Electroporation is a phenomenon based on the change in the permeability of the cell membrane by applying an electric field [1–5]. This electric field produces a strong change of transmembrane potential and, consequently, this induced potential can trigger different mechanisms that change the properties of the membrane [6] (Figure 1).

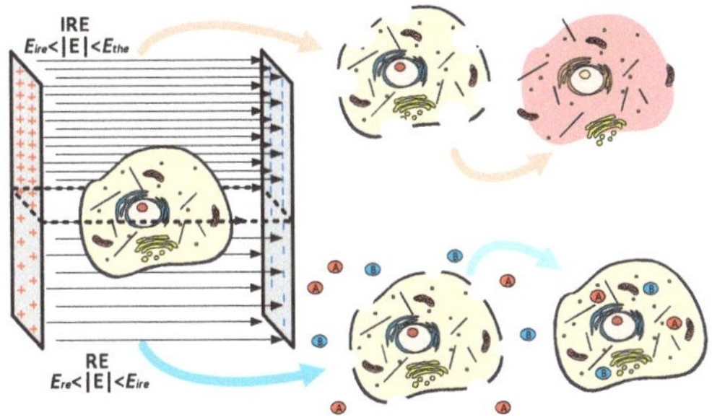

Figure 1. Reversible (RE) and irreversible (IRE) electroporation of an animal cell.

Figure 1 shows how the electroporation has been classified according to the effect produced in the cells depending on the intensity of the applied electric field, in reversible electroporation (RE), and irreversible electroporation (IRE). RE causes temporary permeabilization and is typically used to improve drug absorption, whereas IRE causes cell death, being used as an ablation mechanism with no thermal effect [7].

This paper focuses on the use of electroporation for oncological electrosurgery. In this subject, electroporation has great advantages including:

- The treatment efficacy is not depended on thermal effect, this allows treating high blood perfusion tissues, and it does not produce thermal damage [8,9].
- This technique does not destroy blood vessels structures, this allows the preservation of more healthy tissue [1,3–5].
- Electroporation is compatible with many of the current treatments, such as radiotherapy and chemotherapy, being an effective adjuvant mechanism [6].

In electroporation, an adequate distribution of the electric field is essential to achieve a homogeneous and safe treatment. Electroporation is applied using differential electrodes; at present, the two most used electrodes for the application of the treatment are: parallel-needle electrodes as shown in Figure 2a and parallel-plate electrodes as shown in Figure 2e.

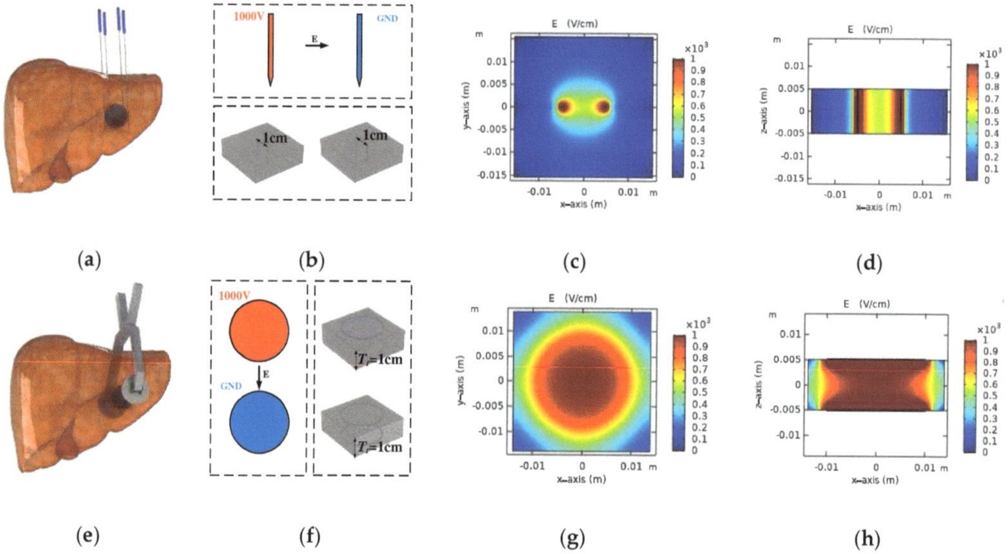

Figure 2. State-of-the-art electroporation electrodes applied to human liver tissue: Needle (**a**) and parallel-plate (**e**) electrodes. Finite element modeling of electroporation: Needle (**b**) and parallel-plate (**f**) electrodes. Electric field modulus in horizontal slice (z = 0): Needle (**c**) and parallel-plate (**g**) model. Electric field modulus in vertical slice (y = 0): Needle (**d**) and parallel-plate (**h**) model.

Currently, the most commonly used electrodes are parallel-needle electrodes [10–12], as they normally allow quick and easy placement. The main limitation of needle-based electrodes is that they create an irregular electric field [13], which is concentrated in several points producing undesired thermal effects [14]. In Figure 2c,d, a simulation of the electric field produced by two needle electrodes has been represented. Also, this type of electrode normally requires several repositioning/application processes to treat large volumes [10,15]. For this reason, this paper is focused on the study of parallel-plate electrodes [3,16–18] to achieve adequate electric field distribution [19,20], as can be seen in Figure 2f–h.

Biological tissues are complex structures [21] and, for this reason, it is complicated to treat a large volume in a homogeneous way by applying a single electric field vector.

In this sense, each spatial point of a tissue has an optimal direction in which the induced transmembrane potential is maximized. This direction changes due to tissue heterogeneity, conductivity variations, and the different cell geometry, among other factors. Nowadays, the most extended method to apply more than one electric field vector is to physically reposition them [22–25]. This procedure is not always available, and in the case of needle electrodes, repositioning the electrodes increases the damage and extend the treatment.

The objective of this paper is to propose and validate a matrix parallel plate electrode (MPPE) [26]. This electrode has the advantages of state-of-the-art parallel-plate electrodes, which allow treating large volumes of tissue and apply more uniform electric field than needle-based electrodes. Also, unlike state-of-the-art parallel-plate electrodes, the proposed structure is composed of different electrically insulated conductive electrode cells, allowing the application of a wide variety of electric field vectors. The rotation of the electric field at each point of the tissue will also allow the transmembrane potential to be applied more efficiently, thus achieving more focused and homogeneous treatment.

Based on a conference publication [26], this paper delves into the analysis of the MPPE, adds experimental validation of the MPPE, and rethinks and expands the optimization of the electrode. Also, for the experimental validation it has been necessary to develop a multi-output generator, considering that MPPE requires as many outputs as the proposed multi-electrode electrode cells.

The remainder of this paper is organized as follows—Section 2 describes the basic aspects of the proposed multi-electrode, and details the model carried out and its evaluation. In Section 3, the main experimental results proving the feasibility of this proposal and an example of optimization are included. Finally, the conclusions of this paper are drawn in Section 4.

2. Materials and Methods

2.1. Proposed Multi-Electrode Architecture

Despite its safety and efficacy, electroporation suffers from several handicaps. One of the challenges of electroporation is to achieve a homogeneous treatment on all types of tissues despite their heterogeneity and anisotropy. To overcome this limitation, this paper proposes the design of a new multi-electrode that allows to apply wide variety of electric field vectors without manipulating the electrode. The MPPE is the proposed multi-electrode to improve electroporation homogeneity.

2.1.1. Proposed Multi-Electrode Geometries

Two geometries of MPPE have been studied, considering square or circular electrodes. These electrodes are made up of nine C_{ij} electrode cells, as it can be seen in Figure 3. Each electrode cell is made of conductive material and it is electrically isolated from the others. The main geometric parameters are: L_I is the side of the square electrode cell and L_E is the total side of the square electrode including the insulation; R_I is the radius of the circular center electrode cell and R_E is the total radius of the circular electrode including the insulation.

Considering that the proposed electrodes consist of 9 electrode cells, it is possible to apply more than 40 different electric field vectors. Increasing the number of electrode cells increases the number of electric field vectors that can be applied, but also leads to a complex generator implementation. The two proposed geometries have been selected with a number of electrode cells that can be controlled in an easy way, varying the inclination of the applied electric field vectors by means of their central electrode cell size. These allow the proposed electrodes to apply electric field vectors in a flexible way.

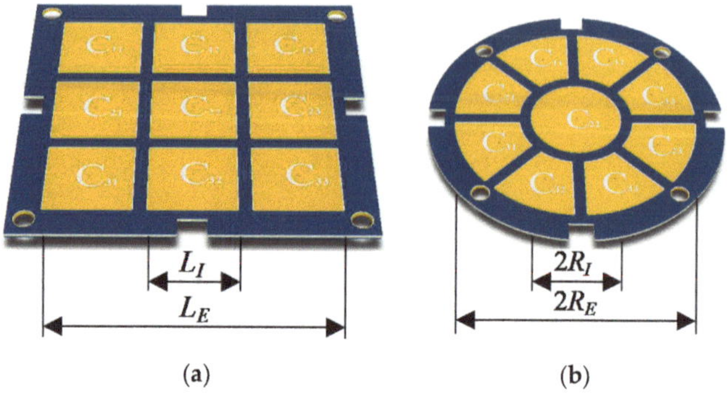

Figure 3. Proposed multi-electrode geometries: (**a**) squared electrode and (**b**) circular electrode.

2.1.2. Operation of Proposed Multi-Electrode

To use the MPPE as well as to analyze its results, different strategies have been proposed. The studied electrodes have been made up of 9 isolated electrode cells, with different applied voltages to obtain the desired electric field distribution. As in state-of-the-art parallel-plate electrodes, two electrodes are used differentially, being necessary to establish and control up to 2×9 different voltages. Electroporation is based on the application of electric field, for this reason, it is necessary to control the applied voltages as a function of the distance between the electrode cells to apply the desired electric field. Moreover, unlike parallel-plate electrodes, multiple electric field vectors are applied at each point, so the application protocols and the evaluation of the results must be different.

In the case of parallel-plate electrodes, the only thing that can be controlled is the applied voltage between their differential plates, and in this way, it is controlled a single electric field vector. Furthermore, the proposed multi-electrode allows to apply voltage between different combinations of the electrode cells that compose it. Controlling the voltage of MPPE electrode cells allows to apply different electric field vectors $\mathbf{A}_n$ at each point of the tissue. To control the different electrode cell settings (CS), a matrix system is proposed that relates the voltages applied to each of the electrode cells with the n electric field vectors that can be applied to each point of the tissue.

Figure 4a shows a schematic of topology proposed to control a circular MPPE of 9 electrode cells. This topology with the designed generator applies a voltage V_C_{nn} to each electrode cell. Two direct voltage sources (Vcc and $-Vcc$) have been used, one positive and other negative, which can be connected by switching devices to both the upper and lower electrode cells and this allows to apply bipolar electric field pulses. Moreover, the electrode cells of the two electrodes have been connected by independent switching devices. So, the voltage is only applied in the selected electrode cells of the electrode. Each of these CS will produce at each point of the tissue $\mathbf{P}_i$ an electric field vector $\mathbf{A}_n$ as can be seen in Figure 4. by sequentially applying of 32 CS in the circular electrode, the different electric field vectors represented in Figure 4b,c can be obtained.

When a uniform electric field is applied in a single direction to a biological tissue, a transmembrane potential is induced in the membranes of its cells. The intensity of this potential at each point of the membrane depends on the applied normal electric field. The main advantage of the proposed electrode over traditional flat plates is that the area of the cell membrane where the transmembrane potential is concentrated can be changed. This is a great advantage, due to the geometry and imperfections of cell membrane, its dielectric strength is not constant over its entire surface and has areas that are more susceptible to electroporation. The proposed electrode allows applying several electric field vectors by means of different CSs that can concentrate the electric field in different areas of the membrane, therefore it is possible to find an electric field vector or a combination that allows electroporating the cells of the tissue in a more homogeneous and easy way.

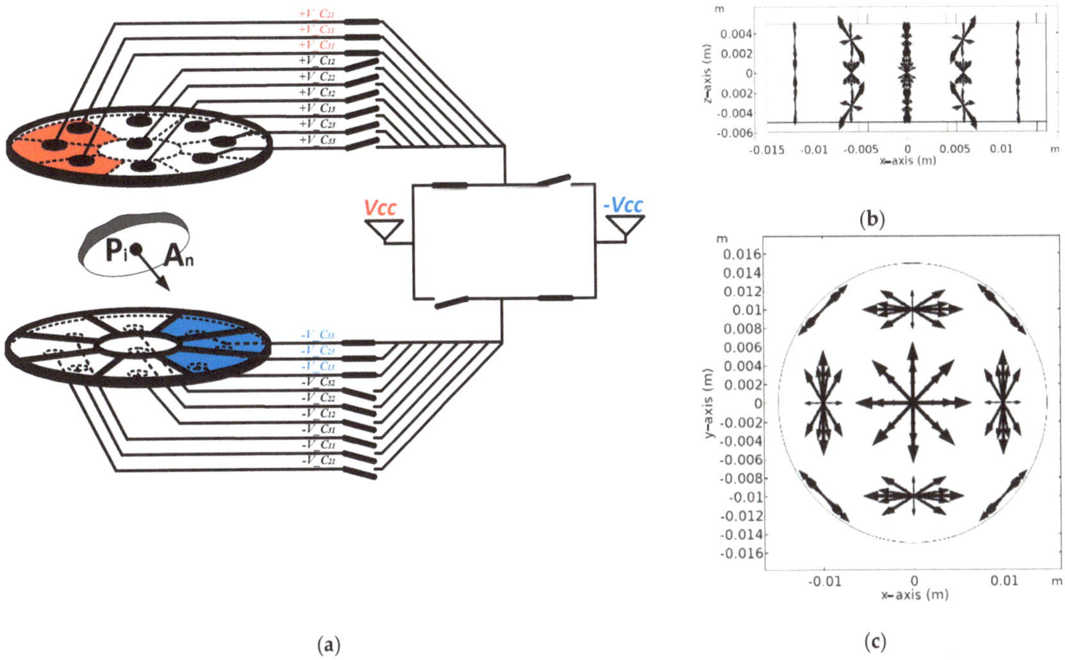

Figure 4. Schematic of the application of voltage in example setting: (**a**) Electric field vectors in the circular model applying 32 voltage settings; (**b**) in vertical slice (y = 0); (**c**) in horizontal slice (z = 0).

With the different systems that have been proposed it is possible to control the operation of the electrode, and it is necessary to establish how to model and evaluate its operation and results.

2.2. MPPE Modeling and Evaluation

Two finite element models have been developed. These types of models are used in electroporation to plan treatments estimating the volumes in which the electric field has exceeded the electroporation threshold. Unlike these, the models have been created focusing on the study of the module and the direction of the applied electric field. In addition, the models do not evaluate a single electric field application, as so it is studied the relationship of the different electric field vectors applied at each point of the tissue by n CSs. This is necessary as the use of MPPE is based on improving the treatment homogeneity applying wide variety of electric field vectors, and not a single vector. This section presents the models developed and how their results have been evaluated.

2.2.1. MPPE Model

With the objective of modelling, the proposed MPPE and improve their design, two finite element models have been created using the software COMSOL Multiphysics. These are stationary models developed with the Electric Currents block of the AC/DC module (Figure 5). The meshed models can be seen in Figure 5a,e. These models allow to analyze the direction and intensity of the electric field at each point of the tissue, assuming ideal conditions, as can be seen in Figure 5b–d,f–h.

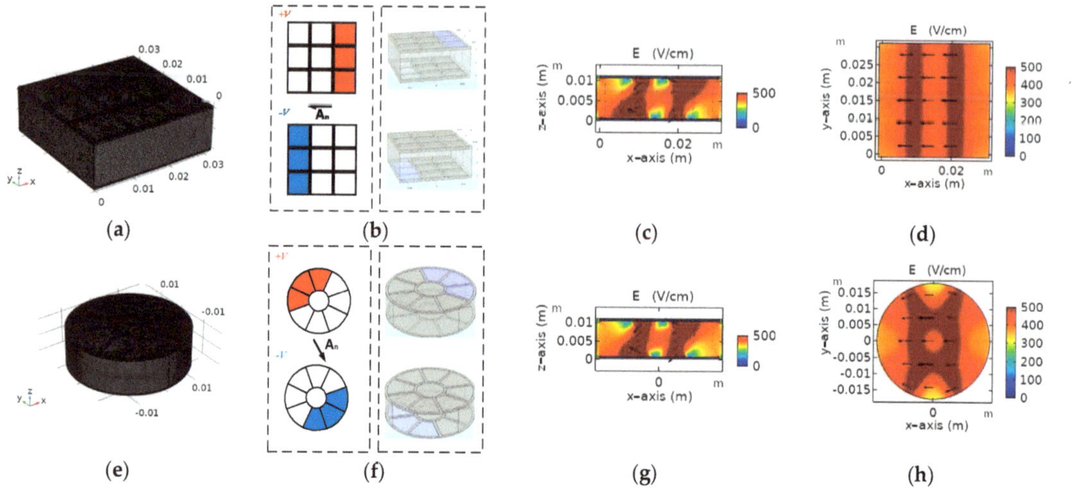

Figure 5. Proposed multi-electrode models meshing: (**a**) squared electrode model, (**e**) circular electrode model. Cell settings (CS)where the voltage is applied: (**b**) squared electrode, (**f**) circular electrode. Module and direction of the electric field in vertical slice (y = 0): (**c**) squared electrode, (**g**) circular electrode. Module and direction of the electric field in horizontal slice (z = 0): (**d**) squared electrode and (**h**) circular electrode.

The model consists of two MPPE made of FR4 as electrical insulation material, and steel as conductive material. A biological tissue of a thickness of T_t is placed in between the electrodes. To model the electrical behavior of the tissue, the electrical conductivity in each point depends on the electric field applied in this point. This relationship depends on measured parameters of the tissue [21,27].

Conductivity behavior of the tissue represented in Figure 6 is described in (1) where σ_0 and σ_f are the initial and final electric conductivities of the tissue, respectively. Field E_{th} models the electric field in which the conductivity reaches half of its maximum value. Finally, the constant K_v controls the slope of the curve.

$$\sigma(|\mathbf{E}|) = (\sigma_0 + \frac{\sigma_f - \sigma_0}{2}(1 + \tanh(K_v(E - E_{th})))). \quad (1)$$

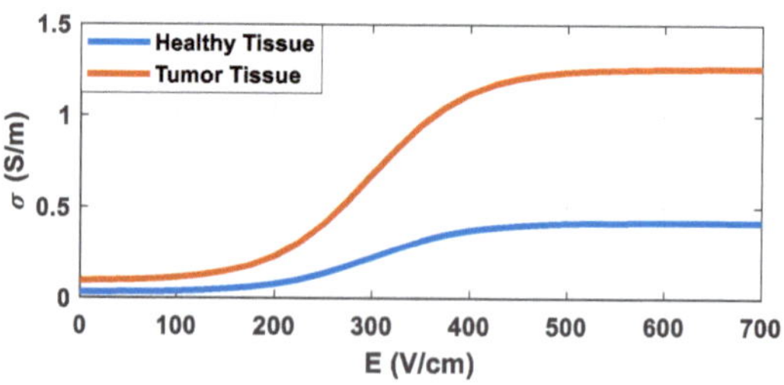

Figure 6. Electrical conductivity of the simulated tissue, as a function of the electric field applied at 10 kHz [21,27].

Table 1 lists the main electrical parameters used in the models presented. All data has been extracted from previous research [27,28]. As it can be seen in Figure 6, two types of tissue have been modeled. One handicap of electrosurgery is the strong differences in electrical conductivity between tissues. Specifically, in tumor electroporation, the electric field is concentrated in healthy tissue as it is a worse electrical conductor than the tumor.

In past research, the conductivity measured in tumors is three times higher than that of healthy tissues [29].

Table 1. Model parameters extracted from [27,28].

Symbol	Parameter	Value (Unit)
σ_i	Insulating Electrical Conductivity	0.0001 (S/m)
σ_s	Steel Electrical Conductivity	300000 (S/m)
σ_{f_h}	Final Healthy Electrical Conductivity	0.4 (S/m)
σ_{0_h}	Initial Healthy Electrical Conductivity	0.03 (S/m)
K_{v_h}	Healthy Slope Constant	0.01
E_{th_h}	Healthy Electric Field Threshold	300 (V/cm)
σ_{f_t}	Final Tumor Electrical Conductivity	1.2 (S/m)
σ_{0_t}	Initial Tumor Electrical Conductivity	0.09 (S/m)
K_{v_t}	Tumor Slope Constant	0.01
E_{th_t}	Tumor Electric Field Threshold	300 (V/cm)

In this model, the external surface is considered to be electrically isolated except for the surfaces where the electric potential is applied, which are the active electrode cells. Isotropic and homogeneous electrical properties in all the volumes are assumed. The tissue has been modeled without taking into account the structure and geometry of the cells, assuming an homogeneous block with equivalent electrical properties; and the effects of temperature on conductivity have been considered negligible because, despite the high instantaneous power applied, the pulses applied in the experiment had a duration of 100 μs.

2.2.2. Evaluation of Independent Electric Field Vectors

The objective of the proposed electrode is to apply the electric field at each point with a modulus above the objective threshold and an appropriate orientation. Due to the difficulty of achieving a homogeneous treatment by applying an electric field in only one direction, the MPPE applies n different vectors to achieve the application of electric field in a direction as optimal as possible. The perfect way to apply the vectors would be a homogeneous distribution, since if the vectors are not sufficiently out of phase with each other they can have the same effect as applying a single vector.

In order to evaluate the MPPE effect, a methodology has been proposed to calculate the number of independent electric field vectors applied at each point. In this subject, two vectors are independent when the angle between them is greater than a set threshold. However, in the case of applying n vectors it is necessary to ensure that each vector meets this condition with the other n-1 vectors. Increasing the number of independent electric field vectors applied at a point, it is more probable to apply the field in the optimal way. Calculating the number of independent electric field vectors applied at each point, it is possible to know the volumes in which the MPPE has been most effective.

Taking into account the above, the first step to evaluate the model is simulating the n electric field distributions produced by the n CS. At each point $\mathbf{P_i}$ of the model, the n vectors of electric field are evaluated as can be seen in Figure 7d. In this example three different configurations have been applied. At each $\mathbf{P_i}$ the modulus of the three applied electric field vectors is evaluated, and to consider that an electric field vector has been effective, its modulus must be greater than the electroporation threshold E_{lim}. In the example, the three vectors applied in $\mathbf{P_1}$, have a modulus above the objective threshold, and therefore it is necessary to calculate if the three vectors are independent. To consider these vectors independent, the angle of each vector with the others must be between established thresholds θ_{min} and θ_{max}. Therefore, the first step is to calculate the angles α_1, β_1, and Ω_1 (Figure 7a–c) which are all the possible relationships between the three electric field vectors applied on $\mathbf{P_1}$ of the example. These three vectors are considered independent if $\theta_{max} > \Omega_1 > \theta_{min}$, $\theta_{max} > \beta_1 > \theta_{min}$ and $\theta_{max} > \alpha_1 > \theta_{min}$, that is, if all angle relationships are above the objective threshold and therefore the three vectors are well distributed. Moreover, if $\theta_{max} > \beta_1 > \theta_{min}$, $\theta_{max} > \Omega_1 > \theta_{min}$, and $\alpha_1 < \theta_{min}$ or $\alpha_1 > \theta_{max}$ the red and yellow vector are considered to be applied almost collinearly, since the angle between them is considered not above the objective threshold. Consequently, they will produce very similar effect and only one can be considered as an independent vector. Moreover, if $\theta_{max} > \Omega_1 > \theta_{min}$, $\alpha_1 < \theta_{min}$ or $\alpha_1 > \theta_{max}$, and $\beta_1 < \theta_{min}$ or $\beta_1 > \theta_{max}$ the angular relationship of the yellow vector with the red and blue vectors is insufficient but between the red and blue vectors the angular relationship is above the objective threshold. Therefore, they can produce different effects at the point $\mathbf{P_1}$ and therefore only the yellow vector is not considered independent. Finally, if, $\alpha_1 < \theta_{min}$ or $\alpha_1 > \theta_{max}$, $\beta_1 < \theta_{min}$ or $\beta_1 > \theta_{max}$ and, $\alpha_1 < \theta_{min}$ or $\Omega_1 > \theta_{max}$ three vectors are considered to create the same effect and, therefore, in the point $\mathbf{P_1}$ only has been applied one independent electric field vector. This process is carried out in the i points of the model and next it is estimated the percentage that has been treated with 1, 2 or 3 independent electric field vectors. This methodology has been implemented in both models to evaluate the effectiveness of the different designs by applying n different CSs.

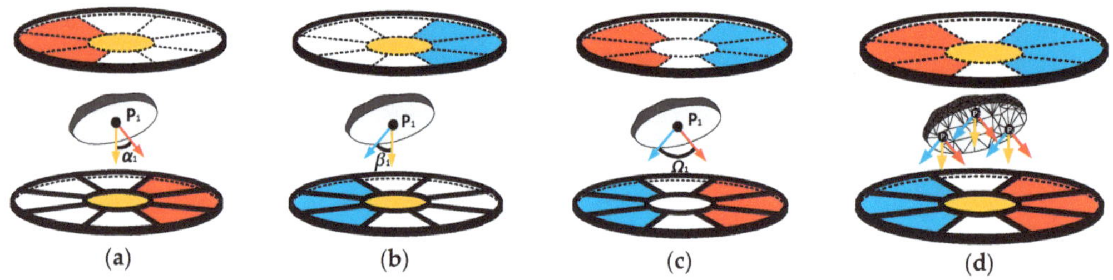

Figure 7. Example of evaluation of independent electric field vectors: (**a**) angle between the electric field vectors produced by red CS and configuration yellow CS, (**b**) angle between the electric field vectors produced by blue CS and configuration yellow CS, (**c**) angle between the electric field vectors produced by blue CS and configuration red CS and (**d**) evaluation of independent electric field vectors in each point of the mesh.

With the methods presented to operate the electrodes, to control the applied electric field, and to evaluate the models carried out, in the following section the results obtained are presented.

3. Experimental and Simulation Results and Discussion

The objective of this section is to prove the feasibility of the proposed MPPE design and to present an example of optimization for a theoretical case.

Both the model and the experiments have been carried out on potato tissue. The behavior of the electroporation of vegetal tissue is documented [21,30], and specifically potatoes allow us to carry out an in-vivo experimentation to validate the proposed models. Figure 8 shows the setup used in the experiments and the prototype under development.

The setup used in the experiment is composed of three parts:

- High voltage generator. This test-bench subset consists of high voltage monitoring/acquiring: an 8-bit LeCroy oscilloscope Wavesurfer 3024, three differential high-voltage probes LeCroy HVD3206, and one Pearson current monitor model 110.
- Electroporation ad hoc multioutput generator, which has been designed to be able to power eighteen outputs and to provide online impedance measurements [27]. The proposed generator is based on a multiple-output structure featuring IGBT (1700-V 100-A 3-phase IGBT power module FS100R17N3E4) an MOSFET (40-V NVMTS0D4N04CTXG MOSFETs) devices [31]. This implementation allows high-performance omnidirectional electroporation treatments.
- Experimentation area, where the potato specimens are carved and placed.

The experimentation developed, and the results obtained by experimentation on potato tissue are described below.

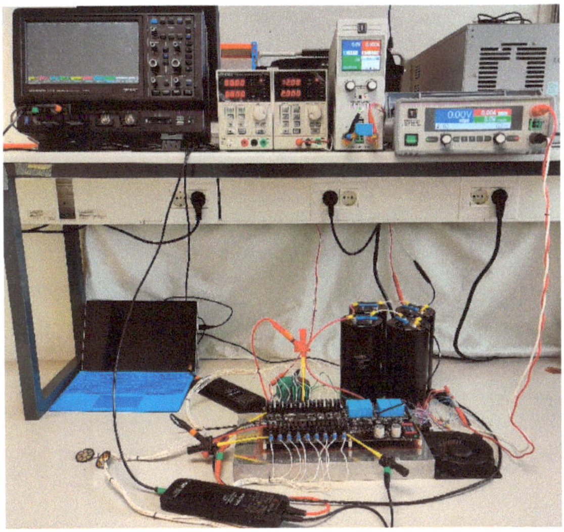

Figure 8. Image of setup used in the experiment.

3.1. Electric Field Distribution

There are direct and indirect methods of studying the effects of electroporation. The most widespread indirect method to observe the effect of electroporation in real time is the study of the change in impedance in the tissue. The problem is that this method only gives a global idea of the state of the tissue, and to compare the results of the model it is not necessary to study the effect in real time. Histopathological methods allow a much more accurate understanding of the state of each part of the tissue. In this subject, different techniques have been developed that allow to know the tissue state in a visual way. In this paper, two different techniques have been used that allow studying the distribution of the electric field by means of the coloration of the tissue. First, dye [32] was used to color the areas most affected by electroporation, the intensity of the dye is proportional to the electric field that has been applied. The other technique used [21] is the removal of the tissue after natural degradation. It has already been described that, if the tissue is left to rest for a few hours, it deteriorates naturally, darkening the areas affected by the treatment in proportion to the intensity of the treatment applied.

To validate these models, the experiments represented in Figures 9 and 10 have been carried out. For these experiments, electrodes with a square electrode cell side of 1 cm spaced 0.8 cm have been used in addition to a 1 mm spacing between electrode cells to ensure electrical insulation, and in the circular case, electrodes with equal interior and exterior surfaces to the square case and an equal spacing have been used.

It can be seen in Figure 9a the CS of electric field that had been applied. In each experiment in Figure 9, 30 pulses of 100 μs separated by 100 ms of 800 V have been applied. The potato variety was young Monalisa and the separation between the electrodes has been established in 8 mm. In Figure 9b–e the simulated electric field, the image of the treated potato, and the superposition on the image of the potato with the contour that delimits the treated area at 300 V/cm or more, this value being the electroporation threshold for the tissue used [27].

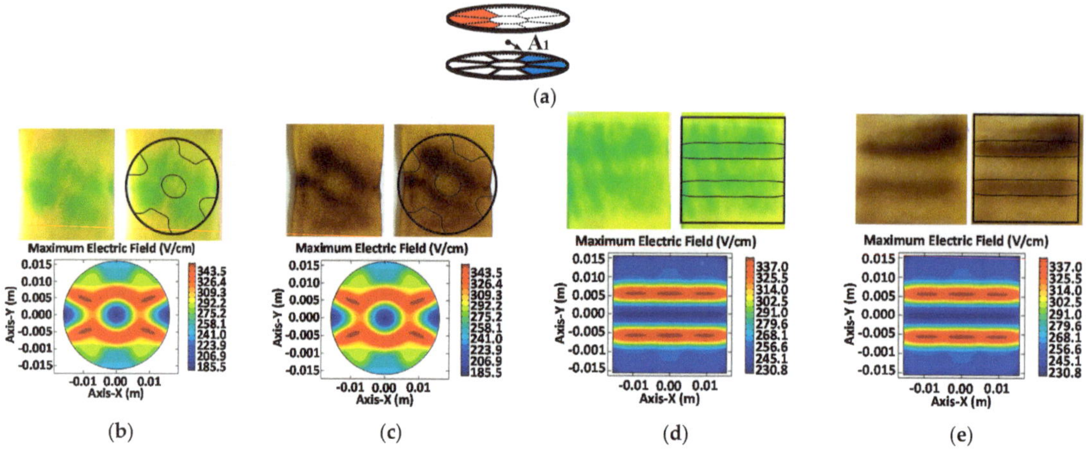

Figure 9. Results of the application in plant tissue of a single CS: (**a**) applied CS, (**b**) photo of the dyed potato and simulated electric field, using circular electrode, (**c**) photo of the degraded potato and simulated electric field, using circular electrode, (**d**) photo of the dyed potato and simulated electric field, using square electrode, and (**e**) photo of the degraded potato and simulated electric field, using square electrode.

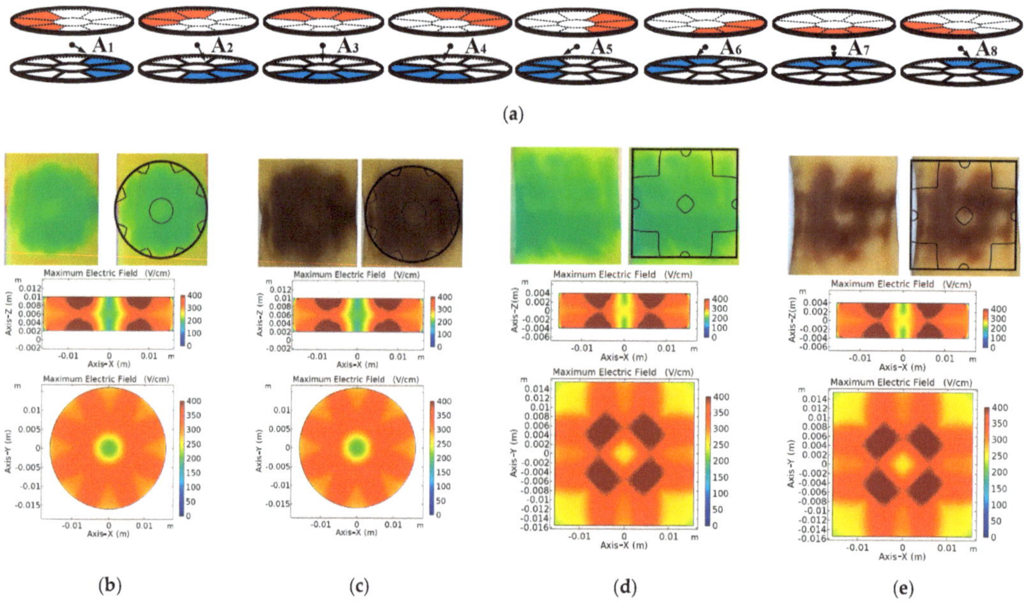

Figure 10. Result of the application in plant tissue of 8 CSs: (**a**) applied CSs, (**b**) photo of the dyed potato and simulated maximum electric field at each point, using circular electrode, (**c**) photo of the degraded potato and simulated maximum electric field at each point, using circular electrode, (**d**) photo of the dyed potato and simulated electric field, using square electrode, and (**e**) photo of the degraded potato and simulated electric field, using square electrode.

In Figure 10a, the CSs that had been applied and their main electric field directions are represented for each combination of active electrode cells. In each experiment, 10 pulses of 100 µs, 800 V, and separated by 100 ms have been applied; and this process has been repeated in the 8 configurations represented in the Figure 10a. Electroporation was applied to 8 mm thick samples of young Monalisa potatoes. In Figure 10b–e, the images of the treated potatoes are shown, and the simulated contour of the area treated at 300 V/cm or more is superimposed on the image. In addition, these figures show a vertical and horizontal slice of the simulated models, where are represented the maximum electric field at each point of the tissue produced by the 8 simulated electric field distributions. These results allow us to see that the models developed enable the prediction of the applied electric field. The next step has been the use of the models developed to achieve the optimization of the electrodes.

3.2. Size and Geometry Optimization

The MPPE prototypes are designed to apply the electric field in the best possible way. One of the advantages of this electrode is that its geometry can be varied and optimized for each case. In this section, a theorical example has been proposed with a spherical tumor of 0.75 mm in diameter placed in the center of a healthy tissue without imperfection, 1 cm thick. The objective is to present a method that allows optimizing the geometry of the electrode based on an initial study of the geometry of the tumor and its positioning. In order to evaluate both geometries, the treated volumes have been estimated in the two models, varying the surface of the electrodes and with an equivalent electrode surface ($L_E = \sqrt{\pi} \cdot R_E \quad \text{and} \quad L_I = \sqrt{\pi} \cdot R_I$). To vary the surface of the electrodes, only the side of the electrode cell of the square electrodes has been varied and then a circular electrode with the same surfaces has been constructed. The data have been analyzed assuming electroporation limit values of 800, 600 and 400 V/cm and an angle between 30° and 150°. In the simulation, 32 different configurations of the electrodes have been applied. There are more CSs that may be useful in other cases in which the tumor is positioned differently, or the electrodes are not positioned evenly. For the example that has been used, the 32 proposed CSs have been studied individually (Figures 11a and 12b). The CSs used are considered to apply a uniform electric field and do not produce many edge effects.

Figure 11a shows the configurations applied to the square electrode, in Figure 12b shows the configurations applied to the circular electrode and in Figures 11b and 12b their results can be seen. It shows the data obtained from the comparison of the proposed electrodes. Firstly, both electrodes allow to treat a large all tumor in 10 directions at 400 V/cm as minimum [27]. It is also possible to cover an important part of the volume in 16 directions. In this example, the optimal geometry within those studied would be the circular electrode with the surface of the square electrode 0.75 mm on a side. In both geometries the best results have been obtained with the same size of central electrode cell. This is because the ideal electrode cell size depends on both the tumor radius and the spacing between the electrodes. Regarding the size of the tumor, the electrode cells must have a comparable size to be able to focus on the treatment. The electrode cells should also have a size comparable to the separation of the electrodes since it has been observed that if the electrode cells are too large, the electric field concentrates on the edges of the electrodes treating the same volume, this separates the outer electrode cells, producing a worse field distribution.

The results make it clear that it is possible to optimize the geometry of the presented electrodes, and that their use allows to obtain a significantly increase of number of directions in which each point of the tissue can be treated, also being able to obtain the advantages of state-of-art electrodes.

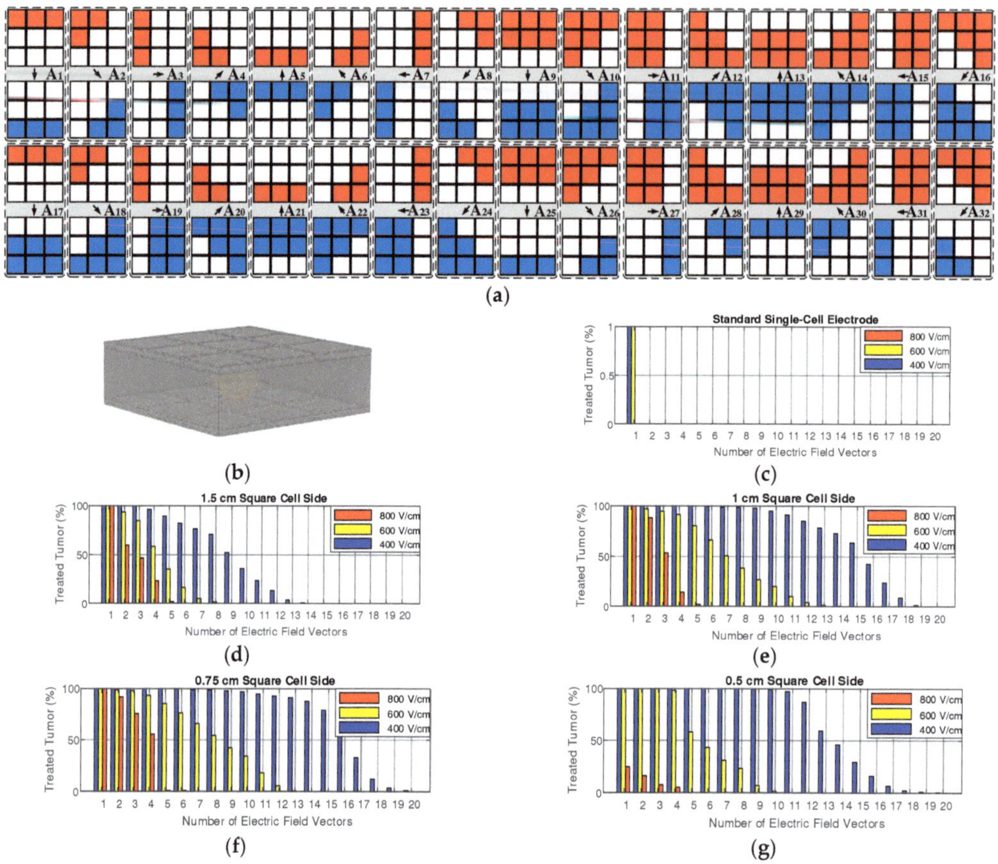

Figure 11. Square electrode optimization results: (**a**) 32 CSs applied, and (**b**) geometry of the square matrix parallel-plate electrode (MPPE) with tumor. Bar graphs with the percentage of tumor treated: standard electrode (**c**), 1.5 cm square electrode cell side (**d**), 1 cm square electrode cell side (**e**), 0.75 cm square electrode cell side (**f**) and 0.5 cm square electrode cell side (**g**).

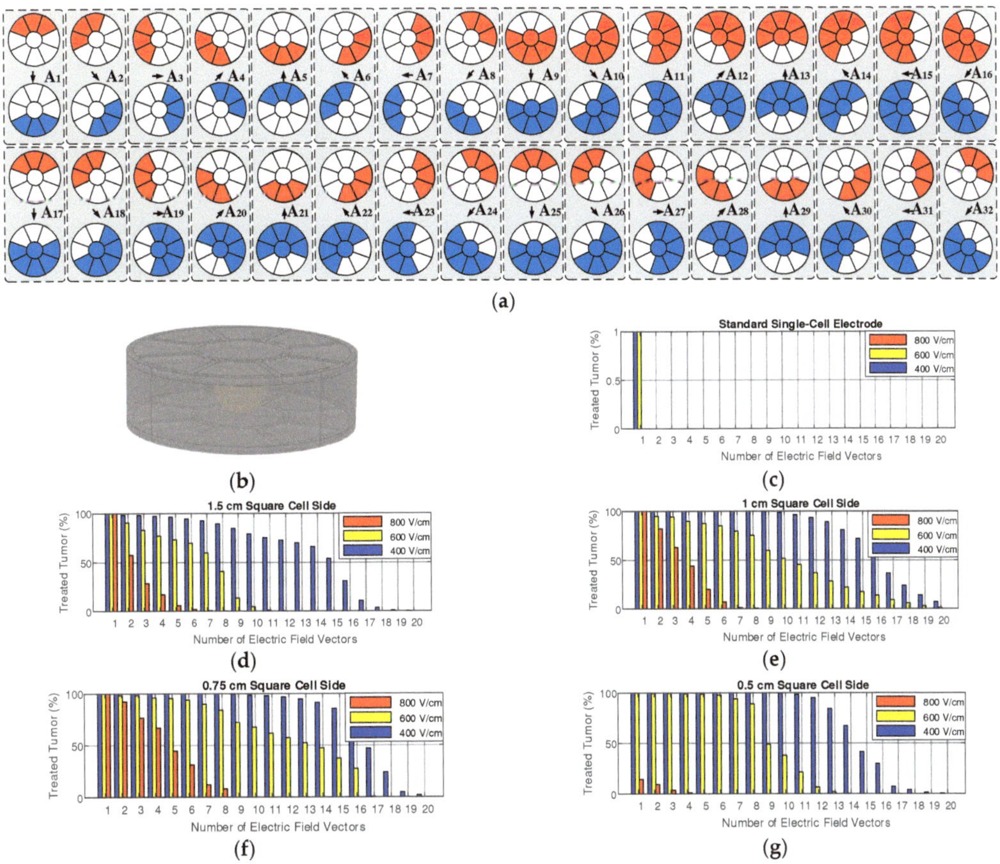

Figure 12. Circular electrode optimization results: (**a**) 32 CSs applied, and (**b**) geometry of the circular MPPE with tumor. Bar graphs with the percentage of tumor treated: standard electrode (**c**), 1.5 cm square electrode cell side (**d**), 1 cm square electrode cell side (**e**), 0.75 cm square electrode cell side (**f**) and 0.5 cm square electrode cell side (**g**).

4. Conclusions

Electroporation is a cancer treatment technique with promising results that needs to use the latest power electronics technologies and electromagnetic designs. To achieve effective treatments, an adequate electric field distribution application, one interesting way is the use of advanced multi-electrode proposals and multi-output power converters.

The proposed structure allows the application of a uniform electric field like the current flat plates but also allows the application of several electric field vectors with different orientations. Applying electric field vectors in different orientations can improve the homogeneity an the efficacy of the treatment [33–35] as is done with needle multi-electrodes [23–25,36,37], but unlike these the proposed electrode does not need to be repositioned. Although the use of flat plate electrodes is not very common in internal tissues, some experiments have been carried out, and it is believed that with a suitable support the proposed multi-electrodes could be better in some cases.

From the obtained results, it can be concluded that it is possible to improve the treatment by means of the proposed multi-electrodes, but their dimensions and design must be adapted in order to optimize the application for each tumor geometry and position. Unlike traditional needle electrodes, the proposed methodology reduces the edge effect in the electric field strength distribution, minimizes thermal effects, and reduces injuries that can occur due to electrode repositioning.

The proposed multi-electrode is an alternative to the current parallel-plates, but it allows applying the electric field in a more flexible way to improve the homogenization of the treatments. It is therefore useful in the treatment of large tumors located in areas accessible without surgery, or even in the treatment of internal organs with high blood perfusion where needles cannot be used.

The proposed electrodes could be useful in the treatment of liver tumors where metastases are common. In this area, electroporation has great advantages since it is not affected by the vascularization of the liver, and also the liver has been documented to be able to regenerate from the effects of electroporation [5], since the electric field does not affect vascular scaffold. Therefore, the proposed methodology could treat large volumes of liver with metastases in a fast and effective way, allowing the regeneration of healthy tissue. Finally, although parallel plates are often difficult to apply because they are usually applied by tweezers, the proposed electrodes can be built with flexible PCBs and could be applied in an easier way avoiding the use of tweezers.

To test the proposed electrodes, a new multi-output generator has been developed. The proposed structures have been analyzed and compared, and finite element analysis models have been developed and validated.

Author Contributions: Conceptualization, B.L.-A., H.S. and O.L.; methodology, B.L.-A., H.S. and O.L.; validation, B.L.-A. and P.B.; formal analysis, B.L.-A.; investigation, B.L.-A., H.S., O.L. and P.B.; writing—original draft preparation, B.L.-A.; writing—review and editing, O.L., H.S., P.B. and J.M.B. All authors have read and agreed to the published version of the manuscript.

Funding: This work was partly supported by the Spanish MICINN and AEI under Project PID2019-103939RB-I00, co-funded by EU through FEDER program, and by the MECD under grant FPU16/03765.

Conflicts of Interest: The authors declare no conflict of interest.

References

1. Junquera, M.C.; Castiella, T.; Gracia-Llanes, J.; López-Alonso, B.; Hernáez, A.; García-Hernández, L.; Güemes, A.; Mejía, E.; Iruzubieta, P.; Naval, A.; et al. Electron microscope evaluation of irreversible electroporation on the liver in a porcine model. *Histol. Histopathol.* **2017**, *32*, 125.
2. Weaver, J.C. Electroporation of cells and tissues. *IEEE Trans. Plasma Sci.* **2000**, *28*, 24–33. [CrossRef]
3. Lopez-Alonso, B.; Hernaez, A.; Sarnago, H.; Naval, A.; Guemes, A.; Junquera, C.; Burdio, J.M.; Castiella, T.; Monleon, E.; Gracia-Llanes, J.; et al. Histopathological and Ultrastructural Changes after Electroporation in Pig Liver Using Parallel-Plate Electrodes and High-Performance Generator. *Sci. Rep.* **2019**, *9*, 12. [CrossRef] [PubMed]
4. Monleón, E.; Castiella, T.; Mejía, E.; Gracia-Llanes, J.; Iruzubieta, P.; Junquera, M.C.; Hernáez, A.; Güemes, A.; López-Alonso, B.; Sarnago, H.; et al. Remodelling of liver tissue after irreversible electroporation in porcine. *Histol. Histopathol.* **2019**, *34*, 132.
5. Iruzubieta, P.; Gracia-Llanes, J.; Meleón, E.; Mejía, E.; Castiella, T.; Hernáez, A.; Güemes, A.; López-Alonso, B.; Sarnago, H.; Lucía, O.; et al. Ultrastructural study of pig liver regeneration following ablation with irreversible electroporation. *Histol. Histopathol.* **2019**, *34*, 131.
6. Geboers, B.; Scheffer, H.J.; Graybill, P.M.; Ruarus, A.H.; Nieuwenhuizen, S.; Puijk, R.S.; Tol, P.M.; Davalos, R.V.; Rubinsky, B.; Gruijl, T.D.; et al. High-Voltage Electrical Pulses in Oncology: Irreversible Electroporation, Electrochemotherapy, Gene Electrotransfer, Electrofusion, and Electroimmunotherapy. *Radiology* **2020**, *295*, 254–272. [CrossRef]
7. Yarmush, M.L.; Golberg, A.; Sersa, G.; Kotnik, T.; Miklavcic, D. Electroporation-Based Technologies for Medicine: Principles, Applications, and Challenges. In *Annual Review of Biomedical Engineering*; Annual Reviews; Yarmush, M.L., Ed.; Palo Alto: Santa Clara, CA, USA, 2014; Volume 16, pp. 295–320.
8. O'Brien, T.; Passeri, M.; Lorenzo, M.; Sulzer, J.; Lyman, W.; Swet, J.; Vrochides, D.; Baker, E.; Iannitti, D.; Davalos, R.; et al. Experimental High-Frequency Irreversible Electroporation Using a Single-Needle Delivery Approach for Nonthermal Pancreatic Ablation In Vivo. *J. Vasc. Interv. Radiol.* **2019**, *30*, 854–862.e7. [CrossRef]
9. Zhang, Y.; Lyu, C.; Liu, Y.; Lv, Y.; Chang, T.T.; Rubinsky, B. Molecular and histological study on the effects of non-thermal irreversible electroporation on the liver. *Biochem. Biophys. Res. Commun.* **2018**, *500*, 665–670. [CrossRef] [PubMed]
10. Mathy, R.M.; Tinoush, P.; da Florencia, R.D.; Braun, A.; Ghamarnejad, O.; Radeleff, B.; Kauczor, H.-U.; Chang, D.-H. Impact of needle positioning on ablation success of irreversible electroporation: A unicentric retrospective analysis. *Sci. Rep.* **2020**, *10*, 21902. [CrossRef] [PubMed]

11. Huang, D.; Zhao, D.; Li, J.; Du, L.; Wei, Z.; Liang, Z.; Li, Z. In A minimally invasive in vivo electroporation method utilizing flexile electrode and microneedle roller. In Proceedings of the 2017 19th International Conference on Solid-State Sensors, Actuators and Microsystems (TRANSDUCERS), Kaohsiung City, Taiwan, 18–22 June 2017; pp. 1684–1687.
12. Neal, R.E., 2nd; Singh, R.; Hatcher, H.C.; Kock, N.D.; Torti, S.V.; Davalos, R.V. Treatment of breast cancer through the application of irreversible electroporation using a novel minimally invasive single needle electrode. *Breast Cancer Res. Treat.* **2010**, *123*, 295–301. [CrossRef]
13. Berkenbrock, J.A.; Machado, R.G.; Suzuki, D.O.H. Electrochemotherapy Effectiveness Loss due to Electric Field Indentation between Needle Electrodes: A Numerical Study. *J. Healthc. Eng* **2018**, *2018*, 6024635. [CrossRef]
14. Davalos, R.V.; Garcia, P.A.; Edd, J.F. Thermal Aspects of Irreversible Electroporation. In *Irreversible Electroporation*; Rubinsky, B., Ed.; Springer: Berlin/Heidelberg, Germany, 2010; pp. 123–154.
15. Biasiolo, L.; Bariani, P.; Marchi, N.; Dughiero, F.; Sieni, E.; Campana, L. An Electrode for the Treatment of Large Surfaces in ECT. In Proceedings of the IECON 2018 44th Annual Conference of the IEEE Industrial Electronics Society, Washington, DC, USA, 21–23 October 2018; pp. 3319–3323.
16. Hansen, H.; Bourke, M.; Stigaard, T.; Clover, A.; Buckley, M.; O'Riordain, M.; Winter, D.; Johannesen, H.; Hansen, R.; Heebøll, H.; et al. Electrochemotherapy for colorectal cancer using endoscopic electroporation: A phase 1 clinical study. *Endosc. Int. Open* **2020**, *8*, E124–E132.
17. Rombouts, S.J.E.; van Dijck, W.P.M.; Nijkamp, M.W.; Derksen, T.C.; Brosens, L.A.A.; Hoogwater, F.J.H.; van Leeuwen, M.S.; Borel Rinkes, I.H.M.; van Hillegersberg, R.; Wittkampf, F.H.; et al. Clinical and pathological outcomes after irreversible electroporation of the pancreas using two parallel plate electrodes: A porcine model. *Hpb Off. J. Int. Hepato Pancreato Biliary Assoc.* **2017**, *19*, 1058–1065. [CrossRef]
18. Panella, C.; Castellvi, Q.; Moll, X.; Quesada, R.; Villanueva, A.; Iglesias, M.; Naranjo, D.; Sanchez-Velazquez, P.; Andaluz, A.; Grande, L.; et al. Focused transhepatic electroporation mediated by hypersaline infusion through the portal vein in rat model. Preliminary results on differential conductivity. *Radiol. Oncol.* **2017**, *51*, 415–421. [CrossRef] [PubMed]
19. Denzi, A.; Strigari, L.; Di Filippo, F.; Botti, C.; Di Filippo, S.; Perracchio, L.; Ronchetti, M.; Cadossi, R.; Liberti, M. Modeling the positioning of single needle electrodes for the treatment of breast cancer in a clinical case. *Biomed. Eng. Online* **2015**, *14*, S1. [CrossRef] [PubMed]
20. Cvetkoska, A.; Pirc, E.; Rebersek, M.; Magjarevic, R.; Miklavcic, D. Towards standardization of electroporation devices and protocols. *IEEE Instrum. Meas. Mag.* **2020**, *23*, 74–81. [CrossRef]
21. Bernardis, A.; Bullo, M.; Campana, L.G.; Barba, P.D.; Dughiero, F.; Forzan, M.; Mognaschi, M.E.; Sgarbossa, P.; Sieni, E. Electric field computation and measurements in the electroporation of inhomogeneous samples %. *J. Open Phys.* **2017**, *15*, 790–796. [CrossRef]
22. Pavliha, D.; Kos, B.; Marcan, M.; Zupanic, A.; Sersa, G.; Miklavcic, D. Planning of Electroporation-Based Treatments Using Web-Based Treatment-Planning Software. *J. Membr. Biol.* **2013**, *246*, 833–842. [CrossRef]
23. Campana, L.G.; Bullo, M.; Di Barba, P.; Dughiero, F.; Forzan, M.; Mognaschi, M.E.; Sgarbossa, P.; Tosi, A.L.; Bernardis, A.; Sieni, E. Effect of Tissue Inhomogeneity in Soft Tissue Sarcomas: From Real Cases to Numerical and Experimental Models. *Technol. Cancer Res. Treat.* **2018**, *17*, 1533033818789693. [CrossRef] [PubMed]
24. Castiello, M.; Dughiero, F.; Scandola, F.; Sieni, E.; Campana, L.; Rossi, C.; Mattei, M.; Pellati, A.; Ongaro, A. A New Grid Electrode for Electrochemotherapy Treatment of Large Skin Tumors. *Dielectr. Electr. Insul. IEEE Trans.* **2014**, *21*, 1424–1432. [CrossRef]
25. Ongaro, A.; Campana, L.; Mattei, M.; Dughiero, F.; Forzan, M.; Pellati, A.; Rossi, C.; Sieni, E. Evaluation of the Electroporation Efficiency of a Grid Electrode for Electrochemotherapy: From Numerical Model to In Vitro Tests. *Technol. Cancer Res. Treat.* **2015**, *15*, 296–307. [CrossRef]
26. López-Alonso, B.; Sarnago, H.; Burdío, J.M.; Lucía, Ó. In Multi-Electrode Architecture Analysis and Modeling for Cancer Treatment using Electroporation. In Proceedings of the IECON 2020 The 46th Annual Conference of the IEEE Industrial Electronics Society, Singapore, 18–21 October 2020; pp. 718–723.
27. López-Alonso, B.; Sarnago, H.; Lucía, O.; Briz, P.; Burdío, J.M. Real-Time Impedance Monitoring During Electroporation Processes in Vegetal Tissue Using a High-Performance Generator. *Sensors* **2020**, *20*, 3158. [CrossRef] [PubMed]
28. López-Alonso, B.; Sarnago, H.; Burdío, J.M.; Lucía, O. Electro-thermal modeling of irreversible electroporation and validation method of electric field distribution. *Int. J. Appl. Electromagn. Mech.* **2020**, *63*, 41–50. [CrossRef]
29. Corovic, S.; Lackovic, I.; Sustaric, P.; Sustar, T.; Rodic, T.; Miklavcic, D. Modeling of electric field distribution in tissues during electroporation. *Biomed. Eng. Online* **2013**, *12*, 16. [CrossRef]
30. Breton, M.; Buret, F.; Krähenbühl, L.; Leguèbe, M.; Mir, L.; Perrussel, R.; Poignard, C.; Scorretti, R.; Voyer, D. Non-Linear Steady-State Electrical Current Modeling for the Electropermeabilization of Biological Tissue. *IEEE Trans. Magn.* **2015**, *51*, 1–4. [CrossRef]
31. López-Alonso, B.; Sarnago, H.; Lucía, O.; Burdío, J.M. In Multiple-Output Generator for Omnidirectional Electroporation and Real-Time Process Monitoring. In Proceedings of the 2021 IEEE Applied Power Electronics Conference and Exposition (APEC21), Phenix, AZ, USA, 9–12 June 2021.
32. Castellví, Q.; Banús, J.; Ivorra, A. 3D Assessment of Irreversible Electroporation Treatments in Vegetal Models. In *1st World Congress on Electroporation and Pulsed Electric Fields in Biology, Medicine and Food & Environmental Technologies*; Jarm, T., Kramar, P., Eds.; Springer: Singapore, 2016; pp. 294–297.

33. Valic, B.; Golzio, M.; Pavlin, M.; Schatz, A.; Faurie, C.; Gabriel, B.; Teissié, J.; Rols, M.P.; Miklavcic, D. Effect of electric field induced transmembrane potential on spheroidal cells: Theory and experiment. *Eur. Biophys. J.* **2003**, *32*, 519–528. [CrossRef]
34. Reberšek, M.; Faurie, C.; Kandušer, M.; Čorović, S.; Teissié, J.; Rols, M.-P.; Miklavčič, D. Electroporator with automatic change of electric field direction improves gene electrotransfer in-vitro. *Biomed. Eng. Online* **2007**, *6*, 25. [CrossRef]
35. Čorović, S.; Županič, A.; Kranjc, S.; Al Sakere, B.; Leroy-Willig, A.; Mir, L.M.; Miklavčič, D. The influence of skeletal muscle anisotropy on electroporation: In vivo study and numerical modeling. *Med. Biol. Eng. Comput.* **2010**, *48*, 637–648. [CrossRef] [PubMed]
36. Leticia Vera-Tizatl, A.; Elizabeth Vera-Tizatl, C.; Vera-Hernandez, A.; Leija-Salas, L.; Rodriguez, S.; Miklavcic, D.; Kos, B. Computational Feasibility Analysis of Electrochemotherapy with Novel Needle-Electrode Arrays for the Treatment of Invasive Breast Ductal Carcinoma. *Technol. Cancer Res. Treat.* **2018**, *17*, 1533033818794939. [CrossRef]
37. Brock, R.M.; Beitel-White, N.; Davalos, R.V.; Allen, I.C. Starting a Fire Without Flame: The Induction of Cell Death and Inflammation in Electroporation-Based Tumor Ablation Strategies. *Front. Oncol.* **2020**, *10*, 1235. [CrossRef]

energies

Article

On-Line Optimization of Energy Consumption in Electromagnetic Mill Installation

Szymon Ogonowski

Department of Measurements and Control, Silesian University of Technology, Akademicka 16, 44-100 Gliwice, Poland; szymon.ogonowski@polsl.pl

Abstract: Milling is one of the most energy consuming stages of the value production chain in many industries. To minimize the specific energy required, new and more efficient devices and circuits are designed and dedicated optimizing control strategies are applied. This research presents the results of innovative electromagnetic mill energy consumption reduction with dedicated supervisory on-line optimizing control algorithm. The paper describes an algorithm that uses the active power measurement and searches for the minimum on the active constraints of the optimization problem. The constraints follow from the product quality, mill supply voltage and magnetic induction requirements. Algorithm performance was tested in simulations, but the main validation was performed on a semi-industrial dry grinding and classification circuit equipped with an electromagnetic mill. The results of the experiments presented in this paper show that the application of the on-line optimization algorithm allows for even a 40% reduction in the electromagnetic mill energy consumption when compared to the nominal operating point.

Keywords: electromagnetic mill; magnetic induction; comminution; energy consumption reduction; on-line optimization; process control

Citation: Ogonowski, S. On-Line Optimization of Energy Consumption in Electromagnetic Mill Installation. *Energies* **2021**, *14*, 2380. https://doi.org/10.3390/en14092380

Academic Editor: Ivica Stevanovic

Received: 17 March 2021
Accepted: 19 April 2021
Published: 22 April 2021

1. Introduction

Comminution is one of the most commonly used processing stages in almost every type of industry, including the mining, mineral processing, chemical, construction, pharmaceutical cosmetics and food industries, among others [1]. It is a process where coarse particles of the feed stream are subjected to various mechanical operations e.g., smashing, breaking, attrition, pealing, cutting or crushing to reduce their average size [2]. In many applications, it also serves as a preparation stage for the following operations. In mineral processing, grinding is a part of the production circuit which aim is to concentrate the valuable minerals contained in raw ores. Minerals are initially liberated from the ore matrix by comminution and particle size separation processes. In the next stage, they are separated from the gangue using processes that select particles based on their physical or chemical properties, such as surface hydrophobicity, specific gravity, magnetic susceptibility, chemical reactivity or color [3,4]. Efficiency of the following operations requires strictly defined features of the comminution product, so the size reduction process is subjected to important technological constraints.

Industry has used comminution for years; however, it is still the most energy consuming operation in the whole value production chain [5]. It is responsible for over 50% of mining companies' total power consumption [6,7], reaching even higher values of 70% for grains' size reduction from 30–50 mm to 20–50 microns [8,9]. Required energy input (often called specific energy) varies depending not only on the size reduction ratio but mostly on the final average size of the product particles. Specific energies observed in the industry can vary significantly: e.g., only 1 kWh/t is required to produce 10 mm particles, while 200 kWh/t is required for 1 μm particles [5]. Such important differences result from longer duration of material processing, less energy-efficient mechanical operations and different machinery required for finer products. Another important issue follows from deteriorating ore grades. To meet the market requirements, the mineral industry is often forced to process low-grade ores with fine-graded valuable minerals, which in turn requires finer milling before further processing is possible [10].

Economic and environmental aspects indicate the relevance of the higher efficiency of the comminution process that can be obtained by applying novel and more effective processing technologies, devices and circuits [11]. It can also be achieved with the operation optimization using dedicated measurements, modeling and control techniques [12,13], or even with new, more environmentally friendly thermoelectric materials [14]. In this research, the energy consumption of electromagnetic mills (EMM) is discussed. The device was designed and patented [15] by the ELTRAF company (Lubliniec, Poland) for fast and efficient ultra-fine milling in batch and continuous processes. One can find only one other commercial solution on the market with a similar design, from GlobeCore [16]; however, it is dedicated only to batch milling. There are several studies reported on EMM application with other inductor designs as well [17,18]. While EMM studies are becoming more popular, one can find discussion on the efficiency of the existing EMM design [19,20]. The authors stress the need to improve the design of existing EMM constructions based on the modeling and simulation results. While such actions are of the high importance, the research reported in this paper is focused on the improvement of the energy efficiency of the existing EMM's operation using control techniques. When dynamic and static requirements for the measurements and actuators are met it is possible to design multilayered control system to optimize the whole system performance according to the chosen criterion [12]. In this approach, the key issue is to stabilize the process using direct control system and then to optimize the chosen criterion for the given operating point. For the dry grinding and classification circuit with EMM used in this research the most important issue is to stabilize the EMM working chamber load by controlling the amount of the material being processed in the mill [21]. Then it is possible to introduce other objectives to the multilayered control system [22] based on technological and economical requirements.

This paper presents research results on reduction of the electromagnetic mill energy consumption with dedicated on-line optimizing control algorithm that takes into account constraints resulting from process technological requirements and EMM construction limitations. Section 2 describes principles of the electromagnetic mill operation and presents the semi-industrial grinding and classification circuit used in the research. Section 3 describes the optimization process: problem formulation, objective function and constraints definition as well as the optimization algorithm design. Simulation results and algorithm validation are also presented in this section. Section 4 describes the experimental verification of the algorithm using semi-industrial circuit with EMM. Practical implementation aspects are discussed, and experimental results are presented and commented on.

2. Electromagnetic Mill

2.1. Construction and Operating Principles of EMM

The electromagnetic mill developed by ELTRAF is a novel grinding device in which a rotating magnetic field directly causes movement of ferromagnetic grinding medium in the working chamber [23]. Figure 1 presents an exemplary D200 EMM with a working chamber diameter equal to 200 mm. The milling process is performed in the working chamber of the mill, which is a non-ferromagnetic tube placed inside the electromagnetic (EM) field inductor. The inductor is composed with three pairs of salient poles placed radially around the working chamber and shifted to each other by an angle of 120° (Figure 1a). After being powered by three-phase grid, they create a rotating magnetic field inside the working chamber, transverse to the chamber length and the direction of the processed material flow. The working chamber is filled with small ferromagnetic rods serving as grinding medium. The size parameters of the rods depend on the mill diameter and the average grain size of the milled material, and are usually 1–3 mm in diameter and 10–15 mm in length. When the rotating magnetic field is inducted in the working chamber, the grinding medium start to move. The processed material is directed to the working chamber, transverse to the grinding medium movement. Material particles collide with the grinding medium when passing through the working chamber what is the main action causing that size reducing of the particles. The rotating magnetic field keeps the grinding medium inside the working chamber, preventing its movement with the processed material stream. Collisions between the rods themselves and with the processed material cause chaotic movements inside the chamber (Figure 1c) and increase the milling effect. At the same time, friction between the rods, material and working chamber walls produces an important amount of heat, further increased by the induced eddy currents. Excessive heat is removed from the system by the air cooling system installed inside the mill's chassis (Figure 1b) and by the stream of processed material with the transport medium (gas or liquid).

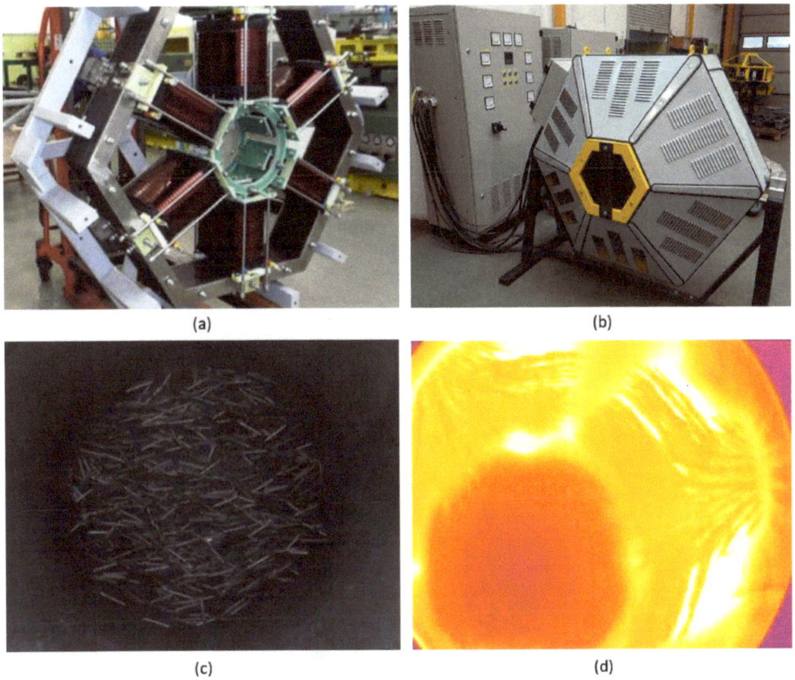

Figure 1. Electromagnetic mill D200: (**a**) inductor; (**b**) mill on the test stand without the working chamber; (**c**) ferromagnetic rods (grinding medium) moving inside the working chamber; (**d**) grinding medium during stop procedure arranging according to the pole lines (thermo-vision image).

The physical parameters of the mill, including working chamber dimensions, distance between the chamber and inductor, number of inductor cores, etc., were optimized to maximize the force exerted by the electromagnetic field and to minimize the electric energy consumption [23]. It makes the electromagnetic mill highly competitive for other milling solutions.

In this research, EMM is a part of the dry grinding and classification circuit, described in more detail in Section 2.2, equipped with dedicated control system based on PLC and SCADA systems. The EM field inductor of the mill is powered by the Mitsubishi Electric FR—F700 frequency inverter with scalar control [24]. Additionally, the user can change the base frequency f_b, which is one of the inverter parameters and corresponds to the rated operating frequency of the motor, if the inverter powers such a motor. Thanks to this, knowing the rated supply voltage, the inverter is able to determine the ratio and change the output voltage in such a way as to maintain its value when the output frequency changes (according to the principle of operation of the scalar control of the inverter) [25].

The frequency inverter allows continuous control and modification of the mill's operating frequency without any technological breaks, and in turn fast reaction, for processing disturbances (milled material parameters variation, flowrate changes, product quality requirements, etc.). It is one of the most important advantage of EMM when compared to the classic milling solutions, where rotation speed of the drum mills is constant, and its change requires technological downtime and mechanical modifications in the mill's installation [26].

2.2. Semi-Industrial Grinding and Classification Circuit

ELTRAF's electromagnetic mill is designed to operate in dry and wet batch and continuous milling process. While initial tests for new materials are usually performed in controlled batch experiments, industrial solutions uses dedicated circuits for continuous processes. In this research D100 EMM with 100 mm working chamber diameter is used in the dry grinding and classification circuit located in the Laboratory of Comminution Processes Control at Silesian University of Technology in Gliwice, Poland (see Figure 2). The circuit is designed according to the patented methodology [27] which assumes vertical positioning of the mill and forced air transported movement of the processed material.

Figure 2. Semi-industrial dry grinding and classification circuit with EMM D100: (**a**) installation with air transport and heat recovery systems; (**b**) control and power supply cabinets.

In the circuit presented in Figure 2a, the mill (no. 1) is placed horizontally, and the feeding system (no. 2) directs the processed material to the mill from the top of the working chamber. Particles of the material falls by gravity into the working chamber and are kept inside by the transport air stream produced by the negative pressure fan (no. 5), directed from the bottom of the working chamber. The amount of material inside the working chamber (load of the mill) is controlled by dedicated algorithm using indirect acoustic measurement [21]. During milling process the material particles are becoming smaller in size and lighter, and in turn, they start to be moved upwards outside the working chamber into the main classifier (no. 3), which creates two material streams: the product stream (particles with diameter $\leq$50 µm) and the recycle stream (particles with diameter $\geq$50 µm) [28]. The product stream is then separated from the air stream in the cyclone (no. 4) and the recycle stream is fed back to the mill from its bottom to ensure additional passing of the particles through the working chamber for the further size reduction.

The whole process is controlled by dedicated power supply and control system presented in Figure 2b. The measurement and control cabinet (no. 6) is equipped with two PLC systems based on SIMATIC S7-300 [29] and SIMATIC S7-1200 [30] controllers with digital and analog I/O modules for communication with sensors and actuators. The system measures transporting air flows, pressures, temperatures and humidities in each supply line of the installation, feed flowrate, electric parameters (voltage, current, active power, etc.), mill load, and mass of the grinding medium inside the working chamber. Research has been performed to improve the system with on-line feed humidity measurements and product quality [31,32]. The system controls the position of damping flaps for the transporting air flow management, process material flowrate and most important—the frequency of the mill operation. S7-1200 controller is connected via MODBUS-RTU communication with Mitsubishi Electric F700 frequency inverter and Lumel ND20 3-phase power network meter [33] installed in the supply cabinet (no. 8). For the purpose of the interaction with the process operator SIMATIC Panel PC [34] was installed and equipped with SCADA system based on GE iFIX platform [35].

The above configuration made it possible to design and implement the multilayered hierarchical control system with dedicated direct control, supervisory control, optimization and production planning layers [36]. Each layer is responsible for specific control tasks, split according to the specific plant sub-systems' dynamics (control horizons), frequency of disturbances to be attenuated and physical range of the control actions in the plant. The research reported in this paper focus only on the optimization layer actions, so for further discussion it is enough to state that other control layers stabilize the process in the chosen operating point. Control set-points for each sub-system are calculated and distributed by the production planning layer based on the technological requirements.

3. Power Optimization Problem

3.1. Objective Function

To maximize profits, one may improve the operation of the distinguished sub-systems of the technological installation. Such division of the global optimization task into separated sub-problems reduces the complexity of the optimization problem; however, the performance of each optimization algorithm should be supervised by an upper-level system that combines all optimization sub-problems into one, and in turn leads to the minimization of the global objective function. This approach is a well-known decomposition of complex optimization tasks, where the coordination of partial optimization tasks takes place by changing the parameters (objective functions and/or constraints) of the partial tasks [37].

In the installation described in Section 3.2, the electromagnetic mill consumes the largest amount of the electrical energy in the system [38]. Thus, it is reasonable to formulate the optimization task as the minimization of the electromagnetic mill active power taking into account quality of the final product and production efficiency.

To formulate the objective function of the optimization problem, a series of tests were required to determine the static active power characteristics of the electromagnetic mill. The dependence of the static characteristic on the mill parameters allows to define decision variables. The mill is supplied by a frequency converter, which output frequency f_o and the base frequency f_b can be changed. Additionally, the mass of the grinding medium m inside the working chamber may be different as well. To create a static active power characteristic of the mill P_o, experiments were carried out in which the output frequency f_o and the base frequency f_b were changed. Additionally, the experiments were carried out for three different values of the grinding medium mass inside the mill's working chamber, within the permissible range of grinding medium mass: m^i = [200,400,600] g. The experiments were conducted for three values of the base frequency f_b^j = [40,50,60] Hz, and for changes of the output frequency f_o according to the schedule presented in Figure 3. Indices i and j represent the succeeding sample numbers. The active power was measured after reaching steady sate, i.e., 20 s after each change of the frequency set-point.

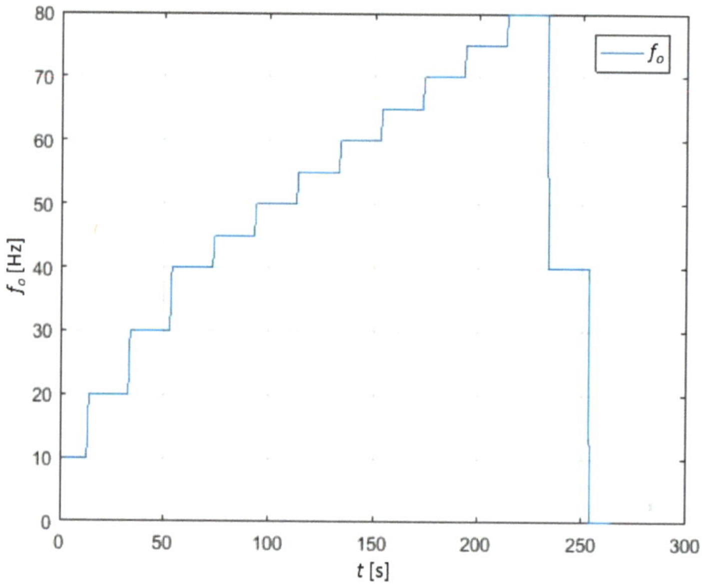

Figure 3. Schedule of the output frequency f_o changes during identification experiments.

The results of the experiments are presented in Figure 4. Note that the maximum active power is achieved for the output frequency f_o equal to the base frequency f_b. Figure 4 also shows obvious dependence on the mass of the grinding medium in the working chamber: power P_o rises with mass m. The control system has the ability to change these quantities, which means that they can become decision variables of the optimization problem. Mass m depends, however, on the required product quality and is controlled by a separate system; thus, it can be excluded from the set of decision variables and may be treated as a disturbance that affects the operation of the optimization algorithm. Finally, the following optimization problem is formulated:

$$min_{f_o,f_b} J(f_o, f_b), \tag{1}$$

where the objective function $J(f_o, f_b)$ is the static characteristic of the mill's active power (P_o). As is further explained, due to the form of the constraints and convexity of J, the analytical model of J is not necessary to be known. For simulation purposes, it is enough to approximate J with non-parametric model, e.g., using Artificial Neural Network (ANN) [39]. There are several ANN models used in this research for approximation of the active power measurements in different working conditions or for approximation of the mill's supply current measurements. In every instance two-layered neural network was used with one hidden layer. Such structure approximates non-linear characteristic more precisely than one-layered structure [39]. The number of neurons in the hidden layer was determined with the mean square error (MSE) value for measurements approximation. Approximation results were always verified with respect to the expected outcome, consistent with the observed behavior of the device, to avoid over-parametrization of the model. In most cases, seven to ten neurons in the hidden layer were used as a result of the ANN structure design.

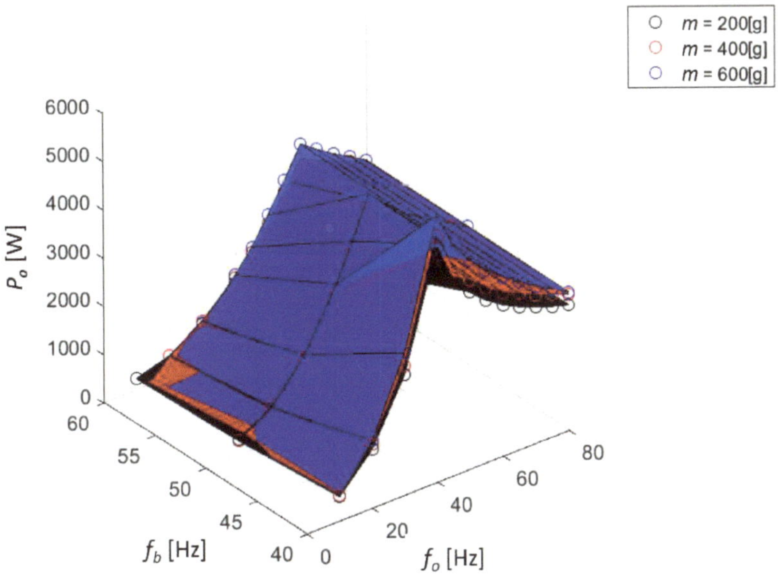

Figure 4. Static characteristic of the active power versus output and basic frequencies obtained for different mass of grinding medium.

3.2. Constraints

In industry, the technological process is always subject to constrains following from product quality requirements, safety condition of the process management, and ensuring failure-free operation. Thus, the allowed area of the J minimum search is always limited by technological constraints.

3.2.1. Product Quality Constraints

The main goal of the control system is to achieve the required production efficiency, expressed by the volume flow of the processed material F_f, and to maintain the assumed quality of the final product. The grinding process product quality is usually assessed by distribution of grain classes, represented by $\boldsymbol{p_p}$ vector of percentage content of grain classes in the product, which is obtained for a given set of grain sizes $\boldsymbol{d}$ expressed in mm [28], as for example in (2).

$$\boldsymbol{d} = [0.05,\ 0.25,\ 0.5,\ 0.57,\ 1,\ 1.5]^T, \tag{2}$$

$$\boldsymbol{p_p} = [25,\ 31,\ 31,\ 33,\ 38,\ 48,\ 100]^T, \tag{3}$$

Many factors determine the quality of the final product (e.g., feed stream, physical parameters of the ground material, etc.) [40], which means that the vector $\boldsymbol{p_p}$ depends on the parameters characterizing the process:

$$\boldsymbol{p_p} = \boldsymbol{p_p}(\boldsymbol{\Omega},\ \boldsymbol{F},\ \boldsymbol{A},\ \boldsymbol{R},\ \boldsymbol{M}), \tag{4}$$

where $\boldsymbol{\Omega}$ is a vector containing the output and base frequencies; $\boldsymbol{F}$ is a vector containing parameters of the processed material, such as volumetric flow, density, humidity, temperature, and a vector characterizing the granulation; $\boldsymbol{A}$ is a vector containing the parameters of the transport air flowing through the working chamber, such as volumetric flow, temperature and humidity; $\boldsymbol{R}$ is a vector containing the parameters of the air-material stream flowing through the recycle, i.e., air volume flow, material mass flow and the material grain size vector in recycling; $\boldsymbol{M}$ is a vector containing parameters of the grinding medium such as weight, size and material. It is assumed that the final product quality requirements will be met if the following inequality is met:

$$p_p^l \leq p_p \leq p_p^h. \tag{5}$$

From a multi-level control system point of view, most of the above-mentioned variables cannot be changed by the optimization layer but they are changed by the direct control layer or serves as a technological parameter (treated in the control system as a disturbance). Additionally, only the output and base frequencies (vector $\boldsymbol{\Omega}$) have an impact on both the product quality and the active power, and therefore they become decision variables of the optimization problem. Other values that affect the quality of the product ($\boldsymbol{F}$, $\boldsymbol{A}$, $\boldsymbol{R}$, $\boldsymbol{M}$) that are not the decision variables should be treated as constant values, depending on the type of material processed or following from the operating point of the plant. The production management layer imposes restrictions on the output frequency f_o and the base frequency f_b (6) and (7), which make it possible to maintain the desired quality of the final product:

$$f_o^l \leq f_o \leq f_o^h, \tag{6}$$

$$f_b^l \leq f_b \leq f_b^h, \tag{7}$$

Constraints (6) and (7) are determined for the certain operating point which is defined by the following set of equations:

$$\boldsymbol{F} = \boldsymbol{F_O}, \boldsymbol{A} = \boldsymbol{A_O}, \boldsymbol{R} = \boldsymbol{R_O}, \boldsymbol{M} = \boldsymbol{M_O}. \tag{8}$$

If constraints (6) and (7) are met at the system operating point defined by the independent variables (8), then the required quality of the final product is reached. The limit values in (6) and (7) depend on the type of material processed and limit values of (5). In practice, all these limits are determined in laboratory tests and analyzes made by technologists. It is assumed that they are known a priori and are given as the constant parameters of the optimization task.

3.2.2. EMM Voltage Constraint

The electromagnetic mill is equipped with six poles. The pole's windings have a limited resistance to electric breakdown. This means that they can be supplied with effective voltage U_o not higher than the breakdown voltage U_{brk}:

$$U_o \leq U_{brk}. \tag{9}$$

The electromagnetic mill is supplied by a scalar-controlled inverter. The inverter changes the output frequency f_o, but it changes the mill winding voltage U_o as well. The principle of the scalar control in the case of electric motor power supply is to maintain a constant value of the magnetic flux amplitude in the stator, equal to its value in the nominal operating conditions. The maintenance of the nominal magnetic flux causes the motor to operate under nominal conditions [25]. For this reason, in the inverters with scalar control, the output voltage is changed according to the following equation:

$$|U_o| = |U_{sup}|\frac{f_o}{f_b}, \tag{10}$$

where U_{sup} is the supply voltage. Considering the mill supply voltage limitation (9) and Equation (10), it is possible to derive the frequency limitation of the inverter:

$$f_o \leq \frac{|U_{brk}|}{|U_{sup}|} f_b. \tag{11}$$

By introducing the limitation (11) to the optimization problem, the maximum value of the voltage supplying the electromagnetic mill is limited to the level of the breakdown voltage U_{brk}.

3.2.3. Magnetic Field Induction Constraints

The basic parameter of the electromagnetic mill, which makes it possible to determine the value of the forces acting on the grinding medium in different areas of the working chamber, is the magnetic field induction B [23]. The induction directly impacts the grinding medium movement, and determines its energy to grind a feed. Also, the induction keeps grinding medium vertically in the working chamber. As field quantity, the induction is characterized by its distribution in space, and in particular determines the grinding medium movement [23].

To evaluate the quality of the electromagnetic mill inductor performance, a set of measurements of magnetic induction is carried out at selected points in the mill's working chamber, not filled with grinding medium or material. Thus, it is possible to measure and model the distribution of induction values in the working chamber space. The measurements are made for different values of the output frequency f_o. Figure 5 shows the spatial distribution of the results obtained for the frequency f_o = 50 Hz. In Figure 5, the x and y axes define the plane of the cross section of the working chamber, while the z axis is parallel to the side of the working chamber and the induced electromagnetic field rotates around it.

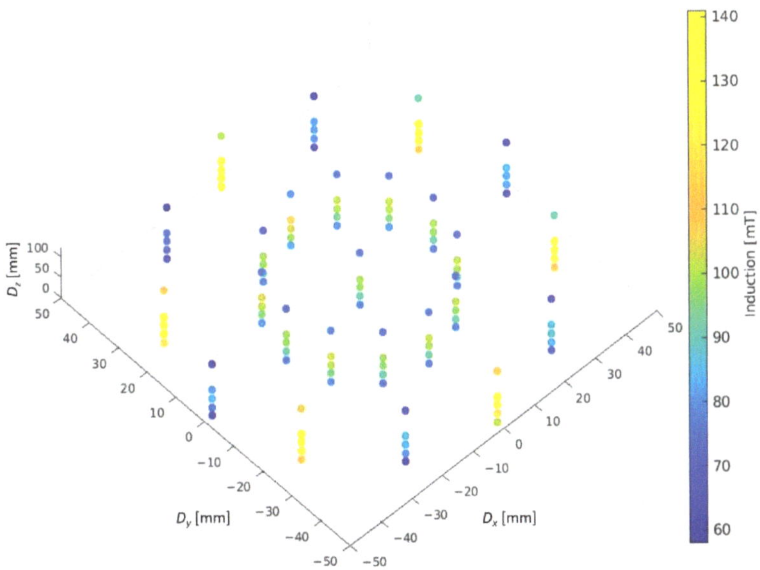

Figure 5. Distribution of the measured value of the magnetic induction inside the mill's working chamber. Dx and Dy are the distance in mm from the center of the working chamber, while Dz is the distance from the bottom (beginning) of the working chamber. The value of the induction in mT is marked with the appropriate color, shown in the legend.

To better illustrate the uniform distribution of induction, it can be presented along a line between opposite poles at an equal distance from the bottom and top of the working chamber (for $Dx = 0$) (Figure 6a) or along a line parallel to the z axis (or the side edge of the working chamber) (Figure 6b). Figure 6a shows the induction values for f_b = 50 Hz and f_o = [40,50,60,70] Hz. Figure 6b shows the measurement results for $f_b = f_o$ = 50 Hz and the respective curves correspond to the points on the cross-section of the working chamber.

Exemplary values of the measured induction and respective current I_o, as in Figure 6a, are presented in Table 1.

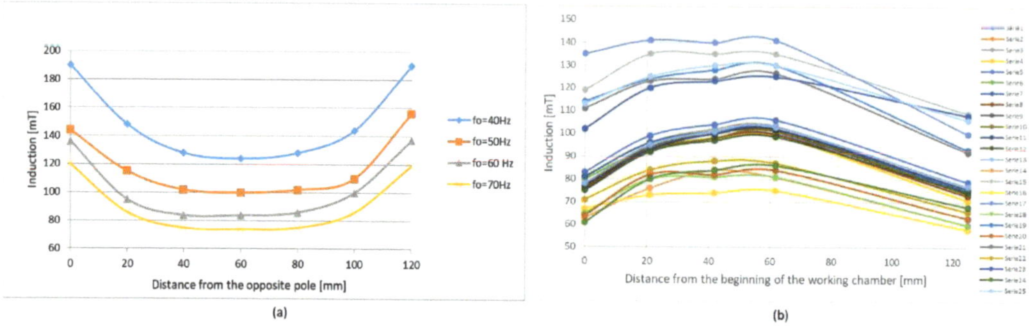

Figure 6. Distribution of the measured value of the magnetic induction inside the working chamber: (**a**) along a line between opposite poles; (**b**) along lines parallel to the side edge of the working chamber (each series represents different measuring point on the working chamber cross-section).

Table 1. Measured values of magnetic induction B for different output frequencies f_o and measured output current I_o.

Distance from Pole		Magnetic Induction B [mT]			
Left	Right	$f_o = 40$ Hz	$f_o = 50$ Hz	$f_o = 60$ Hz	$f_o = 70$ Hz
x [mm]	x′ [mm]	$I_o = 90$ A	$I_o = 68$ A	$I_o = 56$ A	$I_o = 49$ A
0	120	190	144	136	120
20	100	148	115	95	86
40	80	128	102	84	75
60	60	124	100	84	74
80	40	128	102	86	75
100	20	144	110	100	86
120	0	190	156	137	119

The measurements show that the induction decreases with the distance from the pole front and reaches the minimum B_{centr} in the center of the working chamber independently on the frequency f_o. For technological reasons, in order to enable the proper movement of the grinding medium to ensure an effective grinding process, it is important to keep the induction over minimum value, therefore it is enough to require:

$$B_{centr} \geq B_{min}, \tag{12}$$

where B_{min} is the smallest value of magnetic induction for which the grinding medium has sufficient energy to effectively grind the material and keep the grinding medium inside the working chamber. B_{min} depends on the physical properties of the processed material. It is important to note that B_{min} depends as well on the quantity of grinding medium inside the working chamber and it decreases if this quantity grows [23]. In practice, the B_{min} is determined by laboratory experiments or results from the analysis of measurements.

It can be observed in Figure 6 that the induction distribution changes with the change of the output frequency f_o. The value of B_{centr} is also changing and is smaller, the more the output frequency f_o exceeds the value of the base frequency f_b, and inversely, if f_o is smaller than f_b. Table 1 shows that B_{centr} increases with the increase of the windings current, which is also shown in Figure 7. Proportional increase of B_{centr} in relation to I_o is valid only in the range of linear operation of the magnetic circuit. This occurs if the I_o does not saturate the magnetic core [23]. Therefore, to limit the magnetic induction B_{centr} (12), there is no need to introduce an upper limit on I_o of the mill winding. Additionally, the maximum speed of the grinding medium caused by the electromagnetic field is not greater than the frequency of the voltage supplied to the mill [23]. Thus, it is not necessary to introduce an upper limit on the magnetic induction value to keep the upper speed limit of grinding medium (6).

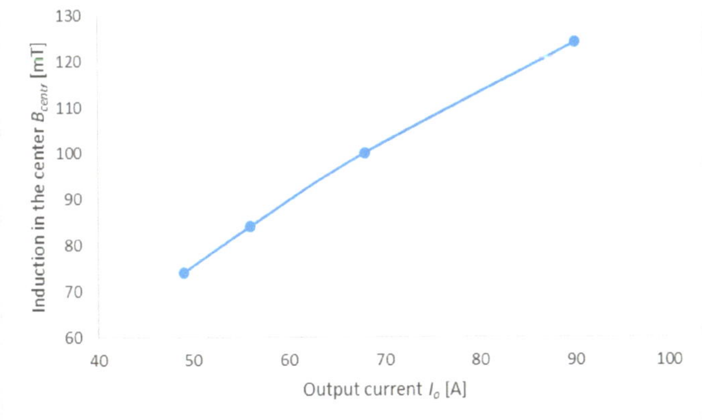

Figure 7. Magnetic induction measured in the center of the working chamber in the function of the windings current.

Due to lack of an upper limit on the induction, only the lower limit (12) is valid. Maintaining the limitation (12) requires the determination of the magnetic induction inside the working chamber during the grinding process. It is, howeverr, impossible, because while working, the chamber is filled with grinding medium and material (see Figure 1c). An indirect method should then be used based on the linear relation of induction B and the current I_o (see Figure 7). Experimental static characteristic of the current was made depending on the output frequency f_o and the base frequency f_b for empty working chamber. For the approximation purposes, a two-layer neural network with one hidden layer was used. The two-layer network allows for a more precise approximation of non-linear characteristic [39] then a single-layer variant. To ensure a sufficiently accurate and smooth approximation, a structure with seven neurons in the hidden layer was finally used. The neural network approximator allows to determine the segment where current I_o reaches a given value. The experiments with several different materials showed that if the current is greater than 55 A, the limitation of the lower value of the magnetic induction B_{min} (12) is kept. The current measurements, as well as the result of the characteristic approximation using ANN and the current limitation, are shown in Figure 8a. The approximation result using linear function of the lower current limitation is shown in Figure 8b.

Finally, the constraints were determined (13) that would make it possible to meet the requirements regarding the minimum value of the magnetic induction in the working chamber (12).

$$\begin{cases} f_o \leq a_l f_b - b_l \\ f_o \geq a_p f_b - b_p \end{cases}. \tag{13}$$

The parameters a_l, b_l, a_p, b_p in (13) are determined and are constant for a certain type of processed material and can act as guidelines for the fineness and the type of grinding medium used. The parameter values are based on the linear approximation of I_o lower limit with the least-squares method [41].

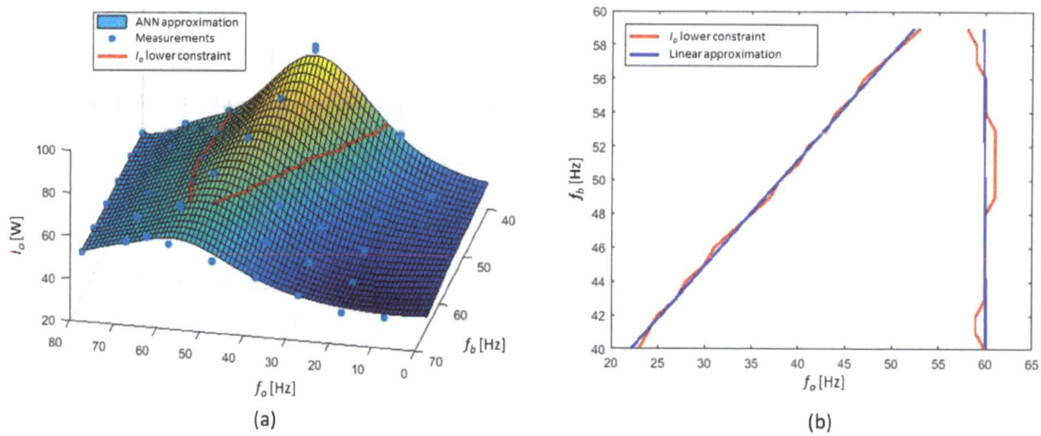

Figure 8. (**a**) The current and its approximation by ANN; (**b**) the current limitation and its linear approximation.

3.3. Optimization Algorithm

Apart from the specification of constraints, the minimization of the mill's active power requires determination of the objective function J for a given value of the decision variables at a given operating point. This may be the phenomenological model or any form of numerical approximation. Depending on the type of model, appropriate, dedicated methods of searching for the optimal solution are used [37]. The use of complex process models, however, can lead to high computational complexity, which in turn may cause unrealizable algorithms in the context of PLC capabilities. In such a situation, on-line algorithms are often used, where after the decision variables change the value of objective function is directly measured. Such an on-line optimization usually employs directional searching which requires some special analysis of the constraints and assumption of unimodality of the objective function [37].

To illustrate the variability of the objective function $J(f_o, f_b)$, a family of static characteristics of the active power was created for different mass of grinding medium inside the working chamber. Figure 9 shows the measurement results and approximations of the objective function $J(f_o, f_b)$. In the discussed case, the objective function is equal to the mill's active power $P_o(f_o, f_b)$.

Figure 9 shows the measured values of active power P_o (represented by the empty circles on the chart), registered by the SCADA system for different set points of base and output frequencies, f_b and f_o, respectively, and for different mass of grinding medium inside the working chamber m. Based on the measurements models approximating the active power P_o were identified for each value of the grinding medium mass m separately. The models were designed as a two-layered ANN with 10 neurons in the hidden layer. Each surface in Figure 9 represents the simulation results for each model separately (e.g., the surface indicating the lowest values of active power is the result of simulation for the model identified for $m = 0$ g, while the surface indicating the highest values of active power is the result of simulations for model obtained for $m = 600$ g). Figure 9 shows that for a fixed value of the base frequency f_b, there is only one extremum (maximum) of active power for the output frequency f_o equal to the base frequency f_b. Additionally, in the direction parallel to the base frequency axis f_b, the static characteristic is monotonic and decreases with increasing value of the base frequency f_b. Therefore, the assumption about unimodality of the static active power characteristic, which is also the objective function of the optimization task $J(f_o, f_b)$ is met.

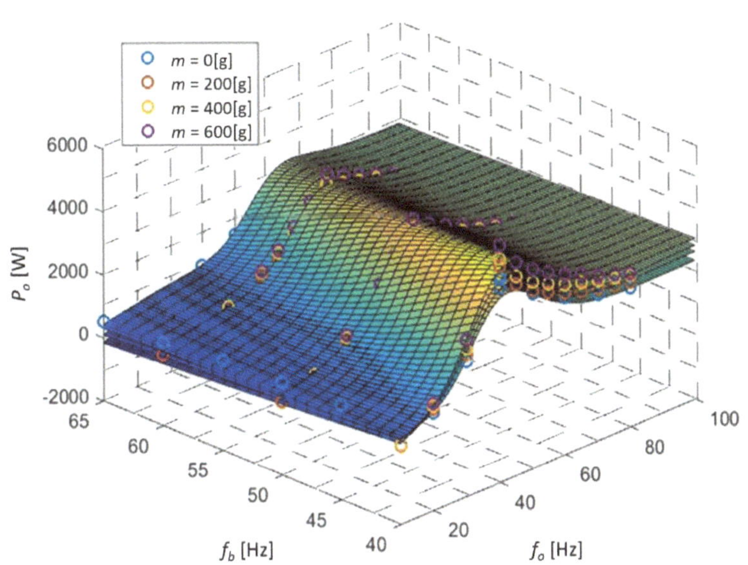

Figure 9. Approximation of the static active power characteristic of the mill depending on the grinding medium mass.

Summarizing the conclusions from Sections 3.1 and 3.2, in particular taking into account (6), (7), (10) and (13), the minimization of the active power of the electromagnetic mill described in Section 2.2 can be achieved by solving the optimization problem written as (1) with respect to the constraints:

$$\begin{cases} 45 \le f_o \le 70 \\ 50 \le f_b \le 55 \\ f_o \le f_b \\ f_o \le 1.6 f_b - 42.4 \\ f_o \ge -0.02 f_b - 60.8 \end{cases} . \quad (14)$$

The coefficients in (14) are parameters of the optimization problem determined by the supervisory control layer for a given operating point of the technological system. The example presented in (14) concerns the D100 mill, product quality (expressed with 80% of product grains with a size <50 μm) and the process efficiency of 150 kg/h of the processed copper ore. Constraints (14) are graphically presented on the active power static characteristic in Figure 10 to indicate the area where optimum solution will satisfy the constraints. Figure 10 combines the active power measurements (dots) and measurements approximation using the ANN model (surface) with graphical representation of constraints (14). Contrary to Figure 9, it presents only one case of the grinding medium mass (m = 200 g) for clarity. Each line in Figure 10 represents the borders of the allowed area indicated by each constraint, e.g., the area between two red lines (including the lines) represents the region where technological constraints on f_o are respected, while the area between two green lines concerns lower limit on B_{centr}. Supply voltage requirements, described by only one constraint in (14) are respected in the region on the right side of the black line, representing this constraint. The intersection of all such regions in Figure 10 represents the search area for the optimization task.

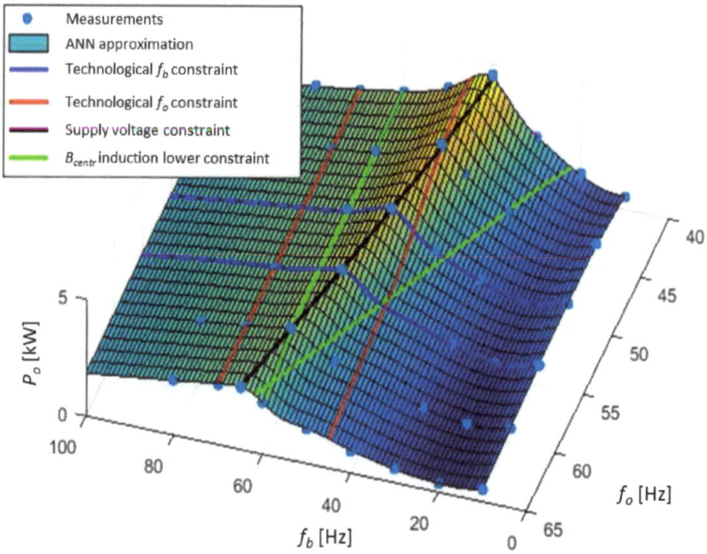

Figure 10. Constraints on the static characteristics of the active power of the inverter.

The analysis of the objective function $J(f_o, f_b)$ allows for the choice of the algorithm of searching for the minimum active power of the mill, which ensures the smallest possible effort needed to determine the optimal point. The convexity of the objective function shows that the active power minimum will be located on one of the active constraints that are presented in Figure 10. It is then enough to search only for the active constraints of the optimization problem.

The constraints imposed on the decision variables (f_o, f_b) are linear (14), which means that a simple one-direction search on the given active constraint may be used to determine the global minimum. There is no need to determine gradient of the objective function, what is the biggest advantage of this approach. Actual value of the objective function can be achieved by changing the output frequency f_o and the base frequency f_b in the frequency converter supplying the electromagnetic mill, and then measuring the value of the mill's active power after reaching steady state.

The general outline of the optimization algorithm is presented in the diagram below (Figure 11). The algorithm starts with determining the active constraints (segments $\overline{AB}$) of the optimization problem. Then, the minimum active power is searched for each segment $\overline{AB}$. The minimum value is stored together with the value of decision variables for which it was obtained. After minimization for each active constraint, the lowest value is selected.

In procedure 1: Determination of active constraints (Figure 11), the active constraints for the i-th step (segments $(\overline{AB}_i)$) are determined by taking the constraint coefficients from the supervisory layer, determining the intersection points of all constraints, selecting points that meet the constraints and creating segments $\overline{AB}_i$.

In procedure 2: Minimizing along the i-th constraint for each of the active constraints, the active power minimum is determined using the golden section method [42]. At the beginning of each iteration of the golden section method, the internal points $F_1(f_o, f_b)$ and $F_2(f_o, f_b)$ are determined at a specific distance from the edge of the segment $\overline{AB}_i$ obtained from the previous block (No. 1 in Figure 11) according to the golden section rule. The decision variables of point F_1 are then set in the inverter. After the active power reaches the steady state, its real value is measured. The whole process is repeated for the point F_2. The active power values obtained for both points F_1 and F_2 are compared and depending on the result of the comparison, one of the constraints of the section $\overline{AB}_i$ (point A or B)

changes and the searching interval is shortened. Each iteration is completed by checking the stop condition (15):

$$\left|\overline{AB_i}\right| < \varepsilon. \tag{15}$$

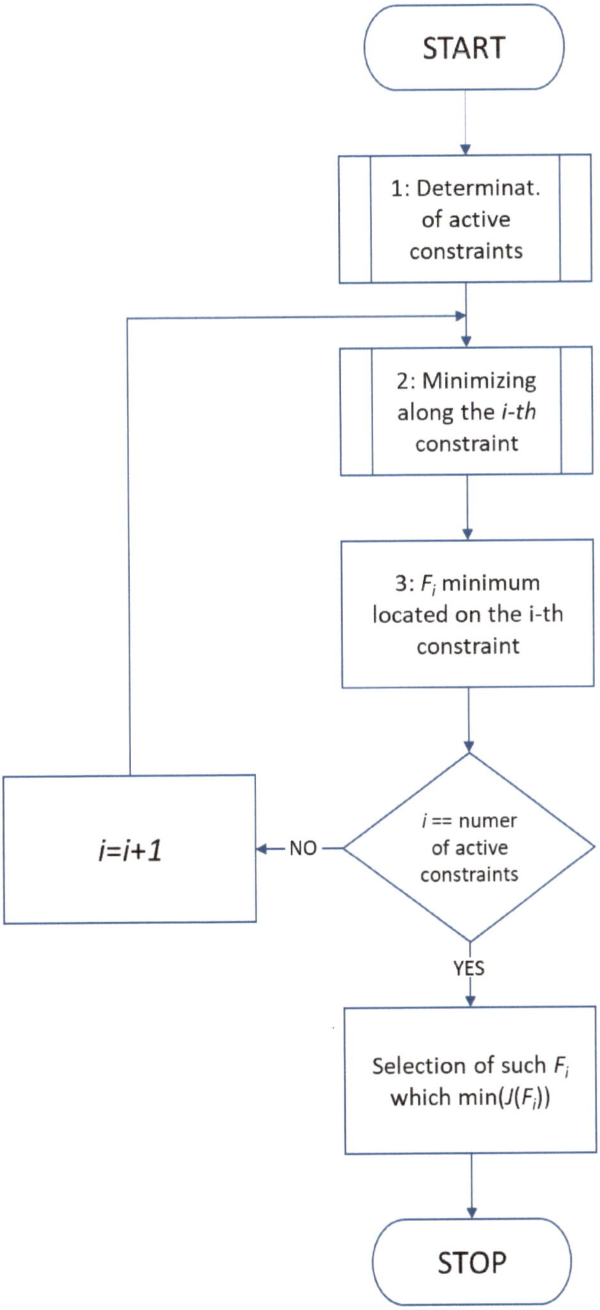

Figure 11. General form of the optimization algorithm.

The maximum number of iterations of Iter [37] is also limited according to Equation (16):

$$\text{Iter} \leq \text{Int}\left[\frac{\ln(\varepsilon) - \ln\left(\left|\overline{A^1B^1}_i\right|\right)}{\ln(0.618)}\right] + 1, \quad (16)$$

where $\left|\overline{A^1B^1}_i\right|$ is the initial length of the interval $\overline{AB}_i$. Meeting the stop condition (15) or exceeding the maximum number of iterations (16) causes the termination of the golden section method.

Each execution of the optimization algorithm results in a new minimum of the objective function that satisfies all constraints of the optimization problem. The proposed optimization algorithm is insensitive to the change of the operating point due to the measurement of the actual value of active power instead of calculation of the value indirectly through a model. However, in order to verify the algorithm in simulations an approximation of the active power was used by means of an ANN as described in Section 3.1 (Figure 10). The simulated process of searching for the minimum is presented in the diagram in Figure 12.

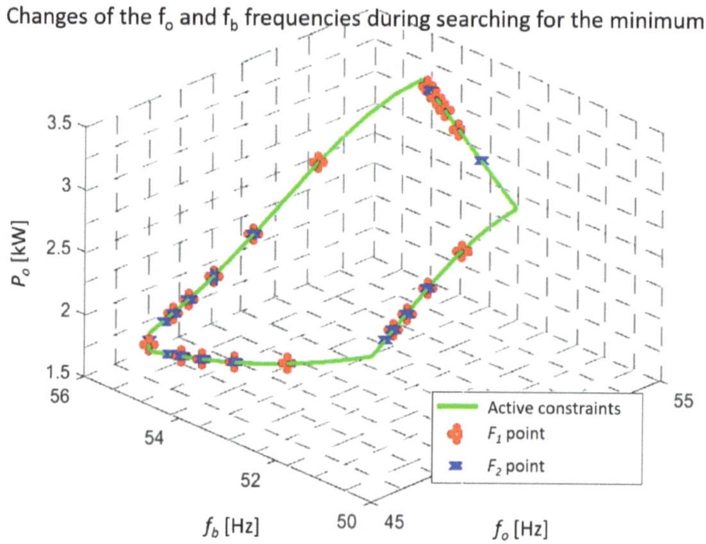

Figure 12. Simulation of searching for the minimum active power by directional optimization on the active constraints.

It can be seen that with each iteration of the golden section method, the decision variables tend to the minimum value of each constraint, narrowing the segment $\overline{AB}_i$. Finally, a minimum is determined for each constraint (Figure 13) and the lowest value is selected. The starting point for the simulated optimization task is the set of nominal frequencies $f_o = f_b$ = 50 Hz resulting in the active power P_o value over 3.6 kW. As Figure 13 shows, the optimization algorithms is calculating the lowest values of the objective function (equal to active power) for each of the active constraints (blue circles). Then, the overall minimum is chosen from all solution candidates (black circle) resulting in the active power reduction to almost 2.1 kW that stands for almost 42% of the power reduction. As follows from the analysis of the objective function model in Figure 13, further reduction of active power is possible, e.g., for lower values of f_o. It would, however, lead to the violation of the imposed technological and quality constraints, hence only the allowed area inside the constrained region can be considered.

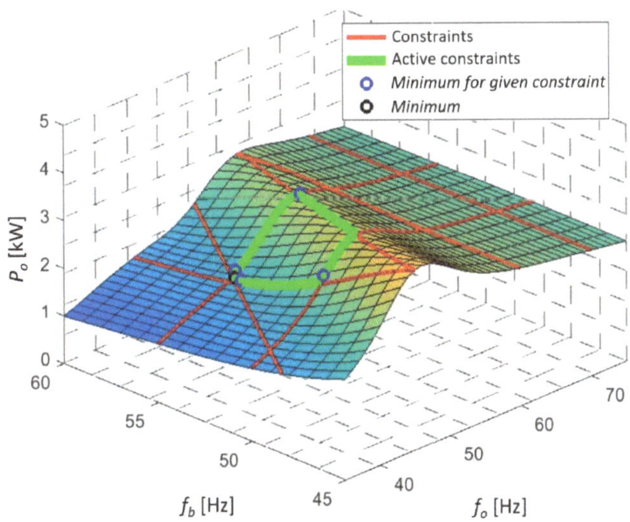

Figure 13. Simulation result of the proposed optimization algorithm.

4. Algorithm Validation in Semi-Industrial Tests

4.1. Algorithm Implementation

Since the algorithm is based on the measurement of active power, the measurement resolution and measurement noise, as well as the resolution of setting the base frequency f_b and the output frequency f_o, should be taken into account. Interactions between these resolutions and the noise range determines the parameter ε choice, which is the stop condition of the optimization task (15). For the tested installation with the D100 mill and Mitsubishi F700 inverter, the value of the stop condition was determined as $\varepsilon = 1$.

Another extremely important aspect is the dynamics of the object, i.e., the dynamics of technological changes in the entire installation, and the dynamics of the control path. The inverter used was parameterized to ensure a frequency change in the range of 0–50 Hz within 3 s. Taking this parameter into account, as well as the settling time of the active power and necessary time to execute the optimization algorithm, the sampling time was set to $T_{opt} = 30$ s. The algorithm was implemented on the PC panel in a SCADA application using the VBA language.

The series of experiments was carried out for various operating points of the installation after implementing the optimization algorithm. From an energy consumption point of view, the key sensitivity of the operating point concerns change of the grinding medium mass in the working chamber. Grinding medium mass follows from technological requirements and depends, inter alia, on the average mill throughput, the type of processed material and requirements on the grinding ratio (average size reduction factor). Such tests made it possible to assess the applicability of the proposed algorithm for a wide range of technological scenarios.

4.2. Results of Experiments

As part of many tests, four experiments E1 to E4 were carried out, in which a mass of grinding medium m_k = [0,200,400,600] g was checked. Figure 14 summarizes the time series of these experiments: the range of changes in the base and output frequencies and the change in active power of the mill.

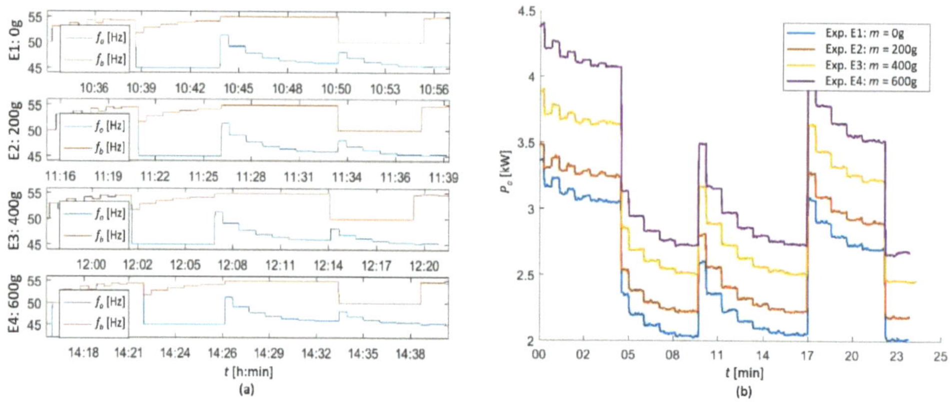

Figure 14. Results of experiments for various mass of grinding medium: (**a**) changes in frequency values; (**b**) changes in the active power of the mill (objective function).

The charts in Figure 14 show that the selected sampling period allows the active power signal to reach a steady state before the succeeding step of the optimization algorithm is performed. The values of the algorithm parameters were also correctly selected to give the algorithm smooth converge. Note that the change of the operating point does not affect the algorithm efficiency. Obviously, the change in the grinding medium mass has an impact on the average value of the active power, which is the expected effect.

Analysis of the results shows that for each of the active constraints, the method determines the minimum of the objective function and then selects the one for which this value is the lowest. The results of the algorithm for four experiments are presented in Figure 15. For each operating point, the method determined the values of the f_b and f_o frequencies, minimizing the energy consumption, which each time is lower by approximately 40% than the value obtained for the nominal settings f_b = 50 Hz and f_o = 50 Hz (compare Figure 14b).

4.3. *Discussion of Results*

The obtained results show that the proposed optimization algorithm enables the minimization of energy consumption by the electromagnetic mill while maintaining the constraints imposed on the process. Similar global minimums follow from the same set of constraints applied in each experiment. In industrial production scenarios, such important changes in the grinding medium mass would usually be combined with different technological requirements for the final product and, in turn, affect the parameters of the imposed constraints. Such modification during experimental verification of the proposed algorithm were not introduced on purpose to allow a clear comparison of results. On-line optimization algorithm was successfully verified; however, the process of searching for the minimum along the active constraint requires many changes of decision variables, which may disrupt the grinding operation.

The reduction of the negative impact of the proposed method can be achieved by carrying out each iteration of the optimization algorithm in an appropriate time window. Time needed to carry out the entire optimization process should be adjusted to the dynamics of disturbances affecting the electromagnetic mill (e.g., changes in the mass of grinding medium, changes in the physical parameters of the feed) so that the determination of the minimum active power follows the change in the operating point of the technological system. The frequency variation can also affect the quality of the final product, whereby in the case of searching and keeping the constraints, the required quality will not be affected.

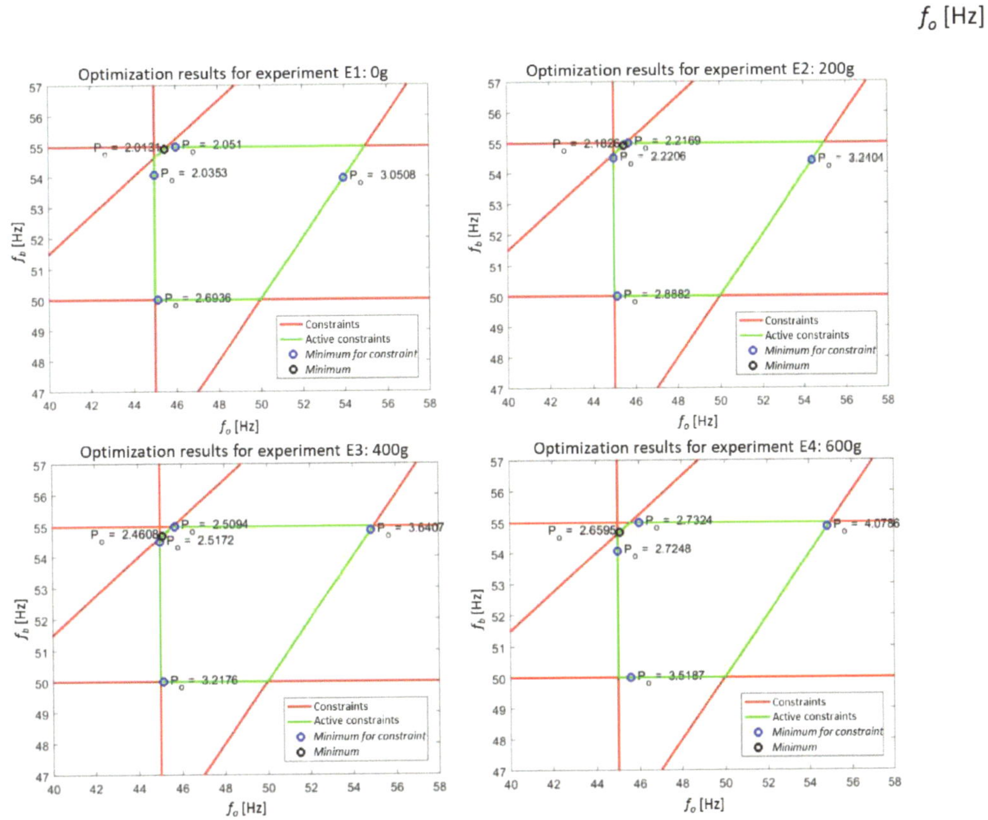

Figure 15. The results obtained from the optimization algorithm for the experiments with different mass of grinding medium.

Frequent changes in the output and base frequencies affect the service life of the actuators. The number of changes in decision variables can be limited by the appropriate modification of the parameter ε. However, the value of this parameter affects the accuracy of the determination of the minimum location. The number of changes also depends on the length of the segment $\overline{AB}$, representing a given active constraint and it becomes smaller the shorter the length of this segment is.

5. Conclusions

The article presents a practical solution to the problem of minimizing electrical energy consumption by the electromagnetic mill. It has been shown that using two variables (output frequency and base frequency) to determine the operation of the inverter makes it possible to change the active power of the electromagnetic mill–inverter system. The search for minimal consumption was shown using the static characteristic of the mill's active power versus the output and base frequency and the mass of grinding medium inside the working chamber. After approximation of the static characteristics with a two-layer neural network, it was possible to analyze the influence of the output frequency, base frequency and grinding medium mass on the active power of the electromagnetic mill. For technological reasons, both frequencies were selected as the two decision variables for the optimization task.

Analysis of the output and base frequencies was carried out, taking into account the quality of the final product. It was noticed that the change in the base frequency caused a change in the supply voltage of the mill coils, which, in the absence of the internal inverter limitation, may lead to their damage. Also shown is the indirect influence of decision variables on the magnetic flux of the working chamber, which determines the ability of the grinding medium to grind the material and keep them inside the working chamber. As a result, the optimization task objective function was designed with the set of technological and construction constraints. Satisfying the constraints makes it possible to keep the required quality of the final product, enables failure-free operation of the mill coils, and finally, creates magnetic field intensive enough to hold the grinding medium inside the working chamber while rotating fast enough to break the feed grains.

Convexity of the objective function was confirmed by experimental measurements and allows to created effective optimization algorithm which searches minima of the active power of the mill on the active constraints. The proposed optimization algorithm finds the global minimum of the active power, regardless of the operating point of the technological system. The efficiency of the proposed on-line optimization algorithm was successfully verified during experiments on a dry grinding installation with an electromagnetic mill D100. Further research will focus on constraint modification, e.g., using additional information on the product quality based on the on-line product grain sizes measurement.

Funding: The research reported in this paper was performed on installation co-financed by the National Centre for Research and Development, Poland, under Applied Research Programme, project number PBS3/B3/28/2015 (acronym SYSMEL). The APC was funded by the rector's habilitation grant, Silesian University of Technology, grant number 02/050/RGH20/0014.

Institutional Review Board Statement: Not applicable.

Informed Consent Statement: Not applicable.

Data Availability Statement: Data sharing not applicable.

Acknowledgments: The author would like to acknowledge a vital role of Markus Swierzy (member of the SYSMEL team) in preparing and conducting the experiments.

Conflicts of Interest: The author declares no conflict of interest. The funders had no role in the design of the study; in the collection, analyses, or interpretation of data; in the writing of the manuscript, or in the decision to publish the results.

References

1. Wills, B.A.; Napier-Munn, T. Comminution. In *Wills' Mineral Processing Technology*; Butterworth-Heinemann: Oxford, UK, 2005; Chapter 5; pp. 108–117.
2. Fuerstenau, M.C.; Han, K.N. *Principles of Mineral Processing*; SME: Chicago, IL, USA, 2003.
3. Mular, A.L.; Doug, N.H.; Derek, J.B. *Mineral Processing Plant Design. Practice and Control Proceedings*; SME: Chicago, IL, USA, 2002.
4. Leistner, T.; Peuker, U.A.; Rudolph, M. How Gangue Particle Size Can Affect the Recovery of Ultrafine and Fine Particles During Froth Flotation. *Miner. Eng.* **2017**, *109*, 1–9. [CrossRef]
5. Wang, Y.; Forssberg, E. Enhancement of energy efficiency for mechanical production of fine and ultra-fine particles in comminution. *China Particuol.* **2007**, *5*, 193–201. [CrossRef]
6. Jeswiet, J.; Szekeres, A. Energy consumption in mining comminution. *Procedia CIRP* **2016**, *48*, 140–145. [CrossRef]
7. Fuerstenau, D.; Abouzeid, A.Z. The energy efficiency of ball milling in comminution. *Int. J. Miner. Process.* **2002**, *67*, 161–185. [CrossRef]
8. Walkiewicz, J.; Clark, A.; McGill, S. Microwave-assisted grinding. *IEEE Trans. Ind. Appl.* **1991**, *27*, 239–243. [CrossRef]
9. Musa, F.; Morrison, R. A more sustainable approach to assessing comminution efficiency. *Miner. Eng.* **2009**, *22*, 593–601. [CrossRef]
10. Hesse, M.; Popov, O.; Lieberwirth, H. Increasing efficiency by selective comminution. *Miner. Eng.* **2017**, *103*, 112–126. [CrossRef]
11. Wang, Y.M.; Forssberg, E. *New Milling Technology. Technical Report*; MinFo: Stockholm, Sweden, 2001.
12. Grimble, M.J. *Industrial Control System Design*; Wiley: Hoboken, NJ, USA, 2001.
13. Hodouin, D. Methods for automatic control, observation, and optimization in mineral processing plants. *J. Process Control* **2011**, *21*, 211–225. [CrossRef]
14. Simon, J.; Guélou, G.; Srinivasan, B.; Berthebaud, D.; Mori, T.; Maignan, A. Exploring the thermoelectric behavior of spark plasma sintered Fe7-xCoxS8 compounds. *J. Alloy. Compd.* **2020**, *819*, 152999. [CrossRef]
15. Lokiec, H.; Lokiec, T. Inductor of Electromagnetic Mill. Polish Patent No. PL 226554 B1, 19 May 2015. (In Polish).
16. Electromagnetic Mill. Available online: https://globecore.com/products/magnetic-mill/ (accessed on 10 March 2021).

17. Styla, S. Laboratory studies of an electromagnetic mill inductor with a power source. *Int. Q. J. Econ. Technol. Model. Process.* **2017**, *6*, 109–114.
18. Styla, S. New Grinding Technology Using an Electromagnetic Mill–Testing the Efficiency of the Process. *Int. Q. J. Econ. Technol. Model. Process.* **2017**, *6*, 81–88.
19. Calus, D.; Makarchuk, O. Analysis of interaction of forces of working elements in electromagnetic mill. *Przeglad Elektrotechniczny* **2019**, *1*, 64–69. [CrossRef]
20. Makarchuk, O.; Calus, D.; Moroz, V. Mathematical model to calculate the trajectories of electromagnetic mill operating elements. *Tekhnichna Elektrodynamika* **2021**, *2*, 26–34. [CrossRef]
21. Ogonowski, S.; Ogonowski, Z.; Swierzy, M.; Pawelczyk, M. Control system of electromagnetic mill load. In Proceedings of the 25th International Conference on Systems Engineering, Las Vegas, NV, USA, 22–24 August 2017; pp. 69–76. [CrossRef]
22. Ogonowski, S.; Ogonowski, Z.; Pawelczyk, M. Multi-objective and multi-rate control of the grinding and classification circuit with electromagnetic mill. *Appl. Sci.* **2018**, *8*, 506. [CrossRef]
23. Sosinski, R. Opracowanie Metodyki Projektowania Trójfazowych Wzbudników z Biegunami Jawnymi Pola Wirującego do Młynów Elektromagnetycznych (Development of Design Methodology for three-Phase Salient-Pole Exciters of Rotating Field for Electromagnetic Mills—In Polish). Ph.D. Thesis, Czestochowa University of Technology, Czestochowa, Poland, 2006.
24. Mitsubishi Electric F700 Inverter Instruction Manual. Available online: https://dl.mitsubishielectric.com/dl/fa/document/manual/inv/ib0600177eng/ib0600177engf.pdf (accessed on 10 March 2021).
25. Kabzinski, J. *Advanced Control of Electrical Drives and Power Electronic Converters*; Springer International Publishing AG: Cham, Switzerland, 2017; ISBN 978-3-319-83362-0.
26. Goralczyk, M.; Krot, P.; Zimroz, R.; Ogonowski, S. Increasing Energy Efficiency and Productivity of the Comminution Process in Tumbling Mills by Indirect Measurements of Internal Dynamics—An Overview. *Energies* **2020**, *13*, 6735. [CrossRef]
27. Pawelczyk, M.; Ogonowski, Z.; Ogonowski, S.; Foszcz, D.; Saramak, D.; Gawenda, T.; Krawczykowski, D. Method for Dry Grinding in Electromagnetic Mill. Polish Patent PL 413041, 6 June 2015.
28. Wolosiewicz-Glab, M.; Ogonowski, S.; Foszcz, D.; Gawenda, T. Assessment of classification with variable air flow for inertial classifier in dry grinding circuit with electromagnetic mill using partition curves. *Physicochem. Probl. Miner. Process.* **2018**, *54*, 440–447. [CrossRef]
29. SIMATIC S7-300. Available online: https://new.siemens.com/global/en/products/automation/systems/industrial/plc/simatic-s7-300.html (accessed on 10 March 2021).
30. SIMATIC S7-1200. Available online: https://new.siemens.com/global/en/products/automation/systems/industrial/plc/s7-1200.html (accessed on 10 March 2021).
31. Buchczik, D.; Wegehaupt, J.; Krauze, O. Indirect measurements of milling product quality in the classification system of electromagnetic mill. In Proceedings of the 22nd International Conference on Methods and Models in Automation and Robotics, Miedzyzdroje, Poland, 28–31 August 2017; pp. 1039–1044. [CrossRef]
32. Budzan, S.; Buchczik, D.; Pawelczyk, M.; Tuma, J. Combining Segmentation and Edge Detection for Efficient Ore Grain Detection in an Electromagnetic Mill Classification System. *Sensors* **2019**, *19*, 1805. [CrossRef] [PubMed]
33. LUMEL 1 and 3-Phase Power Network Meter ND20. Available online: https://www.lumel.com.pl/en/catalogue/product/1-and-3-phase-power-network-meter-nd20 (accessed on 10 March 2021).
34. SIMATIC IPC 477D Panel PC. Available online: https://mall.industry.siemens.com/mall/en/WW/Catalog/Product/?mlfb=6AV7240-6BD17-0PA2 (accessed on 10 March 2021).
35. GE Digital Proficy HMI/SCADA iFIX. Available online: https://www.ge.com/digital/applications/hmi-scada/ifix (accessed on 10 March 2021).
36. Tatjewski, P. *Advanced Control of Industrial Processes: Structures and Algorithms*; Springer: London, UK, 2007.
37. French, M. *Fundamentals of Optimization*; Springer International Publishing AG: Cham, Switzerland, 2018; ISBN 978-3-319-76191-6.
38. Ogonowski, S.; Ogonowski, Z.; Swierzy, M. Power optimizing control of grinding process in electromagnetic mill. In Proceedings of the 21st International Conference on Process Control, Strbske Pleso, Slovakia, 6–9 June 2017; pp. 370–375. [CrossRef]
39. Braspenning, P.J.; Thuijsman, F.; Weijters, A.J.M.M. *Artificial Neural Networks. An Introduction to ANN Theory and Practice*; Springer: Berlin/Heidelberg, Germany, 1995; ISBN 978-3-540-59488-8.
40. Krauze, O.; Buchczik, D.; Budzan, S. Measurement-Based Modelling of Material Moisture and Particle Classification for Control of Copper Ore Dry Grinding Process. *Sensors* **2021**, *21*, 667. [CrossRef] [PubMed]
41. Wolberg, J. *Data Analysis Using the Method of Least Squares: Extracting the Most Information from Experiments*; Springer: Berlin/Heidelberg, Germany, 2006; ISBN 978-3-540-25674-8.
42. Kiefer, J. Sequential minimax search for a maximum. *Proc. Am. Math. Soc.* **1953**, *4*, 502–506. [CrossRef]

energies

MDPI

Article

A New Filter Concept for High Pulse-Frequency 3-Phase AFE Motor Drives

Stefan Hoffmann [1,*], Matthias Bock [2] and Eckart Hoene [1]

[1] Fraunhofer IZM, Gustav-Meyer-Allee 25, 13355 Berlin, Germany; eckart.hoene@izm.fraunhofer.de
[2] Faculty IV—Electrical Engineering and Computer Science, Technische Universität Berlin, Sekr. TIB 4/2-1, Straße des 17. Juni 135, 10623 Berlin, Germany; mail@matthiasbock.net
* Correspondence: stefan.hoffmann@izm.fraunhofer.de

Abstract: The size of back-to-back converters with active front end is significantly determined by the size of the passive filter components. This paper presents a new complete EMC filter concept for this type of converter system that is effective on the input and the output. This involves filtering the main common mode interferences from the grid and motor sides with a single CM choke. Since only the difference of the generated common mode voltage-time areas of both converters is absorbed by this component, the size of the required filter can be greatly reduced compared to conventional filter concepts. The concept is validated on a grid feeding inverter that can be connected to the public distribution network with an output power of 63 kW. The size reduction is demonstrated by means of a design example on a system with the same power and electrical requirements. It is elaborated why, applying the new filter concept, the impedance of the DC link potentials to ground and other electrical potentials should be as high as possible and therefore associated parasitic capacitances should be minimized. From this requirement, rules for the design of the power modules of PFC and motor converters for the application of this filter concept are derived.

Keywords: EMI filter design; electromagnetic compatibility; power electronic converters; motor drives; active front end; filter topology; CM filter; volume minimization

Citation: Hoffmann, S.; Bock, M.; Hoene, E. A New Filter Concept for High Pulse-Frequency 3-Phase AFE Motor Drives. *Energies* **2021**, *14*, 2814. https://doi.org/10.3390/en14102814

Academic Editor: Ivica Stevanovic and Bernhard Wunsch

Received: 22 March 2021
Accepted: 6 May 2021
Published: 13 May 2021

1. Introduction

1.1. Back-to-Back Converters

Back-to-back converters are widely distributed in the industry. They are utilized in variable speed drives connected to the three-phase public distribution network. In order to comply with the requirements regarding the Total Harmonic Distortion (THD) of the mains current and voltage, in many cases not only the motor inverter but also the rectifier is actively controlled. The basic converter system with this so called Active Front End (AFE) is shown in Figure 1 for the standard 2-level topology addressed in this paper.

One goal of research in the field of back-to-back converters is to increase the power density i.e., the power conversion per physical converter volume. There are various approaches for their size reduction. For example, the utilization of Wide Bandgap (WBG) semiconductors can reduce the size of passive components because it enables the switching frequency to be increased in turn permitting the use of smaller DC link capacitors as well as ripple inductors. Another possibility for volume reduction is the development of new filter topologies.

1.2. Filter Design Considerations

Due to the operation of semiconductors in switched mode, Electromagnetic Compatibility (EMC) disturbances are generated in the converters [1], which must be filtered on the mains side due to EMC requirements, e.g. IEC 61800-3. However, filtering the interferences on the motor side is also reasonable for many applications. On the one hand, differential filtering measures can be applied to prevent high voltage slopes on the motor windings, which can damage the winding insulation due to over-voltages [2]. On the other hand, Common Mode (CM) currents in the motor bearings [3,4] can be reduced [5]. Possibly the motor cable no longer requires a screen because the motor filter reduces the radiated interference to such an extent that the applicable EMC limits can be adhered to. Additionally, motor filters can reduce the occurring losses, especially when long motor cables are used. With each switching operation in the motor inverter the parasitic capacitances of motor and cables must be partially reloaded hard without the application of filters, which leads, among other things, to an increase in switching losses [6,7]. Furthermore, Differential Mode (DM) filtering on the motor side results in lower ripple currents in the cable and in the machine, which leads to the reduction of high-frequency copper losses. Although filters also cause losses, the overall loss balance is usually essentially lower [8].

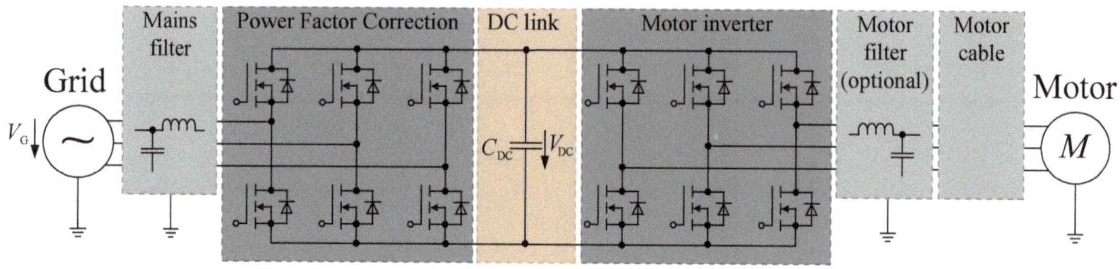

Figure 1. Simplified circuit diagram of a back-to-back converter with AFE in standard 2-level topology segmented into EMC-relevant blocks.

DM filtering is implemented with a ripple filter stage at each of the half-bridge output nodes. They consist of ripple inductors L_r and ripple capacitors C_r, as shown in Figure 2. The inductors absorb almost the entire DM voltage-time area, resulting in a relatively high current ripple in the components of the ripple stage, the semiconductors and the DC link capacitors depending on the ripple inductor's inductance value.

However, both converters also generate a CM voltage that must be reduced for previously described reasons. In [9] it is demonstrated how the CM voltage waveforms in both converters can be made identical by sophisticated voltage pulse generation and thus the CM interference can theoretically be completely eliminated. In fact, in real experiments the CM voltage can be significantly reduced by more than 15 dB, although not completely, due to impediments such as different switching characteristics of the individual semiconductors or dead-time effects (cf. [9]). Another benefit of this measure is that a less saturation-resistant CM choke may be used, since the voltage-time areas, which need to be absorbed, are significantly smaller.

However, for this approach an adjustment of the zero component of the $\alpha\beta 0$ system in the space vector modulation [10] is necessary. This represents a disadvantage as the degree of freedom presented by the zero component is used up by the attempt to eliminate the CM voltage and can hence no longer be used for clamping some transistors at specific moments during the mains period i.e., flat-top modulation [11,12]. When flat-top modulation is applied, not only the number of all switching operations can be reduced by one third. Additionally, the amplitude of the fundamental component can be increased by approximately 15% compared to the conventional modulation scheme [11], enabling the DC link voltage in a motor drive system to be reduced by approximately 13%. As a result of both measures, switching losses can be reduced by more than half. Furthermore, the reduction of the DC link voltage unburdens the inductive components, which have to absorb the main part of the voltage-time areas, with regard to the core and High Frequency (HF) copper losses. It can therefore be concluded that with the approach described in [9] flat-top modulation and the DC link voltage reduction would partially not be possible, leaving even more potential for optimization untapped. This is why it makes more sense to select the degree of freedom of the zero-voltage system for flat-top modulation. The disadvantage that components are required which absorb the HF CM voltage-time area must be accepted. In the following sections, it will be worked out how the volume of these components can be minimized.

1.3. Conventional Filter Concepts

Figure 2 shows the drive system using the most frequently used filter topology with respect to the first filter stage (ripple filter) for the PFC [13] and the motor inverter [5]. The filtering of the DM interferences is achieved by the ripple inductors and capacitors. The respective CM current in the PFC and motor inverter closes in two separate current loops via the ripple inductors, the ripple capacitors and the CM capacitor to the DC link as depicted in red in Figure 2. As a result, the ripple inductors absorb not only the differential voltage-time area but also the respective CM voltage-time area of the corresponding inverter. The increased total voltage-time area results in a higher potential for core losses and HF copper losses in the components. Consequently, the ripple inductors need to be much bigger compared to those that only absorb the DM voltage. It is also possible to distribute the CM and DM components between two inductors connected in series [14]. This would require two additional CM chokes in the back-to-back converter.

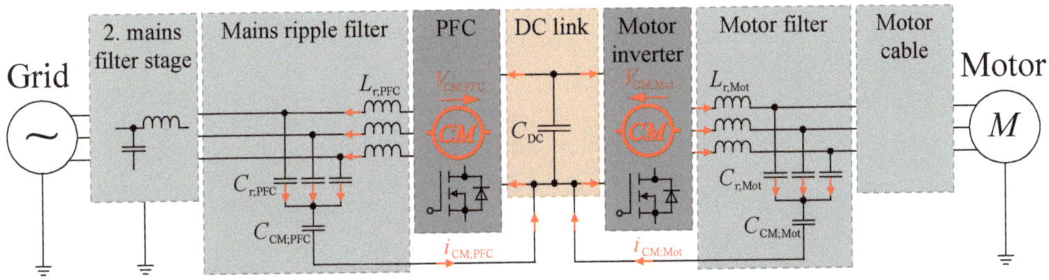

Figure 2. Conventional EMC filter concept for back-to-back converters with AFE, CM current loops for PFC ($i_{CM;PFC}$) and motor inverter ($i_{CM;Mot}$) depicted in red, the generated CM voltages are absorbed by the ripple inductors $L_{r;PFC}$ and $L_{r;Mot}$, respectively.

mention in the text and removed the highlight command.

The low-impedance connection of the capacitor star points to the DC link with the CM capacitors $C_{CM;Mot}$ complicates the implementation of flat-top modulation on the motor side at low output voltages. Due to the high voltage step at the flat-top transitions, when another half-bridge output node is clamped to a DC link potential, filter oscillations with large current amplitudes occur, which can lead to undesirable effects, for example the saturation of the inductors. The oscillation is formed between the inductance of the parallel-connected ripple inductors and the CM capacitor. To overcome the disadvantages a new CM filter concept with a CM choke in the DC link is proposed in this paper.

1.4. New CM Filter Concept with a CM Choke in the DC Link

The first inductive stage of CM filtering can be provided using only a single CM choke for both inverters. In this case, only the difference in the CM voltage-time areas of the PFC and motor inverter must be absorbed by this component. The ripple inductors of the PFC and the motor inverter are unburdened with respect to the potential for core losses because, in contrast to the common concept from Figure 2, they only absorb the DM voltage-time area. The circuitry requirement of this CM filtering concept is that the two DC link potentials must not have any reference to other electrical potentials. The usually existing low-impedance connection of the ripple capacitors to a DC link potential (see Figure 2) and Y capacitors in the DC link cannot be applied.

The entire filter concept is depicted in Figure 3. As shown, the single circuit for the CM current now contains the PFC semiconductors and inductors, the grid-side ripple capacitors, the CM capacitors on the mains and the motor side, the motor-side ripple capacitors and inductors, the motor inverter semiconductors, and the DC link. The CM choke can be placed anywhere in series with the PFC or motor inductors, but also in the DC link. The arrangement in the DC link has the advantages that only two windings are required, the total current is lower due to the higher DC link voltage compared to the two AC voltages and no DM ripple current occur in the component.

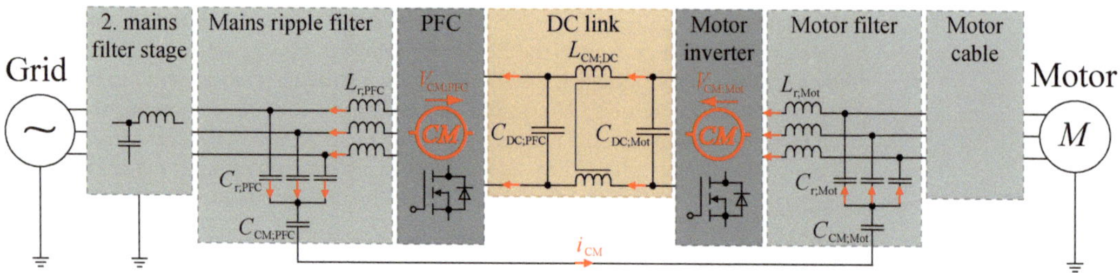

Figure 3. New CM filter concept with one common CM choke for the PFC and the motor inverter, only the main CM current path between the ripple filters is shown, CM currents via parasitic capacitances to the heatsink are disregarded here, the CM voltage time area is reduced to only the difference of both CM voltages.

According to the resulting voltage divider from the inductance values of ripple inductors and CM choke, a low CM voltage also drops across the ripple inductors. However, it is negligible due to the inductance value of the CM inductor, which is usually orders of magnitude higher. The high-frequency voltage drop across the ripple capacitors can also be neglected. Nevertheless, they absorb the low-frequency CM voltage which is unavoidable when using flat-top modulation.

The pulse generation of both inverters should be synchronized and both inverters should be operated with the same switching frequency. For instance, a switching period can start with the turn-on of all High Side (HS) switches of PFC and motor inverter. This ensures that the difference between the CM voltages of the PFC and motor inverter remains minimal without adjusting the zero component in the $\alpha\beta 0$ system. The degree of freedom for the zero component in the $\alpha\beta 0$ system should be utilized for flat-top modulation for the reasons already analyzed. Occurring oscillations due to filter resonances are very low with this concept because they are not excited by the voltage steps at the flat-top transitions but only due to the discontinuity of the pulse patterns.

One disadvantage of this concept is that the CM choke in the DC link introduces an additional leakage inductance between the two DC links. In case of load steps, this can cause undesired oscillations in the DC link voltages. Therefore, this parasitic inductance should be kept very low.

To avoid this problem, it is also feasible that the CM inductor in the DC link is eliminated. In this case, the difference in the CM voltage of the PFC and motor inverter is absorbed by the ripple inductors according to their inductance ratio. Compared to the conventional filter topology, the inductors are also unburdened because they do not absorb the entire CM voltage-time area of the respective converter, but only a part of the difference between the two CM voltages.

1.5. Organization of This Paper

Section 2 contains the applied methods and utilized materials on which the predictions and measurements are based. The prediction methodology for new filter concept introduced in Section 1.4 is experimentally validated in Sections 3.1–3.3 using the example of a grid-feeding inverter with an output power of 63 kW. Subsequently, design rules regarding parasitic capacitances from the power module and the DC link potential to ground are elaborated in Section 3.4. In Section 3.5 the new filter concept is compared to the conventional topology summarized in Section 1.3, in terms of design, using a typical specification catalog with respect to the size of the CM filter components. Section 4 highlights the conclusions of this paper.

2. Materials and Methods

2.1. Experimental Validation Object

Initially, developed models introduced in [15] are compared with measurements. For the validation of the presented filter concept it is not required to operate a complete drive system as shown in Figure 1. It is sufficient to perform the measurements on the power electronic system presented in Figure 4. As necessary for the validation, the CM choke in the DC link absorbs almost the entire CM voltage-time area. One difference compared to the complete motor drive is that the load profile of the DC link CM choke regarding the flux linkage differs when the motor inverter is not operated. Consequently, the design results of this choke optimized with respect to losses would differ. Furthermore, without motor and motor cable a CM interference current path is eliminated due to the high capacitive coupling of the motor and cable to ground, which must be considered in the filter design of the complete motor drive system. Despite these differences the validation object is suitable to qualify the approach presented here. The main CM circuit is marked in red in Figure 4 and runs from the semiconductor modules T_1–T_6 through the ripple inductors $L_{r;U,V,W}$ and capacitors $C_{r;U,V,W}$ as well as the mains-side Y capacitor $C_{Y;AC}$ to ground and closes through the Y capacitor on the DC side $C_{Y;DC}$ and the CM choke $L_{CM;DC}$ to the DC link potentials. Note that the CM capacitors in the ripple stage and the DC link are Y capacitors connected to ground. Independently of this, a CM current can also be generated without a Y capacitor via the parasitic ground capacitances, which are shown in gray in the figure, since the electrical distribution network has a ground reference.

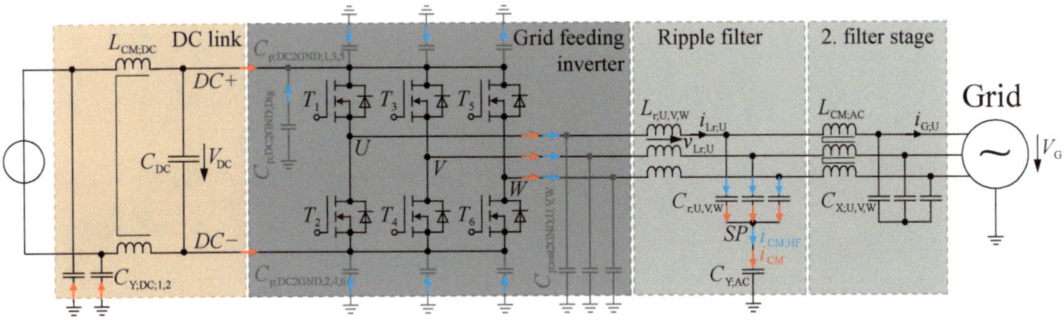

Figure 4. Circuit diagram of the validation object, 3-phase 2-level grid feeding inverter with a DC link CM choke, a ripple stage and one additional EMC filter stage, parasitic ground capacitances are depicted in gray and CM circuit loops in red and blue.

If the predicted voltage and current waveforms, the voltage-time areas of the ripple inductors and the CM choke in the DC link as well as the interference levels at the considered grid feeding inverter correspond to those measured, the methodology can also be applied to the more complex systems considered previously. Once the validity of the models has been demonstrated, a size comparison can be provided between optimized back-to-back converter system using the novel and the conventional filter topology.

The grid feeding inverter is intended for connection to the public distribution network. Table 1 summarizes its electrical requirements and operating parameters. The DC link capacitor, the ripple inductors and the EMC filter are designed for a switching frequency of $f_S = 140\,\text{kHz}$. As a result, EMC limits need only be met from the second harmonic at a frequency of $2f_S = 280\,\text{kHz}$, which reduces the size of the filter components.

Table 1. Requirements and operating parameters of the validation object.

Parameter	Value
Mains voltage	$V_G = 230/400\,\text{V}$
Mains frequency	$f_G = 50\,\text{Hz}$
Maximum output power	$P_{out} = 63\,\text{kW}$
DC link voltage range	$V_{DC} = 610–700\,\text{V}$
Switching frequency	$f_S = 140\,\text{kHz}$
EMC requirement	IEC 61800-3 Class C2

The prototype inverter used in the measurements is shown in Figure 5. The filters are designed in such a way that their components can be placed at the bottom of the device on a PCB with the same base area as the heat sink. Table 2 contains a list of the applied components with the relevant EMC properties. Note that most of the utilized passive components are not volume optimized.

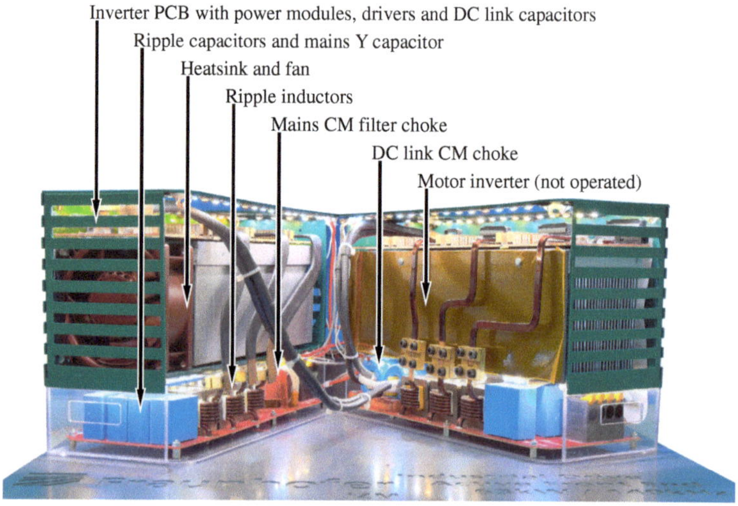

Figure 5. Picture of the public grid connected 63 kW motor drive system with active front end, the motor inverter was not operated.

Table 2. Utilized components and their relevant EMC properties.

Component	Designator	Description	EMC Properties
Semiconductors	T_1–T_6	2 × SiC MOSFET ST SCT100N120G2D2AG	$C_{p;DC2GND;1-6} = 130\,\text{pF}$ $C_{p;DC2GND;Dig} = 320\,\text{pF}$
DC link capacitor	C_{DC}	8 × CeraLink FA10, 2.5 µF, 900 V	$C_{CDC} = 20\,\mu\text{F}$
DC link CM choke	$L_{CM;DC}$	10 × TDK R38.1 × 19.05 × 12.7, N87, $N = 2 \times 2$	$L_{CM;DC} = 144\,\mu\text{H}$, $L_{CM;DC;\sigma} = 350\,\text{nH}$
DC link Y capacitors	$C_{Y;DC;1}, C_{Y;DC;2}$	TDK B32034B4224, MKP	$C_{Y;DC} = 220\,\text{nF}$
Ripple inductors	$L_{r;U}, L_{r;V}, L_{r;W}$	3 × EE4317, KoolMu 60u, $N = 7$	Figure 6, $L_r(I = 0) = 22\,\mu\text{H}$
Ripple capacitors	$C_{r;U}, C_{r;V}, C_{r;W}$	TDK B32926C3565, MKP	$C_r = 5.6\,\mu\text{F}$, $L_{Cr} = 15\,\text{nH}$, $R_{Cr} = 20\,\text{m}\Omega$
Mains Y capacitor	$C_{Y;AC}$	TDK B32926C3565, MKP	$C_{Y;AC} = 5.6\,\mu\text{F}$, $L_{CY;AC} = 15\,\text{nH}$, $R_{CY;AC} = 20\,\text{m}\Omega$
Mains CM filter choke	$L_{CM;AC}$	VAC, T60006-L2050-V565, $N = 3 \times 4$	Figure 6, $L_{CM;AC;\sigma} = 940\,\text{nH}$
Mains filter capacitors	$C_{X;U}, C_{X;V}, C_{X;W}$	2 × TDK B32926C3685, MKP	$C_X = 13.6\,\mu\text{F}$

The impedance of the CM choke on the mains side $L_{CM;AC}$ cannot be described with frequency-independent parameters because the employed nanocrystalline core material has a strong dependence on frequency with respect to the complex permeability due to increasing eddy currents at higher frequencies. The impedance curve used in the simulations was measured with an impedance analyzer (see below) and is shown in the right diagram of Figure 6.

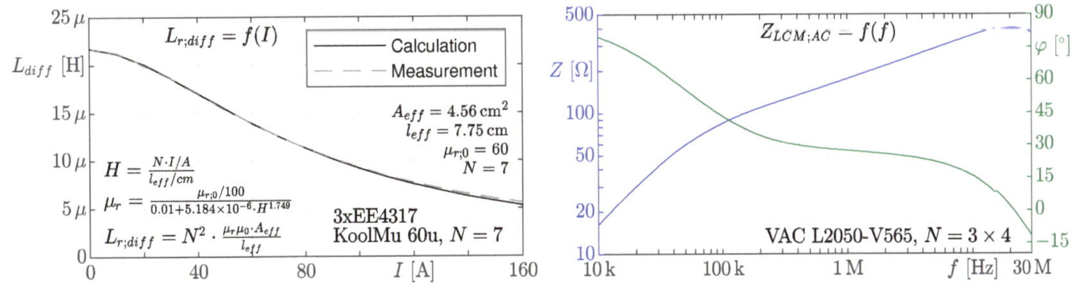

Figure 6. Current dependency of the differential inductance of the ripple inductor and its calculation approach (**left**) and frequency dependent CM impedance and phase angle of the CM choke on the mains side (**right**).

The frequency dependency of all other specified parameters is not as significant (e.g., L_{Cr}, L_{CY}) or can be neglected for the modeling (e.g., R_{Cr}, $R_{CY;AC}$), allowing them to be described with frequency-independent values. The parameters were determined by measurements with the impedance analyzer.

The frequency dependence of the ripple inductors' permeability is neglected, since it drops by only 5% at the maximum frequency used in the EMI simulations (3 MHz) due to the distributed air gaps in the core material. Furthermore, the winding capacitance of the ripple inductors can also be neglected in the simulations because the first parallel resonance only occurs at a frequency above 10 MHz.

However, it must be considered that the inductance value of the ripple inductors is current-dependent due to the utilization of powder core material KoolMu 60u from Magnetics (https://www.mag-inc.com/Products/Powder-Cores/Kool-Mu-Cores (accessed on 19 July 2019)). On the one hand, their voltage-time area and the CM voltage-time area in the DC link CM choke are affected as a function of the load condition. In the left diagram of Figure 6, the partially empirical equations and parameters for calculating this dependency are provided and the result is compared with the measurement performed with a power choke tester. The current dependent inductance can be predicted with very high accuracy. On the other hand, the current dependent inductance value is reflected in higher interference emissions with increasing load currents as described in the following. the company and must therefore be written with a capital M: "Magnetics".

The interference measurements are performed at an output power of approximately $P_{out} = 1.5\,\text{kW}$ and a DC link voltage of $V_{DC} = 610\,\text{V}$. At the operating point with the highest mains current of approximately $i_G = 130\,\text{A}$, their inductance value drops from $L_{diff} = 22\,\mu\text{H}$ at zero current to 7 µH. Due to the shifted phase currents of the three inverter stages the emissions does not increase in a proportional manner. Even when one ripple inductor momentarily carries a high current leading to a strongly reduced inductance, the other two inductors will at the same time carry a lower current and thus, relatively seen, maintain a higher inductance. Analyses show that the PK value increases by a maximum of 7 dB and the AV value by 5 dB at full load compared to the low load case.

Special attention is paid to the design of the DC link CM choke, which is designed having the complete motor drive system in mind. Since it must absorb the difference in CM voltage-time areas from both converters, which is relatively high in certain operating conditions, it must be ensured that the choke does not saturate at any operating point and that it is designed for a minimum of losses. The selected core material is a loss-optimized

MnZn ferrite for the applied switching frequency with a relatively high saturation flux density of $B_{sat} = 390\,\mathrm{mT}$ at a temperature of 100 °C (N87 from TDK). To keep the leakage inductance as low as possible, two stacks of 5 cores R38.1 × 19.05 × 12.7 are wound with 2 windings of 2 turns each as shown in the right picture of Figure 7. The resulting low leakage inductance of $L_{CM;DC;\sigma} = 350\,\mathrm{nH}$ per winding including the connections to the two DC links resonates with the DC link capacitors at a frequency of approximately 43 kHz. Due to the damping caused by the losses in the CM choke and especially in the DC link capacitors, which have a relatively high Equivalent Series Resistance (ESR) at this frequency, the oscillation decayed to below 10% of its initial amplitude after 5 oscillations following a voltage step in one of the two DC link voltages. Since the controller operates with a frequency of 12.5 kHz, no impact on the control is to be expected. However, a load step on the motor side will cause an oscillation with the resonance frequency on the mains side. At this frequency, however, no requirements exist regarding EMC or THD. As the switching frequency is substantially higher than the resonance frequency, no oscillations due to the switching operations of the semiconductors are to be expected. This topic will be subject to further studies in the future.

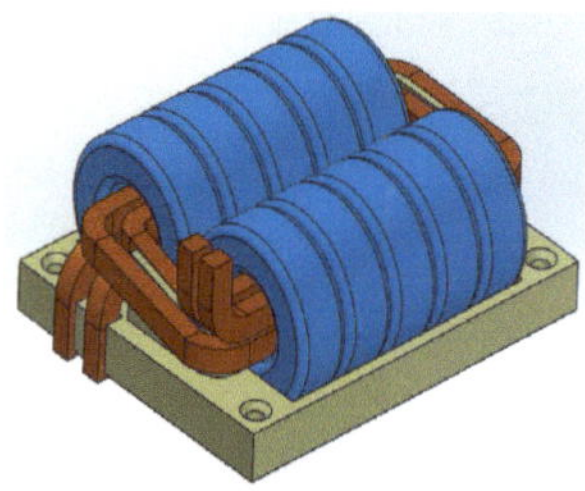

Figure 7. Populated DCB of the EMC optimized power module used for validation, equipped with four SiC MOSFETs, each in CSP, and additional thermal planes on the LS (**left**) and 3D model of the CM choke in the DC link with low parasitic stray inductance (**right**).

For the validation object, power modules depicted in Figure 7 have been EMC-optimized as described in [16]. The basis for the low-inductive design are Chip-Scale Packages (CSPs) of silicon carbide (SiC) Metal-Oxide-Semiconductor Field-Effect Transistor MOSFET dies, which are also used for the component SCTW100N120G2AG from ST in HiP247 package (https://www.st.com/en/power-transistors/sctw100n120g2ag.html (accessed on 7 October 2020)). The semiconductor dies are each embedded in a small circuit board. One module features four such SiC switches, two in parallel for both High Side (HS) and Low Side (LS).

In the first assembly step, the drain contact of the chip is sintered onto a copper foil. The contacting of the gate and the source is realized with microvias in PCB technology. The LS switches are flipped with respect to their relative position to the Direct Copper Bond (DCB). This is done in order to reduce the parasitic capacitance between the LS drain i.e., the half bridge output potential and the heat sink $C_{p;out2GND;U,V,W}$. To avoid imbalances with respect to the occurring parasitic capacitances, the copper areas of the $DC+$ and $DC-$ potentials on the top side of the DCB are of equal size. In first approximation the capacitance between top and bottom copper areas can be simply determined using the equation for the capacitance of a plate capacitor and amounts to $C_{p;DC2GND,1-6} \approx 130\,\mathrm{pF}$ (area 33.9 mm × 13.25 mm, the ceramic material silicon nitride Si_3N_4 is a dielectric with $\epsilon_r = 8.0$ and has a thickness of 250 μm).

Thermal planes, visible in the picture near the LS switches, are added in order to reduce the thermal resistance between the LS chips and the heat sink: Since the LS switches are flipped, their sinter connection, which have a much better thermal conductivity than the microvias, is not facing the DCB but the inverter PCB. The negative impact this has on the thermal properties is accepted in this case for the benefit of a decreased parasitic capacitance. The additional thermal planes are thermally highly conductive cuboids designed to alleviate thermal problems by transporting heat away from top of the CSPs via copper paths on the inverter PCB back down to the DCB.

The top side of the module is not wire bonded, but soldered onto a PCB. Therefore, the commutation and gate inductances are extraordinarily low allowing for extremely fast switching. Using the loss simulation approach in [17] the switching losses can be predicted neglecting the parasitic module inductances. Hence, it can be concluded that the switching speed is limited exclusively by the semiconductor properties (internal gate resistance, capacitances and transfer characteristic).

2.2. Utilized Measurement Devices

The inverter system described above was operated and monitored during operation using the following instruments:

- Oscilloscope with 5 GHz bandwidth Tektronix MSO58;
- 2 DC voltage sources Delta Elektronika SM1500-CP-30;
- High-voltage, isolated, differential voltage probes PMK BumbleBee;
- Current clamps Tektronix TCP0150;
- Line impedance stabilization network 3 Rohde & Schwarz ESH3-Z6.

An Agilent 4294A impedance analyzer was used for impedance measurements. Furthermore a DPG10 power choke tester was applied for inductance measurements in dependence on the current. The evaluation of the interference emissions was performed with PK detector measurements using an EMC test receiver Rohde & Schwarz ESU-8.

2.3. Filter Design Assumptions

After providing experimental evidence of the accuracy of the existing prediction methodology, conventional and novel filter topologies are compared using computational methods. As already mentioned, the analytical calculation tool presented in [15] is used for the calculations. It enables a large number of design possibilities to be calculated in a short time.

The models predict the operating behavior of back-to-back converter systems with AFE according to Figures 2 and 3 respectively, assumed to have a rated power of 63 kW and being connected to the public distribution network. Mains phase Root Mean Square (RMS) voltage is $V_G = 230\,\text{V}$, mains frequency is $f_G = 50\,\text{Hz}$. The maximum DC link voltage is set to $V_{DC} = 610\,\text{V}$, a switching frequency of $f_S = 140\,\text{kHz}$ and flat-top modulation is considered.

For thermal modeling it is assumed, that the ambient temperature is $T_a = 60\,^\circ\text{C}$ maximum and that the winding temperature of the inductors and CM chokes must not exceed $T_L = 120\,^\circ\text{C}$. Furthermore, it is assumed that the entire surface of the inductors can dissipate a power of $k = 50\,\text{W}/(\text{m}^2\text{K})$ to the environment via forced air cooling.

The ripple inductors on the PFC and motor side as well as the DC link CM choke can be designed independently of all other passive components because they are configured according to the lowest losses and not according to EMC attenuation characteristics. Since PFC ripple inductors always absorb the difference between mains and DC link voltage, losses are maximum at full load. The output voltage on the motor side can be varied over a wide range, resulting in the voltage-time area of the associated ripple inductors being much higher than that of the PFC inductors. It makes sense to analyze and compare designs at the operating point of maximum losses. With the use of the analytical calculation tool mentioned above a motor voltage of 145 V and a motor current of 91.3 A is therefore chosen.

With respect to possible core materials, different powder materials are compared, namely the powder material KoolMu from Magnetics with relative permeability of 40, 60 and 90 and the amorphous iron based alloy material 2605SA1 Metglas (https://metglas.com/magnetic-materials/ (accessed on 20 February 2021)). Their parameters were taken from their respective datasheets and used to calculate the core losses applying the improved generalized Steinmetz equation [18–20]. The Steinmetz parameters are listed in Table 3. The factor for calculation with a sinusoidal excitation k_p is converted for use in the generalized Steinmetz equation as described in [20]. The dependence of the core losses on the premagnetization [21] is neglected for simplicity. The core shapes considered for the design are limited to the standard cores available from the manufacturers. E cores are considered for the KoolMu cores and AMCC cores for the Metglas cores.

Table 3. Steinmetz parameters of the considered core materials.

Material	Sinus Excitation Proportional Factor k_p	Frequency Exponent α_{Fe}	Flux Density Exponent β_{Fe}
KoolMu 40u	$5.6 \times 10^{-6}\,\mathrm{W/cm^3}$	1.45	2.09
KoolMu 60u, KoolMu 90u	$5.0 \times 10^{-6}\,\mathrm{W/cm^3}$	1.36	1.77
Metglas 2605SA1	$1.36 \times 10^{-6}\,\mathrm{W/cm^3}$	1.51	1.74
Vitroperm 500F	$7.15 \times 10^{-9}\,\mathrm{W/cm^3}$	1.82	2.08

Since the permeability of the core materials strongly changes with the magnetic field strength, the dependence of the differential inductance of the inductors on the current is taken into account by the calculation tool (cf. left hand diagram of Figure 6). To simplify the calculation of the current ripple, the average differential inductance during a switching period corresponding to the instantaneous low-frequency load current is used. The error that occurs is negligible.

A single-layer rectangular winding with wires bent over the thinner edge is used as winding, which ensures a high copper fill factor and low HF copper losses in the components. Such rectangular wires are available in a large geometric variety. Therefore, the width and height of the winding wire are automatically calculated by the calculation tool, taking into account the necessary insulation distances and manufacturability. Those are assumed to be 1 mm between copper and core where the wire is surrounded by core material, 5 mm where the winding runs outside the core and 300 µm from one turn to the next. The frequency-dependent resistance of the winding is calculated with a 2D FEM simulation at an assumed winding temperature of $T_L = 120\,^{\circ}\mathrm{C}$. The results are used to determine the Low Frequency (LF) and HF copper losses. In this process, just the resistance in the winding part enclosed by core material is calculated. Since the HF copper losses are only a small part of the inductor losses, it is assumed that the current density distribution is identical in the winding part outside the core. To account for the copper losses in the winding part that is not enclosed by the core material, the resistance calculated with the FEM is multiplied by the winding factor. This is calculated analytically previously and specifies the ratio of the actual length of the winding wire to the length considered in the simulation.

For the design of the CM choke in the DC link it is important to note that there is no operating point at which all designed chokes have the maximum losses. In the case of motor start-up, core losses occur almost exclusively, because the current in the DC link is very low due to the low output voltage. At this operating point, the core losses are at a maximum. A CM choke optimized for this operating point would have a large number of turns and a winding wire with a low copper cross-section. In contrast to this, however, at maximum output power the copper losses are maximum, but the potential for core losses is not as pronounced. Therefore, several operating points must be considered for the design of the CM choke in the DC link. Table 4 shows the operating points used for the choke design.

Table 4. Considered operating points for the CM choke design.

Phase Motor Voltage and Current V_{Mot}, I_{Mot}	Output Fre-Quency f_{Mot}	Phase Shift between Mains and Motor Voltage
0, 91.3 A	0	-
230 V, 91.3 A	50 Hz	60°
150 V, 91.3 A	50 Hz	60°

Standard ring cores made of nanocrystalline core material from Vacuumschmelze (https://vacuumschmelze.com/products/Inductive-Components-and-Cores/Amorphous-and-Nanocrystalline-Cores (accessed on 17 February 2021)) are used in the design. Since the geometric dimensions are not very fine-grained, not as many variants as for the ripple inductor design can be calculated. In the loss modeling, the HF copper losses are neglected because the switching-frequency current is very low due to the high inductance value of the CM choke. The core losses are calculated using the generalized Steinmetz equation with the parameters specified in Table 3. The low-frequency copper losses result from the determined DC resistance of the flexible copper winding wire at a temperature of $T_L = 120\,°C$ and the DC link current. The latter and the voltage-time area that the CM choke must absorb are determined using the calculation tool presented in [15]. Furthermore, it must be ensured that the core material is not saturated at any operating point. It is also taken into account that the CM voltages are load dependent due to the current dependence of the differential inductance of the ripple inductors (cf. Sections 3.1 and 3.2).

For the EMC filter design the following assumptions are made. Since the switching frequency is selected with $f_S = 140\,\text{kHz}$ and EMC limits values exist from a frequency of 150 kHz, the filters are designed for the interference levels at the second harmonic of the switching frequency. The interference excitation is considered to be a square wave voltage with a voltage step corresponding to the DC link voltage V_{DC} and the highest possible interference voltage value for the harmonic under consideration. The Fourier transformed value for these assumptions results in the PK interference voltage in the frequency domain V_{Noise}:

$$V_{Noise} = \frac{\sqrt{2}V_{DC}}{\pi k} \tag{1}$$

where k is the number of the harmonic which is 2 for the design frequency of 280 kHz.

In order to comply with the EMC regulation limit values on the mains side according to IEC 61800-3 Class C2, further filter components are required in addition to the ripple filter. For the frequency of 280 kHz the PK limit value is 79 dBµV. To enable the cable shield to be eliminated on the motor side, the PK interference voltage must not exceed 80 dBµV in the frequency range 150–500 kHz in accordance with the CISPR 14 standard. For this purpose, the interference voltage on the motor side is measured at an input impedance of the measuring device of 1500 Ω.

The EMC filters are designed according to the PK value. Although the PK and AV values differ by only about 5 dB for a LF sinusoidal excitation, the limits according to IEC 61800-3 Class C2 differ by 13 dB. However, the AV value of the interference can be reduced to the desired difference to the PK value using frequency dithering according to [22,23]. Since the measurement bandwidth of 9 kHz is much smaller than the average switching frequency, the AV value can be reduced widely by more than 10 dB. The PK value cannot be influenced by frequency dithering.

For the size comparison, only the CM filter volume is considered. It is assumed that EMC optimized power modules are utilized. For the conventional topology, this means a minimum ground capacitance of the half-bridge output nodes $C_{p;out2GND}$ according to the design rules in [16]. In contrast, in the case of using the CM filter concept with the CM choke in the DC link, the parasitic capacitances of the DC link potentials $C_{p;DC2GND}$ must be minimized. In the simulations, it is assumed that the module layout is optimized to reduce these capacitances to 5 pF per phase. For the other parasitic ground capacitances of the semiconductor module, a total capacitance of 6 × 130 pF is considered.

As described in [14] the overall capacitance value of all Y capacitors on the mains side must be limited due to safety requirements to a total value of 35 nF. Behind the ripple inductors on the mains and motor side respectively, X capacitors with non-restricted capacitance value can be used for CM filtering because they do not have to be connected to the ground potential. The CM capacitors $C_{CM;PFC}$ and $C_{CM;Mot}$ in Figures 2 and 3 can be omitted. The three capacitors $C_{r;PFC}$ and $C_{r;Mot}$, respectively, are used on each side, each connected to one of the three phases. Thus they have an effect for filtering the CM and DM interferences. Their connection between the mains and motor sides when using the new filter topology or to the DC link potential when using the conventional filter topology must be very low inductive. Assuming three X capacitors with a capacitance value of 2.2 μF each and a parasitic inductance of 20 nH, the impedance of the CM current path is determined exclusively by the inductance from the fourth harmonic on.

The required size of the DM filter is not influenced by the components of the CM filter in this case. Although leakage inductances of the CM chokes have an influence on the DM attenuation, but in this case they have such a minor influence with values in the range of only 100–200 nH that they can be neglected. To minimize the DM filter, additional DM filter chokes are needed to reduce the size of the X capacitors.

The design is done with simple circuit simulation with the simulation program Portunus (https://www.adapted-solutions.com/en/home-englisch/ (accessed on 28 February 2021)). Due to the limited number of available cores made of nanocrystalline core material from Vacuumschmelze, no calculation with the optimization tool for the CM filter is required.

As almost the entire voltage-time area is absorbed by the CM choke in the DC link or by the ripple inductors, and outside the ripple stages the ripple currents are negligible, only the LF copper losses are considered for further required CM chokes. The inductance and resistance values of the CM chokes are calculated according to the cross-section and path length of the utilized core, the complex permeability of the nanocrystalline core material at the design frequency of 280 kHz, as well as the number of turns.

3. Results and Discussion

3.1. Comparison of Predicted and Measured Current and Voltage Waveforms

Since the CM voltage is absorbed by the DC link, the current and voltage waveforms differ from those using the conventional filter topology. To increase the understanding of the system, current and voltage waveforms in the ripple stage are initially analyzed. The left diagram of Figure 8 shows the comparison of measurement (gray) and prediction (red) of the ripple inductor and mains current at an output power of $P_{out} = 25$ kW and operation with flat-top modulation over one mains period in phase U. For better comparability, the predicted inductor current waveform is shown with the envelope of minimum and maximum current $i_{Lr;min}$, $i_{Lr;max}$.

In the measured waveforms of the inductor and mains currents, small inhomogeneities can be seen at multiples of $\Delta\omega t = 60°$. The cause can be explained by the fact that at these points in time the switch, which is permanently turned on, is changed (at $\omega t \approx 60°$ from T_6 to T_1). However, the requirements regarding the THD are still met.

Slight differences can be recognized between the positive and negative inductor current half-waves with regard to the minimum and maximum values. They are caused by the fact that always all top transistors are turned on simultaneously at the beginning of a switching period.

The diagram on the right in Figure 8 shows the curves of inductor current and voltage over a switching period at the arbitrarily chosen mains angle of $\omega t = 20.8°$. The representation of the control signals shows that the half bridge output of phase W is currently clamped to the DC minus potential (T_6 is permanently turned on). The control signal value of 0 indicates that the bottom semiconductor is activated. If the value of 1 is assigned to the control signal, the respective top switch is turned on. Occurring dead times ($t_{dead} = 125\,\text{ns}$) could be assigned to one of both possible states depending on the inductor current direction, but are neglected by the calculation tool, resulting in small time deviations from the measured curves of the inductor voltage (here at $\omega t = 20.83°$). As can be seen, the developed analytical calculation tool [15] enables a precise prediction of all current and voltage characteristics.

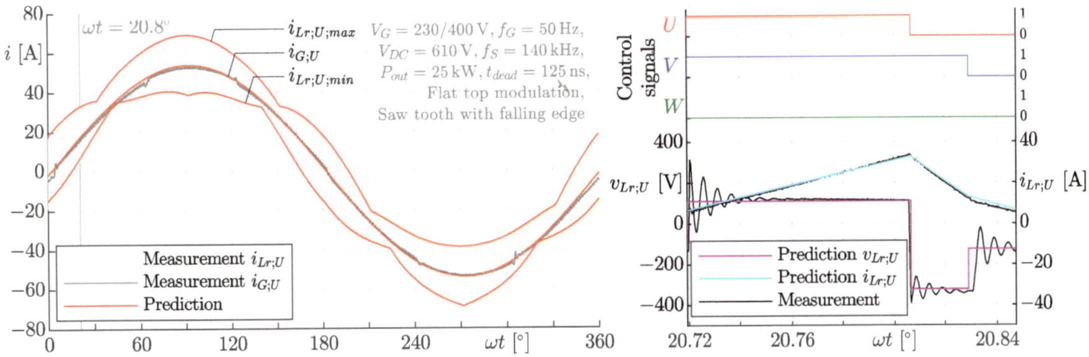

Figure 8. Comparison between predicted and measured waveforms in the ripple inductor during one mains period (**left**) and one switching period (**right**).

It can be seen that the voltage across the ripple inductor changes not only when the switching state of the associated half-bridge changes, but also when that of the other two half-bridges changes. The reason for this is the absence of a low-impedance connection of the filter to the DC link potentials. As a result, the CM voltage drops across the DC link potentials. Assuming current-independent inductance values of the ripple inductors, the CM voltage changes by one third of the DC link voltage for each change in switching state. However, since the inductance value of the ripple inductors is current-dependent and different mains currents flow in the three inductors due to the phase shift of 120°, the CM voltage steps and thus also the inductor voltage steps differ from each other when the switching state of the three half-bridges changes. In the example shown at $\omega t = 20.8°$, the current in phase U is the smallest compared to the other phases, the inductance value of the ripple inductor is the highest and consequently the CM voltage step when switching phase U is the lowest. At a DC link voltage of $V_{DC} = 610\,\text{V}$, the CM voltage step at switching in phase U at this operating point is only 176 V instead of 203 V and the inductor voltage changes by 434 V instead of 407 V. Consequently, for the accurate prediction of the voltage-time areas in the ripple inductors and the DC link CM choke, the current dependence of the inductance value of the ripple inductors $L_r = f(I)$ must be taken into account.

3.2. Voltage Time Area of the Ripple Inductor and CM Choke

For the design of inductive components, the determination of the losses is of central importance. Only if the voltage-time areas in the inductors are correctly determined, the core losses can be calculated with high accuracy. For the validation of the design results regarding the core losses from Section 3.5, the predicted voltage-time areas are compared with those measured.

Figure 9 presents the voltage-time areas absorbed by the ripple inductor $\Psi_{Lr} = v_{Lr} \cdot t$ for different configurations as a function of the mains angle ωt for 5 different DC link voltages and an output power of $P_{out} = 11\,\text{kW}$ in reactive power neutral operation $cos\varphi = 1$ at a switching frequency of $f_S = 140\,\text{kHz}$ and a mains voltage of $V_G = 228\,\text{V}$ in each case. The flux linkage Ψ_{Lr} corresponds to the area below the inductor voltage-time curve when the inductor voltage is positive during a switching period $1/f_S$. Illustratively, the voltage-time area for the switching period shown in the right hand diagram of Figure 8 can be calculated with the time interval when the top semiconductor of phase U is activated as $\Psi_{Lr} = 110\,\text{V} \times 4.84\,\mu\text{s} = 0.532\,\text{mVs}$. This value can also be found in the bottom left diagram at a mains angle of $\omega t = 20.8°$ and a DC link voltage of $V_{DC} = 610\,\text{V}$. The magnitude of the negative voltage-time area is almost the same. The low-frequency component and the occurring losses can be neglected.

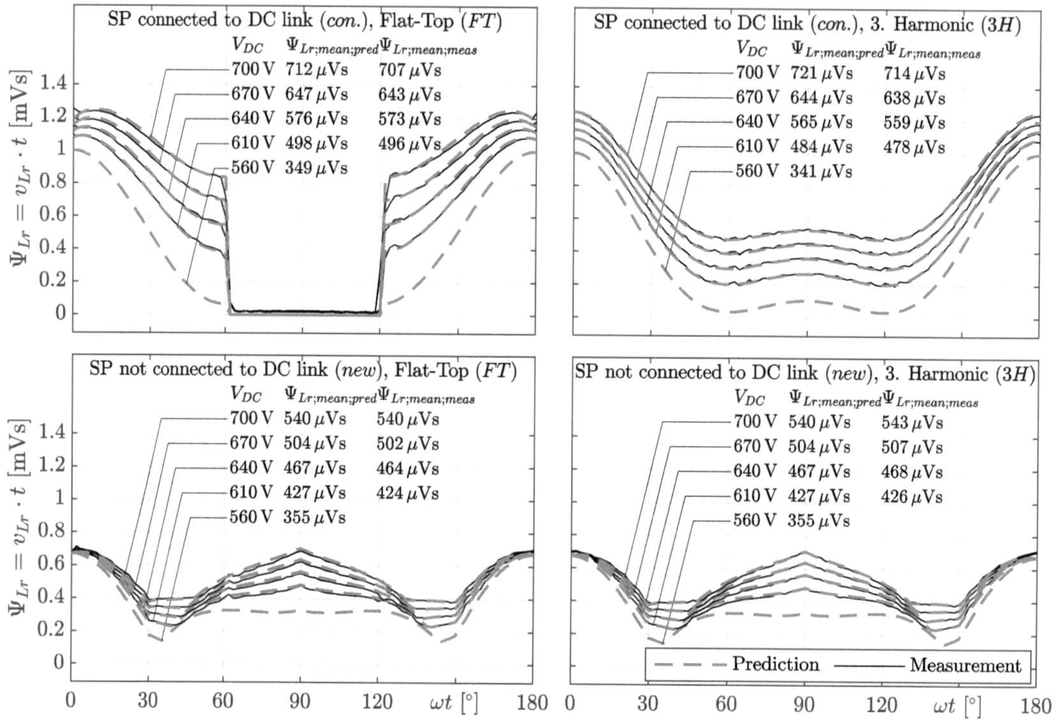

Figure 9. Comparison between predicted and measured voltage time areas absorbed by the ripple inductors in dependence on the mains angle at five different DC link voltages with and without connection of the capacitor star point SP and DC minus potential applying flat-top modulation or imprinted third harmonic.

Since the waveforms are identical from $\omega t = 180°$ to $360°$, the plots are limited to a half mains period. The capacitor star point SP is either connected to the DC minus potential (*con.*) which is related to the conventional filter topology or not connected (*new*) what corresponds to the operation with the new filter topology. In addition, the inverter is optionally operated with the injection of a 3rd harmonic (*3H*) and with flat-top modulation (*FT*).

The flux linkages calculated with the analytical model are shown by the gray dashed lines. It can be seen that the flux linkages can be predicted with high accuracy in all operating conditions. More substantial deviations between measurement and prediction occur in operation with flat-top modulation shortly after another half-bridge transistor is clamped. The filter-frequency oscillations excited during this process slightly increase the voltage-time area in the inductor for a short time. Additionally, there are small dead-time related calculation inaccuracies.

The voltage-time area waveforms for operation with an injected 3rd harmonic and flat-top modulation are identical when the filter is not connected to the DC link (*new*). The reason for this is that the adjusted voltage vectors in the space vector diagram are the same for these two operating conditions and the voltage-time area generated by the zero component of the $\alpha\beta 0$ system (CM) does not occur in the ripple inductor but in the DC link CM choke in this setup configuration. In contrast, in the case of a low-impedance connection between the ripple capacitor star point and the DC link (*con.*), the ripple inductor must absorb the switching-frequency CM voltage. Therefore, its voltage-time area is considerably higher which leads to higher core loss potential. The low frequency CM voltage drops across the ripple capacitors $C_{r;U}, C_{r;V}, C_{r;W}$. Since the CM voltage differs in operation with imprinted 3rd harmonic and flat-top modulation, the voltage-time-area waveforms are also different.

The disproportionate increase of the voltage-time area with the DC link voltage $\Psi_{Lr;mean} = f(V_{DC})$ is caused by the disproportionate increase of the difference between DC voltage and mains voltage. However, it is noticeable that the increase in the mean value of the voltage-time area in the ripple inductors rises with an exponent of approximately 3.2 when the conventional filter topology is used (*con.*), whereas this exponent is only approximately 1.9 when the filter topology with DC link CM choke is applied (*new*). It can be concluded from this that the ripple inductors are unburdened particularly regarding core loss potential at high DC link voltages when the new filter topology is used.

As mentioned earlier, when using ripple inductors with current-dependent inductance applying the new filter topology, the CM voltage changes as a function of the load condition and thus also influences the voltage-time area that the DC link CM choke absorbs. Figure 10 shows the voltage-time areas over half a mains period for three different load conditions. It can be predicted with high accuracy.

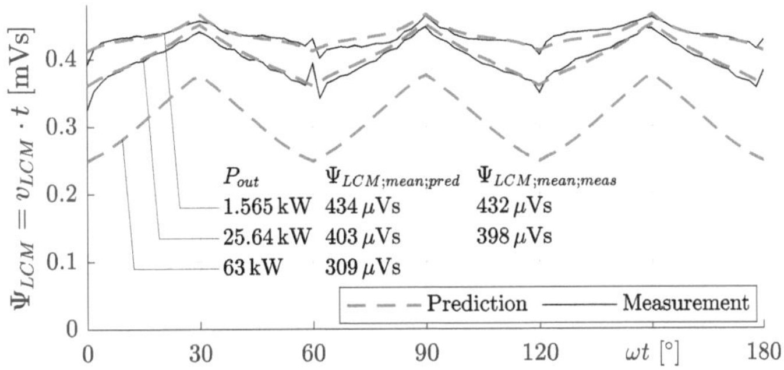

Figure 10. Comparison between predicted and measured voltage time areas of the DC link CM choke in dependence on the mains angle for three load conditions. Prediction and measurement are in very good agreement demonstrating the validity of the underlying model, due to experimental limitations the 63 kW curve is only predicted.

As the load increases, the voltage-time area becomes lower, reducing the potential for core losses. This has a positive influence on the design of the choke because the copper losses increase with the load and thus the overall loss profile is more balanced as a function of the load condition.

3.3. EMC Measurements

The CM emissions of the system with the new filter concept described in Section 2.1 are modeled and measured. In Figure 11 the circuit model is depicted for the prediction of CM emissions. As can be seen in the diagrams, the evaluation of the interference levels measured by the test receiver with the measurement bandwidth of 9 kHz is already included in the voltage source model of the interference excitation VCM_PK. For this purpose, the equation

$$\Delta V = e^{-\left(\frac{\sqrt{ln2}\cdot\Delta f}{\frac{f_{BW}}{2}}\right)^2} \quad (2)$$

given in [24] is implemented in the circuit simulation. For the CM interferences, the ripple inductors, the resistors of the Line Impedance Stabilization Network (LISN), the X capacitors and the parasitic capacitances of the DC link potentials to ground are each connected in parallel. This is taken into account in the simulation by a factor of three or one third (see Figure 11). In the right-hand diagram the red curve depicts the predicted emissions, while the black curve shows the result of the interference measurement, in the frequency range between 100 kHz and 3 MHz. It can be recognized that only the switching-frequency harmonic at 140 kHz exceeds the prediction for CM emissions. According to the inductance and capacitance values of the used components listed in Table 2, these interference levels can be clearly attributed to DM interferences, which will not be analyzed further here.

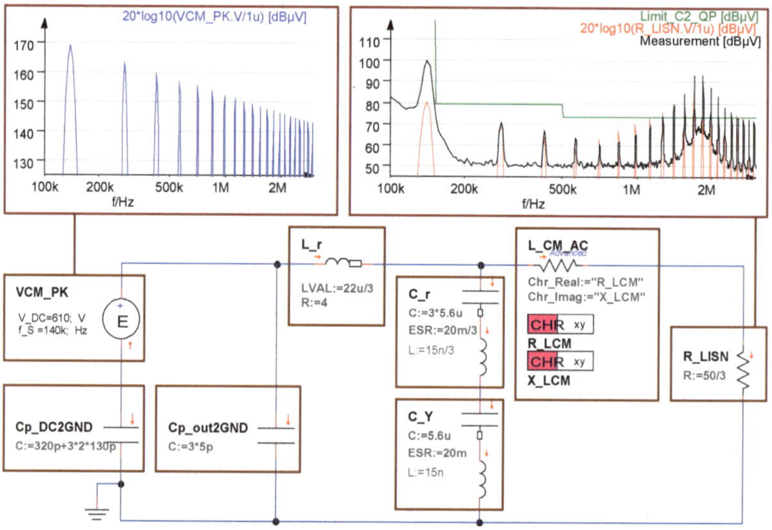

Figure 11. Simplified model for HF CM currents $i_{CM;HF}$ (see also blue arrows in Figure 4) with interference excitation (VCM_PK), the PK spectrum of the excitation is depicted in blue on the top left, the predicted and measured PK interference levels in red and black, respectively, on the top right, prediction and measurement are in very good agreement except for the interference level at 140 kHz.

The interference emissions can be reproduced from the second harmonic onwards with the presented circuit simulation. This only requires consideration of the CM path shown in blue in Figure 4. It can be concluded, that the ground capacitances of the DC link potentials $C_{p;DC2GND}$ play a key role for the modeling these results. These parasitic capacitances are so high that, above a frequency of about 400 kHz, their impedance is lower than that of the CM current path via the DC side Y capacitor and the DC link CM choke, which is marked in red in Figure 4.

The limit exceedance by about 20 dB at a frequency of approximately 1.7 MHz arises due to the resonance of the CM inductance of the ripple inductors $L_r/3$ with the parasitic capacitance of the DC link potentials to ground $C_{p;DC2GND}$. Due to the strong capacitive coupling of the DC link potentials to digital ground caused by the PCB design and the DC link voltage measurement, a portion of $C_{p;DC2GND;Dig} = 320\,\text{pF}$ is added to the parasitic ground capacitance caused by the semiconductor modules $C_{p;DC2GND;1-6}$.

This oscillation can also be seen in the inductor voltage in the right-hand diagrams of Figure 8. However, in this measurement the semiconductor modules were removed from the heat sink. As a result, the resonant frequency increases to approximately 3.3 MHz corresponding to the lower ground capacitance of the DC link potentials ($C_{p;DC2GND} = C_{p;DC2GND;Dig}$) and the interference emissions are reduced by approximately 10 dB. The related EMC interference measurement results are shown in the left diagram of Figure 12.

In contrast to the design rules worked out in [16], it can be concluded from these results that when using the introduced filter concept the ground capacitance of the DC link potentials should ideally be minimized.

The right diagram of Figure 12 shows the interference emissions when wired capacitors with a capacitance value of $C_{p;out2GND;1-3} = 480\,\mathrm{pF}$ each are additionally connected between the half-bridge output nodes and ground. It can be seen that with this measure the resonance at 3.3 MHz which exceeds the limit value can be suppressed. Due to their low impedance, these capacitors return the CM interference current back to the source and can thus be regarded as the first CM filter stage. Since the capacitors and especially their connections with wires have a very high parasitic inductance of about 90 nH, there is a significant resonance at approximately 24 MHz which exceeds regulation limits. However, the interference emissions at the frequencies in the MHz range can be expected to be below the limit values if the ground capacitances of the half-bridge output nodes are not generated by wired capacitors, but by parasitic module capacitances.

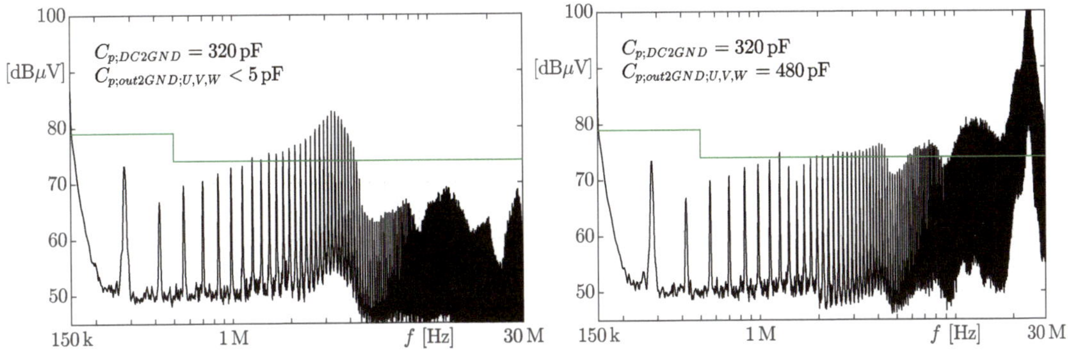

Figure 12. EMC measurement results reducing the parasitic ground capacitance of the DC link potentials to $C_{p;DC2GND}=C_{p;DC2GND;Dig}=320\,\mathrm{pF}$ (**left**) and increasing the parasitic ground capacitance of the half bridge output nodes $C_{p;out2GND} = 1.44\,\mathrm{nF}$ (**right**).

3.4. Derivation of Design Rules for an EMC Optimized Power Module

From these results, design rules for the semiconductor modules when using the new filter topology can be derived. In contrast to the design rules from [16], when using the new filter concept presented here, not the ground capacitances of the half-bridge output potentials $C_{p;out2GND}$ but those of the DC link potentials $C_{p;DC2GND}$ must be minimized in order to optimize EMC. Therefore, not the bottom but the top semiconductors in the power module have to be flipped. In addition, parasitic capacitive couplings of the DC link potentials to all other electrical potentials, such as digital supply and ground potentials as well as measurement lines, need to be minimized. Furthermore, the parasitic capacitances of the half-bridge output nodes should be high and must be the same in all three half-bridges to avoid the generation of CM currents. With each switching operation, the associated ground capacitance of the half-bridge output node is reloaded by two thirds of the DC link voltage. The amount of charge required for this is exactly the same as the amount of charge required for reloading the other two capacitances that change their voltage by one third of the DC link voltage, assuming all half-bridge outputs have the same ground capacitance. Due to the fact that the directions of the currents are distributed in such a way that the current flowing out of the half-bridge outputs is exactly as high as the current flowing in, the change of the switching states of the half-bridges does not cause a CM interference.

3.5. Design Comparison

After verifying the models, the back-to-back converters when using the conventional and the new filter topology can now be optimized according to to assumptions in Section 2.3 and compared with each other in terms of the required volumes.

Figure 13 illustrates the design results for the ripple inductors of the PFC. The left diagram shows the losses and the right diagram the calculated temperature as a function of the inductor volume. From the large number of more than 600 calculated inductor variants, a Pareto front is obtained for each of the assumptions made, which is shown with the solid lines. Based on this, the respective loss and volume optimized inductors can be determined (new filter concept: 3xEE4317, KoolMu 60u, $N = 7$, $A_{Cu} = 6.0\,\text{mm} \times 2.5\,\text{mm}$, $L_{nom} = 22\,\mu\text{H}$, $V = 80.4\,\text{cm}^3$, $P_{Lr} = 32.9\,\text{W}$, conventional filter concept: 4xEE4317, KoolMu 60u, $N = 7$, $A_{Cu} = 6.0\,\text{mm} \times 2.5\,\text{mm}$, $L_{nom} = 29\,\mu\text{H}$, $V = 97.7\,\text{cm}^3$, $P_{Lr;Motor} = 40.8\,\text{W}$).

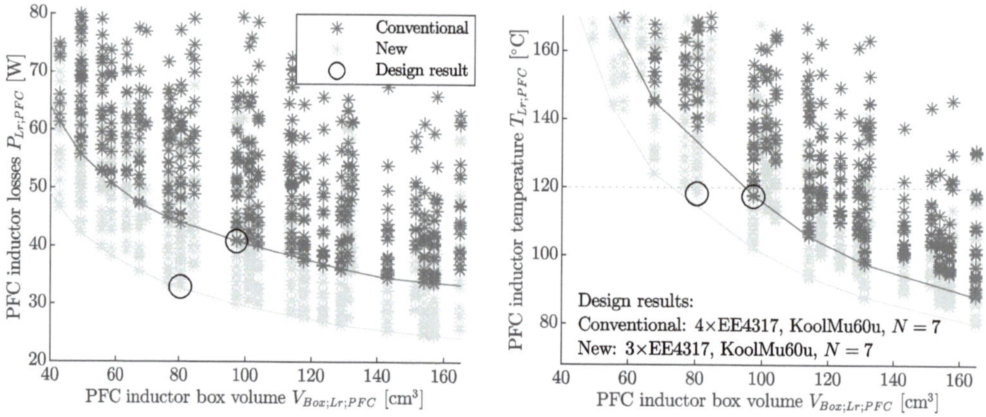

Figure 13. Design results of the PFC ripple inductors for the conventional filter structure and the introduced new filter concept, power losses (**left**) and temperature (**right**) as a function of the box volume, operating parameters $V_G = 230\,\text{V}$, $f_G = 50\,\text{Hz}$, $S = 63\,\text{kVA}$, $\cos\varphi = 1$, $V_{DC} = 610\,\text{V}$, $f_S = 140\,\text{kHz}$, flat-top modulation.

For the motor inductor, only the dependence of the volume on the temperature is shown in the left diagram of Figure 14. Due to the higher loss potential, the size difference of the motor ripple inductors is larger than that of the PFC inductors. This results in an inductor with 6 stacked EE4317 cores (6xEE4317, KoolMu 60u, $N = 7$, $A_{Cu} = 6.0\,\text{mm} \times 2.5\,\text{mm}$, $L_{nom} = 44\,\mu\text{H}$, $V = 131\,\text{cm}^3$, $P_{Lr} = 53.9\,\text{W}$) for the conventional filter topology and an identical inductor but with only 4 stacked EE4317 cores (4xEE4317, KoolMu 60u, $N = 7$, $A_{Cu} = 6.0\,\text{mm} \times 2.5\,\text{mm}$, $L_{nom} = 29\,\mu\text{H}$, $V = 97.7\,\text{cm}^3$, $P_{Lr} = 39.4\,\text{W}$) for the design with CMC in the DC-link.

The design result of the CM choke in the DC link with respect to the calculated temperatures is shown in the right-hand diagram of Figure 14. Based on the set conditions in Section 2.3, the optimal choke has the core L2045-W101 and a winding with a cross-section of approximately $16\,\text{mm}^2$. The predicted winding diameter of 6 mm, which corresponds to a cross-section of approximately $28\,\text{mm}^2$, is rather large, which is understandable, because flexible winding wire has a lower copper fill factor than solid wire. The maximum power dissipation is $P_{L;CMC;DC} = 35\,\text{W}$ and occurs at full load (second operating point in Table 4).

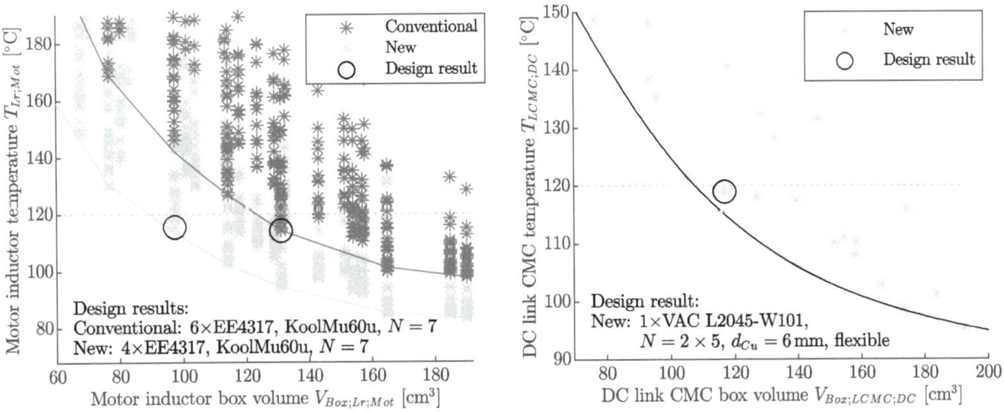

Figure 14. Design results of the motor ripple inductors (**left**) and CM choke in the DC link for the conventional filter structure and the introduced new filter concept, operating parameters $V_G = 230\,\text{V}$, $f_G = 50\,\text{Hz}$, $S = 63\,\text{kVA}$, $\cos\varphi = 1$, $V_{DC} = 610\,\text{V}$, $f_S = 140\,\text{kHz}$, flat-top modulation.

Table 5 compares the required net volumes for the different subassemblies when using the conventional and the new filter topology. The real assembly requires a larger volume because, on the one hand, the PCB and the ventilation duct requires additional space and, on the other hand, the components have to be placed at a distance from each other due to the required heat dissipation and, in some cases, the minimization of electromagnetic couplings. Since it can be assumed that the actually required volume must be scaled by the same factor in each case, the net volumes for both setup variants are comparable.

Table 5. Comparison of the required net filter component box volume for the conventional and the new filter topology, assumed PK limit value on the mains side is 79 dBμV and on the motor side 80 dBμV.

Component	Conventional Filter Topology	New Filter Topology
PFC ripple inductor	$3 \times 97.7\,\text{cm}^3$	$3 \times 80.4\,\text{cm}^3$
Motor ripple inductor	$3 \times 131.0\,\text{cm}^3$	$3 \times 97.7\,\text{cm}^3$
CM choke in the DC link	-	$116.9\,\text{cm}^3$
CM filter on the mains side	$132.8\,\text{cm}^3$	$63.3\,\text{cm}^3$
CM filter on the motor side	$96.4\,\text{cm}^3$	$41.4\,\text{cm}^3$
Total volume	$914.1\,\text{cm}^3$	$754.7\,\text{cm}^3$

It can be seen that already the total net volume of the ripple inductors and the CM choke in the DC link when using the new filter topology is lower than the total volume for the ripple inductors when using the conventional filter topology (sum of the first three lines in the table, $651\,\text{cm}^3$ compared to $686\,\text{cm}^3$). The main reason for this is that only the difference between the two CM voltages has to be absorbed by the CM choke in the DC link.

In addition, when a DC link CM choke is employed, the CM interference voltages across the ripple capacitors on the mains and motor sides are significantly lower compared to the conventional filter topology because the impedance of the CM choke is factors higher than the CM impedance of the PFC or motor inductors. This enables further filter volume to be saved.

Figure 15 shows exemplarily the design result for the CM filter on the mains side when using the conventional filter topology. The interference excitation is modeled with the voltage source on the left side *V_Noise* which is parameterized according to Equation (1).

The remaining interference voltage on the LISN is reproduced with the resistance R_m, whose value is only $50\,\Omega/3$ for CM interference due to the parallel connection of the resistors of all three phases. The ripple stage consists of the previously designed ripple inductors and the ripple capacitors, which are the same on the mains and motor sides for both filter topologies (X capacitors, MKP, $3C_{r;PFC} = 3C_{r;Mot} = 3 \times 2.2\,\mu F$, $V_{Box} = 3 \times 11.3\,cm^3$) and are also required to provide DM attenuation. For the conventional filter topology an LCL CM filter stage is additionally required (two CM chokes VAC L2030-W514, $N = 3 \times 3$, $V_{Box} = 48.4\,cm^3$ and a Y capacitor MKP, $C_{Y;PFC} = 33\,nF$, $V_{Box} = 2.0\,cm^3$), whereas if the filter topology with CM choke in the DC link is applied, only one more small CM choke (VAC L2025-W344, $N = 3 \times 2$, $V_{Box} = 25.3\,cm^3$) is necessary.

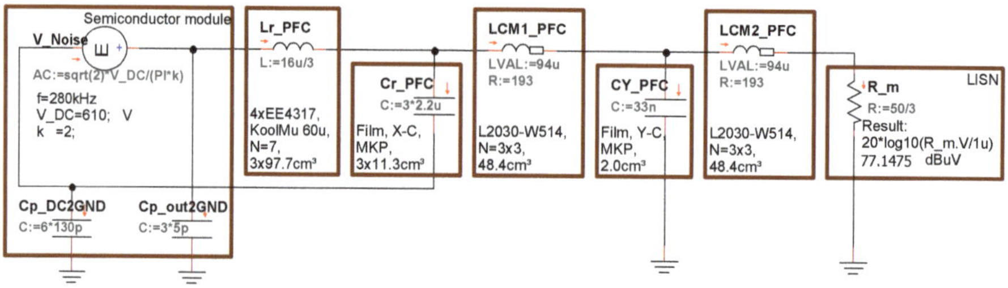

Figure 15. Design result of the CM filter on the mains side using the conventional filter concept, interference excitation is the voltage source V_Noise, design frequency is $2f_S = 280\,kHz$, assumption of an EMC optimized power modules with $C_{p;DC2GND} = 780\,pF$ and $C_{p;out2GND} = 15\,pF$, voltage drop across the resistor R_m reproducing the behavior of the LISN and the test receiver must not exceed the limit value of $79\,dB\mu V$.

On the motor side, only Y capacitors with a total capacitance value of $C_{Y;Mot} = 2 \times 100\,nF$ (MKP, $V_{Box} = 2 \times 3.7\,cm^3$) is required behind the ripple stage to comply with the limit values (cf. Section 2.3) when the new filter topology is applied. When using the conventional filter topology, the CM interference voltage at the ripple capacitors $C_{r;Mot}$ is still so high that a CM choke (VAC L2030-W514, $N = 3 \times 3$, $V_{Box} = 48.4\,cm^3$) and Y capacitors (MKP, $C_{Y;Mot} = 2 \times 220\,nF$, $V_{Box} = 2 \times 7.0\,cm^3$) must be inserted.

The calculations in Table 5 show that the total CM filter volume in this example can be reduced by 17.5% when using the filter topology with the CM choke in the DC link.

4. Conclusions

discussion is unusually long or complex.

In this paper a novel filtering concept for back-to-back converter systems was presented. The main CM current path is designed in such a way that only a single CM choke in the DC link absorbs the entire CM voltage-time area from both converters. Furthermore, this concept has the advantage that the component only absorbs the difference between the CM voltages from the mains and motor side and thus its core losses are very low at most operating points if all semiconductors controlled synchronously.

The CM filter attenuation of the first filter stage is higher by multiples compared to the conventional filter topology, because the inductance of the CM choke is typically orders of magnitude higher than that of the ripple inductors, which usually absorb the CM voltage. On a validation object with a power of 63 kW it was shown that the total CM filter volume can be reduced by approximately 17.5%. The concept is particularly beneficial if unshielded cables are used on the motor side and EMC requirements therefore must be met, or if the cable shields can thus be eliminated. It is also advantageous that the ripple inductors on the mains and motor side are unburdened with regard to the core loss potential, because they now only absorb the DM voltage-time area, and their size can therefore be reduced. Due to the lower dependence of the voltage-time area absorbed by the ripple inductors on the DC link voltage when using the new filter topology, the concept is particularly beneficial for high DC link voltages.

Furthermore, the filter concept allows the application of flat-top modulation in both converter stages almost without the excitation of filter-frequency oscillations, even when the motor output voltage is very low.

It was also shown on the validation object that at high parasitic capacitances of the DC link potentials to ground, the dominant CM current does not flow via both ripple stages but via these parasitic capacitances and only one ripple stage. Together with the CM inductance of the ripple inductors, they cause a pronounced resonance, which leads to increased interference emissions in the frequency range of the resonant frequency.

From this finding it can be derived that when using the filter topology presented in this paper, EMC optimized semiconductor modules should be designed in such a way that parasitic ground capacitances of the two DC link potentials are minimal. Therefore, not the LS but the HS chips have to be flipped. The parasitic capacitances of the half-bridge output potentials to ground have a filtering effect for high-frequency CM noise and must therefore not be limited. However, it must be ensured that they are as equal as possible in all half bridges.

In the future, further studies and EMC-optimized setups are necessary to validate the effectiveness of the presented concept also at high frequencies.

Author Contributions: Conceptualization, S.H. and E.H.; methodology, S.H.; hardware and software for control signal generation, M.B.; experiments, S.H. and M.B.; validation, S.H.; investigation, S.H.; writing—original draft preparation, S.H.; supervision, E.H.; funding acquisition, E.H. All authors have read and agreed to the published version of the manuscript.

Funding: This research was funded by the European Center of Power Electronics (ECPE) within the Lighthouse Program grant numbers 2018.II/PC13 and 2018.II/PC14.

Institutional Review Board Statement: Not applicable.

Informed Consent Statement: Not applicable.

Data Availability Statement: Measurement data and datasets generated during and/or analysed in the presented study are available from the corresponding authors upon reasonable request.

Conflicts of Interest: The authors declare no conflict of interest.

Abbreviations

The following abbreviations are used in this manuscript:

AFE	Active front end
AV	Average
CM	Common mode
CSP	Chip scale package
DCB	Direct copper bond
DM	Differential mode
EMC	Electromagnetic compatibility
ESR	Equivalent series resistance
MOSFET	Metal–oxide–semiconductor field-effect transistor
HF	High frequency
LF	Low frequency
LISN	Line impedance stabilization network
LS	Low side
MKP	Polypropylene (dielectric of film capacitor)
PK	Peak
RMS	Root mean square
SiC	Silicon carbide
THD	Total harmonic distortion
WBG	Wide band gap

References

1. Ran, L.; Gokani, S.; Clare, J.; Bradley, K.; Christopoulos, C. Conducted electromagnetic emissions in induction motor drive systems. I. Time domain analysis and identification of dominant modes. *IEEE Trans. Power Electron.* **1998**, *13*, 757–767, doi:10.1109/63.704152.
2. Finlayson, P. Output filters for PWM drives with induction motors. *IEEE Ind. Appl. Mag.* **1998**, *4*, 46–52, doi:10.1109/2943.644886.
3. Erdman, J.; Kerkman, R.; Schlegel, D.; Skibinski, G. Effect of PWM inverters on AC motor bearing currents and shaft voltages. *IEEE Trans. Ind. Appl.* **1996**, *32*, 250–259, doi:10.1109/28.491472.
4. Chen, S.; Lipo, T.; Fitzgerald, D. Source of induction motor bearing currents caused by PWM inverters. *IEEE Trans. Energy Convers.* **1996**, *11*, 25–32, doi:10.1109/60.486572.
5. Akagi, H.; Tamura, S. A Passive EMI Filter for Eliminating Both Bearing Current and Ground Leakage Current From an Inverter-Driven Motor. *IEEE Trans. Power Electron.* **2006**, *21*, 1459–1469, doi:10.1109/tpel.2006.880239.
6. Matheson, E.; von Jouanne, A.; Wallace, A. Evaluation of inverter and cable losses in adjustable speed drive applications with long motor leads. In Proceedings of the IEEE International Electric Machines and Drives Conference, Seattle, WA, USA, 9–12 May 1999; IEMDC'99, Proceedings (Cat. No.99EX272), doi:10.1109/iemdc.1999.769058.
7. Yi, P.; Murthy, P.K.S.; Wei, L. Performance evaluation of SiC MOSFETs with long power cable and induction motor. In Proceedings of the 2016 IEEE Energy Conversion Congress and Exposition (ECCE), Milwaukee, WI, USA, 18–22 September 2016, doi:10.1109/ecce.2016.7855344.
8. Kampen, D.; Burger, M. Drive system loss reduction by allpole sine filters. In Proceedings of the PCIM Europe 2016; International Exhibition and Conference for Power Electronics, Intelligent Motion, Renewable Energy and Energy Management, Nuremberg, Germany, 10–12 May 2016; pp. 144–147.
9. Videt, A.; Messaoudi, M.; Idir, N.; Boulharts, H.; Vang, H. PWM Strategy for the Cancellation of Common-Mode Voltage Generated by Three-Phase Back-to-Back Inverters. *IEEE Trans. Power Electron.* **2017**, *32*, 2675–2686, doi:10.1109/tpel.2016.2573831.
10. Duesterhoeft, W.C.; Schulz, M.W.; Clarke, E. Determination of Instantaneous Currents and Voltages by Means of Alpha, Beta, and Zero Components. *Trans. Am. Inst. Electr. Eng.* **1951**, *70*, 1248–1255, doi:10.1109/t-aiee.1951.5060554.
11. Taniguchi, K.; Ogino, Y.; Irie, H. PWM technique for power MOSFET inverter. *IEEE Trans. Power Electron.* **1988**, *3*, 328–334, doi:10.1109/63.17951.
12. Kolar, J.; Ertl, H.; Zach, F. Influence of the modulation method on the conduction and switching losses of a PWM converter system. *IEEE Trans. Ind. Appl.* **1991**, *27*, 1063–1075, doi:10.1109/28.108456.
13. Wyss, J.; Biela, J. Volume optimization of a 30 kW boost PFC converter focusing on the CM/DM EMI filter design. In Proceedings of the 2017 19th European Conference on Power Electronics and Applications (EPE'17 ECCE Europe), Warsaw, Poland, 11–14 September 2017, doi:10.23919/epe17ecceeurope.2017.8099251.
14. Boillat, D.O.; Kolar, J.W.; Muhlethaler, J. Volume minimization of the main DM/CM EMI filter stage of a bidirectional three-phase three-level PWM rectifier system. In Proceedings of the 2013 IEEE Energy Conversion Congress and Exposition, Denver, CO, USA, 15–19 September 2013, doi:10.1109/ecce.2013.6646954.
15. Hoffmann, S.; Hoene, E.; Zeiter, O.; Kuczmik, A.; Lang, K.D. Effective method for filter design and semiconductor and inductors loss calculation. In Proceedings of the 2016 18th European Conference on Power Electronics and Applications (EPE'16 ECCE Europe), Karlsruhe, Germany, 5–9 September 2016, doi:10.1109/epe.2016.7695505.
16. Domurat-Linde, A.; Hoene, E. Analysis and Reduction of Radiated EMI of Power Modules. In Proceedings of the 2012 7th International Conference on Integrated Power Electronics Systems (CIPS 2012), Nuremberg, Germany, 6–8 March 2012.
17. Kuczmik, A.; Hoffmann, S.; Hoene, E. Double pulse vs. indirect measurement: Characterizing switching losses of integrated power modules with wide bandgap semiconductors. In Proceedings of the 2020 10th International Conference on Integrated Power Electronics Systems (CIPS), Berlin, Germany, 24–26 March 2020.
18. Reinert, J.; Brockmeyer, A.; Doncker, R.D. Calculation of losses in ferro- and ferrimagnetic materials based on the modified Steinmetz equation. *IEEE Trans. Ind. Appl.* **2001**, *37*, 1055–1061, doi:10.1109/28.936396.
19. Li, J.; Abdallah, T.; Sullivan, C. Improved calculation of core loss with nonsinusoidal waveforms. In Proceedings of the Conference Record of the 2001 IEEE Industry Applications Conference, 36th IAS Annual Meeting (Cat. No.01CH37248), Chicago, IL, USA, 30 September–4 October 2001, doi:10.1109/ias.2001.955931.
20. Venkatachalam, K.; Sullivan, C.; Abdallah, T.; Tacca, H. Accurate prediction of ferrite core loss with nonsinusoidal waveforms using only Steinmetz parameters. In Proceedings of the 2002 IEEE Workshop on Computers in Power Electronics, 2002. Proceedings, Mayaguez, PR, USA, 3–4 June 2002, doi:10.1109/cipe.2002.1196712.
21. Muhlethaler, J.; Biela, J.; Kolar, J.W.; Ecklebe, A. Core losses under DC bias condition based on Steinmetz parameters. In Proceedings of the The 2010 International Power Electronics Conference-ECCE ASIA, Sapporo, Japan, 21–24 June 2010, doi:10.1109/ipec.2010.5542385.
22. Johnson, S.; Zane, R. Custom Spectral Shaping for EMI Reduction in High-Frequency Inverters and Ballasts. *IEEE Trans. Power Electron.* **2005**, *20*, 1499–1505, doi:10.1109/tpel.2005.857565.

23. Gonzalez, D.; Balcells, J.; Santolaria, A.; Bunetel, J.C.L.; Gago, J.; Magnon, D.; Brehaut, S. Conducted EMI Reduction in Power Converters by Means of Periodic Switching Frequency Modulation. *IEEE Trans. Power Electron.* **2007**, *22*, 2271–2281, doi:10.1109/tpel.2007.909257.
24. Wang, Z.; Wang, S.; Kong, P.; Lee, F.C. DM EMI Noise Prediction for Constant On-Time, Critical Mode Power Factor Correction Converters. *IEEE Trans. Power Electron.* **2012**, *27*, 3150–3157, doi:10.1109/tpel.2011.2182059.

energies

Article

Broadband Circuit-Oriented Electromagnetic Modeling for Power Electronics: 3-D PEEC Solver vs. RLCG-Solver

Ivana Kovacevic-Badstuebner [1,*], Daniele Romano [2], Giulio Antonini [2], Jonas Ekman [3] and Ulrike Grossner [1]

1 Advanced Power Semiconductor Laboratory, ETH Zurich, 8092 Zurich, Switzerland; ulrike.grossner@ethz.ch
2 UAq EMC Laboratory, Università degli Studi dell'Aquila, 67100 L'Aquila, Italy; daniele.romano@univaq.it (D.R.); giulio.antonini@univaq.it (G.A.)
3 Department of Computer Science, Electrical and Space Engineering, Luleå University of Technology, SE-971 87 Luleå, Sweden; jonas.ekman@ltu.se
* Correspondence: kovacevic@aps.ee.ethz.ch

Abstract: Broadband electromagnetic (EM) modeling increases in importance for virtual prototyping of advanced power electronics systems (PES), enabling a more accurate prediction of fast switching converter operation and its impact on energy conversion efficiency and EM interference. With the aim to predict and reduce an adverse impact of parasitics on the dynamic performance of fast switching power semiconductor devices, the circuit-oriented EM modeling based on the extraction of equivalent lumped R-L-C-G circuits is frequently selected over the Finite Element Method (FEM)-based EM modeling, mainly due to its lower computational complexity. With requirements for more accurate virtual prototyping of fast-switching PES, the modeling accuracy of the equivalent-RLCG-circuit-based EM modeling has to be re-evaluated. In the literature, the equivalent-RLCG-circuit-based EM techniques are frequently misinterpreted as the quasi-static (QS) 3-D Partial Element Equivalent Circuit (PEEC) method, and the observed inaccuracies of modeling HF effects are attributed to the QS field assumption. This paper presents a comprehensive analysis on the differences between the QS 3-D PEEC-based and the equivalent-RLCG-circuit-based EM modeling for simulating the dynamics of fast switching power devices. Using two modeling examples of fast switching power MOSFETs, a 3-D PEEC solver developed in-house and the well-known equivalent-RLCG-circuit-based EM modeling tool, ANSYS Q3D, are compared to the full-wave 3-D FEM-based EM tool, ANSYS HFSS. It is shown that the QS 3-D PEEC method can model the fast switching transients more accurately than Q3D. Accordingly, the accuracy of equivalent-RLCG-circuit-based modeling approaches in the HF range is rather related to the approximations made on modeling electric-field induced effects than to the QS field assumption.

Keywords: partial element equivalent circuit; finite element method; quasi-static electromagnetic modeling; fast switching power semiconductor devices

Citation: Kovacevic-Badstuebner, I.; Romano, D.; Antonini, G.; Ekman, J.; Grossner, U. Broadband Circuit-Oriented Electromagnetic Modeling for Power Electronics: 3-D PEEC Solver vs. RLCG-Solver. *Energies* **2021**, *14*, 2835. https://doi.org/10.3390/en14102835

Academic Editor: Nicu Bizon

Received: 20 March 2021
Accepted: 12 May 2021
Published: 14 May 2021

1. Introduction

Fast switching power semiconductor devices have a great potential to further increase the performance of advanced power electronic systems (PES), which represent a key enabler for highly efficient generation, distribution, and use of electrical energy. During fast switching transients, the system dynamic performance is strongly influenced not only by the device design but also by the layout parasitics. Accordingly, design optimization of circuit and/or package layouts is required in order to fully utilize the fast switching capabilities of new generations of silicon (Si) power metal–oxide–semiconductor field-effect transistors (MOSFETs), and especially of emerging wide-band gap (WBG) power semiconductor devices, i.e., gallium nitride high-electron-mobility transistors (GaN-HEMTs) and silicon carbide (SiC) power MOSFETs. The prediction of layout parasitics and layout optimization using electromagnetic (EM) modeling evolves as a constitutive step of the PES design. The main aim of using EM modeling tools is to estimate the electrical system performance before actual hardware prototyping. Commonly investigated aspects are conduction and switching power losses, conducted and radiated electromagnetic interference (EMI), over-voltages, over-currents, current- and voltage-slopes, and stability. Power electronics (PE) applications involve the current and voltage signals in a wide frequency range, i.e., line frequency (50 Hz), switching frequency (from tens of kHz up to MHz range) and high frequency associated to switching transients (from ≈50 MHz up to ≈1 GHz), implying the need for accurate and computationally affordable broadband electromagnetic characterization. The broadband measurements of parasitics in multichip power modules lead to extraction of simplified equivalent circuits used for modeling the package EM behavior [1–3], frequently neglecting mutual inductive and capacitive coupling effects. The equivalent electrical circuits of discrete packages can be extracted based on impedance measurements [4], S-parameter measurements [5] or Time Domain Reflectometry (TDR) [6]. Additionally, to achieve a good match between measurements and EM simulations, it is highly important to understand the measurement setup in terms of calibration, excitations ports and measurement accuracy in different frequency ranges [7]. Accordingly, the parasitics extraction of power modules and PE circuit layouts based on experimental measurement procedures is quite limited and computational electromagnetics represents the key enabler for broadband electromagnetic characterization enabling a more accurate virtual prototyping of PES [8].

With the tremendous increase of computational power available on personal computers and server machines, solving Maxwell's equations numerically on a large domain with hundreds of thousands unknowns becomes feasible. As EM modeling at the system level gains more and more attention in recent decades, powerful EM modeling tools dedicated for PE applications have been developed. Selecting the right EM tool for the specified application should be based on selecting the right numerical technique, which in turn depends on the layout geometry and the modeling accuracy requested in different frequency ranges. For example, for the EM modeling of printed circuit board (PCB) layouts with surface mounted devices (SMD), typically the so-called 2.5D (or 3D planar) EM simulators such as, e.g., ANSYS HFSS 3D Layout and KEYSIGHT Momentum are used. These simulators are based on the assumption of dominant two-dimensional (2-D) current distribution, such as in PCB multilayer structures, including the current through PCB vias in a computationally efficient way. On the other hand, in PE applications, the modeling domain often cannot be simplified and the current distribution has to be calculated in arbitrary three-dimensional (3-D) geometries, e.g., PCBs with discrete TO-packages and/or power modules. Hence, EM simulators which can accurately model 3-D current flow are required.

For the dimensions of typical power modules (several cm to some low tens of cm) with fast switching power semiconductor devices, i.e., with current/voltage transients in the range of several ns, the quasi-static (QS) EM field assumption is still valid, whereas the full-wave EM field has to be taken into account to represent longer structures such as, e.g., motor cables, or to simulate radiated EMI. In the state-of-the-art broadband EM modeling for PE applications, two types of numerical techniques are frequently employed for solving numerically the Maxwell's equations: Finite Element (FE)-based methods and equivalent-circuit-based methods. The FE modeling typically does not apply any simplifications on the Maxwell's equations and calculates the solutions for broadband EM field distribution by discretizing the 3-D modeling volume in small mesh elements. As EM modeling considering the low- and high-frequency effects with the same accuracy is a very challenging task, FEM EM field solvers are commonly specialized for either high frequency (HF) or low-frequency (LF) EM simulations. The main benefit of using the equivalent-circuit-based methods is a typically lower computational cost for extracting the layout parasitics in a wide frequency range, i.e., from dc to hundreds of MHz, however, at the price of lower modeling accuracy. The idea behind all equivalent-circuit based methods comes from the Partial Element Equivalent Circuit (PEEC) method. While the PEEC method is a stand-alone numerical technique, the equivalent-circuit-based modeling tools employ several numerical techniques to derive equivalent circuits of layout parasitics. For example, the QS EM simulation tool, ANSYS Q3D, employs the FEM and the Method-of-Moments (MoM), and hence, it is related in the literature to both 3D-PEEC method [9,10] and Finite Element Analysis [11,12]. The Q3D derives equivalent circuits of layout parasitics in a form of R-L-C-G lumped elements, where R-L and C-G are calculated by decoupling the electric and magnetic field components in the Maxwell's equations. A similar approach is also used in [13], where the capacitive and inductive effects are calculated by two independent solvers, a 3-D MoM electrostatic and a 3-D MoM magnetostatic solver, respectively. In [11], decoupled modeling of capacitive and inductive layout behavior was treated as a QS EM field assumption. In particular, a less accurate HF modeling using the RLCG-based modeling tools has been attributed to limitations of the QS field assumption and engineers resort to full-wave FEM EM modeling tools for modeling HF EM field effects more accurately. However, the capacitive layout effects can become dominant in the frequency range above hundreds of MHz for which the coupling between the magnetic and electric field components can start to have a higher impact and the QS assumption is still valid. Accordingly, the simplifications adopted in the RLCG-based EM modeling tools are often not well investigated in terms of their accuracy for accurate capturing of HF effects under the QS EM field assumption. The differences between the RLCG EM modeling approach and a QS 3D-PEEC solver for the fast switching PE applications have not been addressed in a comprehensive way so far. The aim of this paper is to provide a comprehensive overview of these numerical engines in terms of modeling accuracy and overall capability for the extraction of broadband EM macromodels of PCB circuit and power semiconductor package layouts.

2. The State-of-the-Art Device-Circuit Layout Coupled Modeling

The simulations of PE circuits are commonly based on modeling power semiconductor devices using a switch model with look-up tables describing the switching and conduction energy losses [14–16]. This guarantees fast simulations of power converters with a high number of power semiconductor devices, but lacks in accurate modeling of the respective switching transients. The switching transients are strongly influenced by non-linear C-V and I-V device characteristics as well as the circuit layout parasitics [17], which in turn determines the device energy losses and electromagnetic interference (EMI) noise levels. Energy losses and EMI noise become the key design aspects of power converters implementing fast switching power semiconductor devices. The virtual prototyping of power converters, especially the accurate prediction of the generated EMI noise, the device stress, the current and voltage slopes and the ringing, requires equivalent electrical models of circuit layout parasitics and Spice-based compact device models. Therefore, the computational cost is increasing in comparison to the standard PE circuit simulators [16].

The state-of-the-art device-circuit layout coupled modeling for PE applications can be described by a diagram showed on Figure 1. The EM modeling starts from the three-dimensional (3D) model of the circuit layout and the selection of a numerical technique used for solving Maxwell's equations. The most known numerical techniques used in computational electromagnetics for PE are the FEM, the MoM and the PEEC method. Solving numerically Maxwell's equations leads to an EM macromodel in a form of a multi-port linear network, where the ports represent the selected nodes in 3-D layout structure, whose values can be then extracted from the circuit simulation. The multi-port linear network describes the input-output response of the circuit layout either in the time or the frequency domain. To consider the influence of circuit layouts on the system performance, the EM macromodel has then to be simulated together with the electrical models of PE circuit components such as magnetic components, capacitors, gate drivers, and power semiconductor devices. A circuit simulator is used to calculate the respective current and voltage transient waveforms.

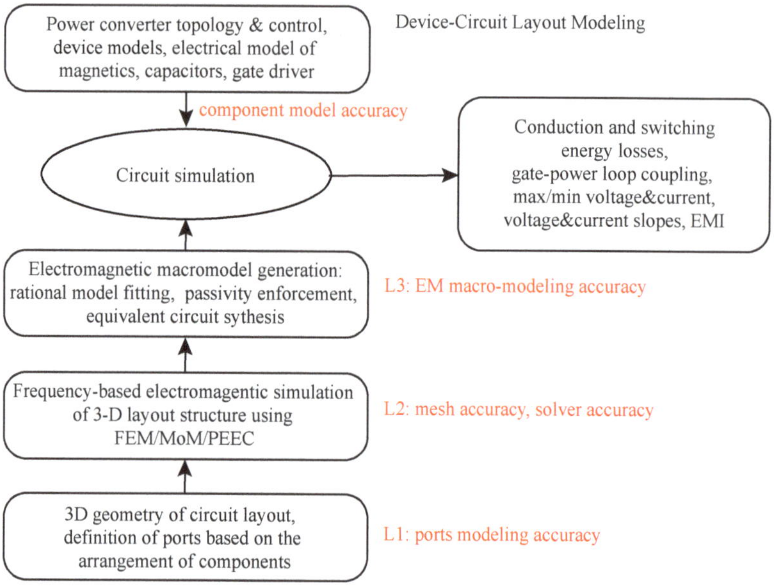

Figure 1. Diagram showing the state-of-the-art device-circuit layout modeling.

2.1. Time Domain EM Macromodeling

Time domain macro-modeling refers to representing the EM performance of circuit layout in time domain by means of the impulse or step response at each port or the time-domain state-space model. The impulse response, i.e., derivative of step response, of the system describes in time-domain all the information contained in the poles and residues of a frequency domain macro-model. When the impulse response is available, it is convolved with the external signals at the ports allowing the circuit simulations coupling time-domain EM macro-models and (non)linear models of power devices and other electrical components. Time-domain impulse response can be directly extracted from Network Analyzer and TDR measurements, or from a transient analysis performed using commercial EM modeling tools typically based on, e.g., Finite Difference Time Domain (FDTD) method or Transmission Line Method (TLM). The time-domain state-space model can also be calculated by numerical techniques such as the PEEC method. As measurements of multi-port networks in a wide frequency range can be very challenging and the standard measurement equipment is typically designed for $Z_0 = 50\ \Omega$, time-domain computational electromagnetics seems to be attractive. However, directly importing an impulse response matrix or a time-domain state-space model to circuit simulators is not supported by any of commercial circuit simulators. Therefore, in PE, frequency domain EM modeling is typically used for extracting the layout parasitics and generating the equivalent electrical models suitable for the time-domain simulations. This means that the frequency-to-time transformation is inherently needed and stability and passivity have to be enforced if not originally satisfied by the frequency-domain dataset. As a consequence, the modeling accuracy may be affected. Namely, a passive model matches a non-passive dataset up to a specified threshold accuracy, which is related to the amount of passivity violation in the data [18].

2.2. Frequency Domain EM Macromodeling

The frequency domain EM field solution at the ports is calculated in N_f frequency points in the form of S, Z, or Y network parameters. In the case of S-parameters, $\mathbf{S} = [S_{ij}(f_m)]$, where $i, j = 1 \ldots N_p$, N_p is number of ports, and $m = 1 \ldots N_f$, a characteristic impedance

Z_0 has to be defined. Such a multi-port linear network represents the broadband frequency behavior of the modeled circuit layout, which is by nature a passive circuit. In order to simulate the frequency domain EM macromodel in the time-domain circuit simulations, an equivalent electrical circuit (netlist) has to be generated. For the $S_{n \times n}$ matrix, either a state-space model or a rational-model using the poles-residues representation is extracted fitting to the calculated samples $S_{ij}(f_m)$, where $i, j = 1 \dots n$ [19]. The most common technique for the extraction of pole-residue rational model is the Vector Fitting (VF) technique [20]. If the passivity of the rational-model is not satisfied, an enforcement passivity procedure is then applied in order to preserve the physical properties of modeling layout [18]. Finally, an equivalent circuit (netlist) can be synthesized from the pole-residue model or the state-space model [19]. For an accurate broadband EM macro-model, it is mandatory to perform a frequency sweep over a wide frequency range resulting in a large number of frequency samples. Circuit simulators such as, e.g., KEYSIGHT ADS and ANSYS Nexxim implement algorithms for direct conversion of S-parameter to time-domain based on either a conversion to impulse response (e.g., Inverse Fast Fourier Transformation) and convolution or via the state-space model generation. Within this paper, we refer to this method as quasi-time-domain macro-modeling, since first S-parameters have to be provided in a large number of frequency points covering a wide frequency range.

3. Circuit-Oriented 3-D Electromagnetic Modeling

Electromagnetic tools used in PE for the extraction of circuit layout parasitics are typically based on the numerical techniques such as the FEM, MoM, and the PEEC method. Hereby, the numerical technique of choice determines the definition of the ports representing the placement of electrical components in the circuit layout, the structure mesh properties, and the broadband EM modeling accuracy.

An accurate extraction of layout parasitics can be performed by using powerful numerical engines based on the FEM. The FEM-based EM solvers resort to the meshing of the entire volume of the modeling domain and approximating the EM fields on this volume mesh. On the other hand, the methods derived from the integral form of Maxwell's equations such as MoM and the PEEC method adopt a volumetric mesh based on the concept of the Green's function [21]. Differently from the FEM-based modeling, the latter avoid the use of absorbing boundary conditions to approximate unbounded media. Another advantage is the possibility to extract impedance of non-closed current paths in a wide-frequency range by defining the input and output ports, while in the three-dimensional (3-D) FEM-based EM solvers, only closed current paths can be defined [7]. In PE applications it is often challenging to define common return current paths, and specialized, therefore the equivalent-RLCG-circuit-based solvers [11,13,22] have been frequently employed in engineering practice for the parasitic extraction and generation of EM macromodels. The RLCG solvers are often related to the 3-D PEEC-based modeling, but there are major differences between these two numerical techniques.

3.1. 3D-PEEC Solver

The 3-D PEEC method [23] leads to a large equivalent circuit coupling magnetic and electric field effects which can be described by the PEEC system matrix (1) using the modified nodal analysis [24],

$$\begin{bmatrix} j\omega\mathbf{P}^{-1} & -\mathbf{A}^T \\ \mathbf{A} & \mathbf{R} + j\omega\mathbf{L}_\mathrm{p} + \mathbf{Y}_\mathrm{d}^{-1}(\omega) \end{bmatrix} \cdot \begin{bmatrix} \mathbf{\Phi}(\omega) \\ \mathbf{I}(\omega) \end{bmatrix} = \begin{bmatrix} \mathbf{I}_s(\omega) \\ \mathbf{V}_s(\omega) \end{bmatrix} \tag{1}$$

where $\mathbf{I}(\omega)$ and $\mathbf{\Phi}(\omega)$ represent the unknown distribution of currents and electric potentials in the discretized 3-D modeling domain, respectively, $\mathbf{P}$ accounts for the coefficients of potential, $\mathbf{L}_\mathrm{p}$ is the partial inductance matrix, and $\mathbf{A}$ is the connectivity matrix, the matrix $\mathbf{R}$ and $\mathbf{Y}_\mathrm{d}(\omega)$ are diagonal, representing the resistances of elementary conductor cells, and the impedances of dispersive and lossy dielectric cells, respectively, [25]. Namely, $\mathbf{Y}_\mathrm{d}(\omega)$ can be written in the form of real and imaginary part, $\mathbf{G} + j\omega\mathbf{C}_\mathrm{e}$, which reduces to only $j\omega\mathbf{C}_\mathrm{e}$

for non-lossy dielectrics. The effects of electric and magnetic field components are taken into account by **P** and $\mathbf{L}_\mathrm{p}$, respectively. The size of **P**, $\mathbf{L}_\mathrm{p}$, **R** and $\mathbf{Y}_\mathrm{d}(\omega)$, **A** are defined by the PEEC mesh. Namely, the PEEC mesh leads to N volume elementary cells and N_p elementary cell nodes, so that **P** is $N_\mathrm{p} \times N_\mathrm{p}$ matrix, **L**, **R** and $\mathbf{Y}_\mathrm{d}$ are $N \times N$ matrices, **A** is $N \times N_\mathrm{p}$ matrix. Here, it should be noted that the inter-dependency of the electric and magnetic fields, i.e., coupling between electric and magnetic field components, are taken into account in a mathematically rigorous way by using a common PEEC mesh for calculating **P**, $\mathbf{L}_\mathrm{p}$ and **R** matrices. The full-wave PEEC modeling is also described by (1), however **P** and $\mathbf{L}_\mathrm{p}$ are then complex matrices. Furthermore, the QS and full-wave PEEC formulation can be extended to include magnetic materials [26]. In the QS PEEC formulation without dielectrics, i.e., the (R,L,P) PEEC method, the PEEC system matrix reduces to (2),

$$\begin{bmatrix} j\omega\mathbf{P}^{-1} & -\mathbf{A}^T \\ \mathbf{A} & \mathbf{R} + j\omega\mathbf{L}_\mathrm{p} \end{bmatrix} \cdot \begin{bmatrix} \mathbf{\Phi}(\omega) \\ \mathbf{I}(\omega) \end{bmatrix} = \begin{bmatrix} \mathbf{I}_s(\omega) \\ \mathbf{V}_s(\omega) \end{bmatrix}. \quad (2)$$

The computational cost of the PEEC EM modeling is defined by the order of the system matrix, $(N_\mathrm{p} + N) \times (N_\mathrm{p} + N)$, where $(N_\mathrm{p} + N)$ can reach more than 100 k unknowns for the actual modeling structures in PE applications. For extracting EM model of the layout, the access to only small number of internal nodes is needed, and Model-Order-Reduction (MOR) techniques [27,28] can be employed to reduce the system complexity and speed-up the system solution in a large number of frequency points. Further simplifications of the 3-D PEEC modeling can be performed by neglecting electric field induced effects, which leads to the QS (R,L) PEEC formulation, implemented in the well-known EM tools for inductance extraction, FastHenry [29] and Inca3D [30]. Similarly, a simulation tool FastCap [31] based on the charge-potential relation, was used for independent extraction of capacitances, which can be associated to the **P** matrix in the (R,L,P) PEEC method. Combining **R** and **L** matrices extracted from, e.g., FastHenry or Inca3D, and the **C**-matrix extracted analytically or using, e.g., FastCap, an equivalent circuit-oriented EM modeling, referred in this paper to the equivalent-RLCG-circuit-based EM modeling, has been frequently adapted for PE applications [32–34].

3.2. RCLG-Solvers

The calculation of electric and magnetic field effects in the QS RLCG solvers are based on two independent solvers, a RL solver and a CG solver, for the extraction of inductive and capacitive circuit behavior, respectively, [11,13]. The EM field distribution to be simulated is approximated on a 3-D modeling structure using the equivalent lumped R-L-C-G circuit representation. By enforcing the Kirchhoff laws to independent nodes and loops, the RLCG system matrix can be written in the form of (3),

$$\begin{bmatrix} \mathbf{G} + j\omega\mathbf{C} & -\mathbf{A}^T \\ \mathbf{A} & \mathbf{R} + j\omega\mathbf{L} \end{bmatrix} \cdot \begin{bmatrix} \mathbf{\Phi}(\omega) \\ \mathbf{I}(\omega) \end{bmatrix} = \begin{bmatrix} \mathbf{I}_s(\omega) \\ \mathbf{V}_s(\omega) \end{bmatrix}, \quad (3)$$

where **C** matrix represents the capacitances between conductors, **G** models the conductances of lossy dielectrics, **R** and **L** model resistance and inductances of conductors, respectively. In particular, using the RLCG solvers, the user first defines the points of interest in 3-D modeling structures, which are then used as input modeling nodes N_c. Each input modeling node is associated to a surface cell (patch) on a conductor, where $\mathbf{Y}_\mathrm{C}$ ($\mathbf{G} + j\omega\mathbf{C}$)-matrix elements model the electric-field induced effects, i.e., the complex admittances of the surface cells. Volume cells are defined between two input modeling nodes, so that the matrix $\mathbf{Z}_\mathrm{L}$ ($\mathbf{R} + j\omega\mathbf{L}$) model the magnetic-field induced effects, i.e., the self- and mutual-complex impedances of the volume cells. The computation of $\mathbf{Z}_\mathrm{L}$ and $\mathbf{Y}_\mathrm{C}$ is based on two independent EM solvers implementing different types of mesh. Namely, in a comparison to the 3-D PEEC solver, surface and volume cells are further discretized by smaller mesh elements to calculate $\mathbf{Z}_\mathrm{L}$ and $\mathbf{Y}_\mathrm{C}$ matrices. This leads to structured differences between the

3D-PEEC (2) and 3D-RLCG modeling (3). While in the 3-D PEEC methods, **R** is diagonal matrix, in a RLCG modeling approach, non-diagonal **R** elements model the real part of the non-zero mutual impedance between two volume cells. In addition, in the RLCG method, the presence of dielectric is included within the $\mathbf{Y}_C$ matrix modeling the Joule losses by **G** and the capacitive effects by **C**, while the 3-D PEEC method includes also the inductive behavior of current paths through dielectric material. Accordingly, the equivalent circuit is formed based on a smaller number of input modeling nodes, i.e., ports, so that the size of the original RLCG system matrix is significantly reduced in comparison to the 3D-PEEC system of equations. In particular, adding more input nodes, the size of equivalent circuit is increasing and the complexity of RLCG-solvers increases towards the complexity of 3-D PEEC solvers. In the Q3D, a RLCG matrix element can be represented an equivalent Transmission Line Model (TLM), i.e., distributing **R**, **L**, **G**, and **C** uniformly across a system of coupled transmission lines. This makes the RLCG solver of Q3D computationally efficient, however not mathematically rigorous. Furthermore, in the presence of non-lossy dielectrics or without dielectrics, the system matrix of RLCG solvers reduces to (4),

$$\begin{bmatrix} j\omega\mathbf{C} & -\mathbf{A}^T \\ \mathbf{A} & \mathbf{R}+j\omega\mathbf{L} \end{bmatrix} \cdot \begin{bmatrix} \mathbf{\Phi}(\omega) \\ \mathbf{I}(\omega) \end{bmatrix} = \begin{bmatrix} \mathbf{I}_s(\omega) \\ \mathbf{V}_s(\omega) \end{bmatrix}. \tag{4}$$

Comparing (4) and (2), it can be seen that the 3-D PEEC system of equations for the QS PEEC formulation without dielectrics has the same form as the RLCG system matrix without dielectrics and with non-lossy dielectrics. Due to this resembling, the RLCG EM modeling is sometimes misinterpreted as 3-D PEEC modeling in the literature, e.g., [9]. In [10], the inaccuracy of HF modeling in Q3D was addressed only for an example of a long transmission line for which it was assumed that the QS assumption becomes invalid above 100 MHz. It was shown that the accuracy of modeling capacitive effects in Q3D can be improved by using TLM and dividing the structure in smaller parts, which corresponds to adding more internal nodes as previously discussed. Here in this paper, the RLCG-based equivalent circuit modeling in the Q3D is investigated with respect to modeling the capacitive HF effects to show that the inaccuracies are not related to the QS field assumptions as often stated in the literature [10,11] but rather to the approximation of the 3-D PEEC modeling approach.

The 3D-PEEC method has been recognized as a numerical technique with a high potential for broadband EM modeling in PE [35–37]. It has been often used assuming a regular mesh with parallelepipeds as basic elements, which allows PEEC matrices, **R**, $\mathbf{L}_p$, and **P**, to be computed fast based on closed-form analytical formulas [38]. However, this prevents using the standard meshing algorithms and thus, modeling of more complex 3-D geometries. With the increasing requirements to accurately capture the skin and proximity HF effects over a wide frequency range for modeling complex geometries without any limitations, the 3-D FEM-based EM simulations have recently received more and more attention [7].

3.3. Comparison between FEM-, RLCG- and PEEC-Based EM Modeling

A comprehensive analysis on the differences between the 3-D PEEC method, the RLCG method with two independent RL and CG solvers, and the 3-D FEM with respect to predicting the fast switching transients of power semiconductor devices is based on a double pulse test (DPT) circuit with a non-optimized layout developed so that it promotes the high parasitic effects and can be accurately modeled by all three numerical techniques. As shown in Figure 2, the DPT circuit layout includes PCB tracks, two TO packages of a free-wheeling diode, and a power MOSFET. The EM frequency-domain macromodel of the layout is extracted by three solvers: (1) the well-known RLCG solver of ANSYS Q3D Extractor [22], (2) the powerful FEM solver of ANSYS HFSS [7], and (3) a 3-D PEEC solver developed in-house [37].

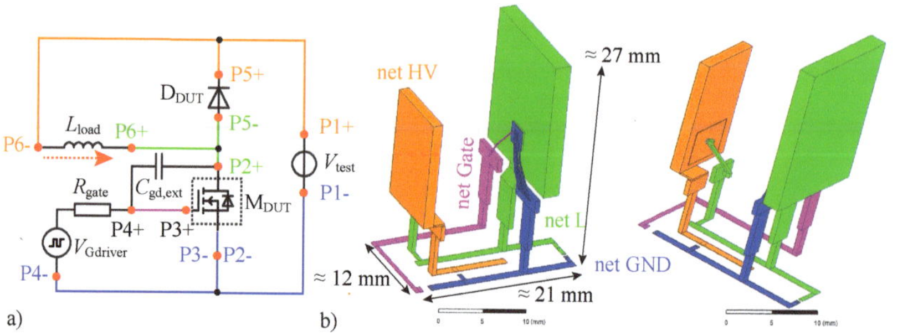

Figure 2. EM modeling of fast switching transients based on a DPT circuit: (**a**) the DPT circuit schematic with marked connected areas and 12 ports, (**b**) 3D circuit layout including PCB tracks and two TO-packages for a free-wheeling diode and a power switch.

The accuracy of EM modeling can be defined at three levels as shown in Figure 1. The first level (L1) is the geometry import and definition of ports, the second level (L2) is the meshing and solver accuracy, and the third level (L3) is extraction of EM macro-model for time-domain simulations. The modeling example under test is designed so that further geometry simplifications are not required for any of the three numerical techniques. Namely, the bond-wires of TO-packages are modeled by non-orthogonal parallelepipeds [39] and the other parts are modeled by orthogonal PEEC cells.

3.3.1. Ports

The length of ports is kept small in order to minimize the effects of different port definitions between the solvers [7,40]. Moreover, once the S-parameters are extracted in the frequency range [0 GHz, 1 GHz] in N_f points, they are imported in the same way in the ANSYS Nexxim circuit simulator, which internally uses the state-space modeling to generate the quasi-time-domain EM macro-models. Therefore, the same modeling accuracy at the level L3 is assumed for all three numerical techniques in the time-domain, and only the differences due to different numerical techniques affecting the modeling accuracy at the level L2 are emphasized.

The DPT circuit defines four connected areas, referred to as nets, and six ports, P$i+$ and P$i-$, where $i = 1 \ldots 6$, marked in Figure 2 by different colors. In comparison to the ports P1-4 with the references (P$i-$) to the ground (GND) plane, the ports P5 and P6 are differential ports with the internal nodes P5$-$ and P6$-$, respectively, which do not belong to the GND net. Using the HFSS FEM solver, only 6×6 S-parameters can be extracted, while both the PEEC solver and the RLCG solver of Q3D return 12×12 S-parameters directly. For the calculation of S-parameters, the characteristic impedance $Z_0 = 50\ \Omega$ is used. In the next step, $S_{6\times6}$ parameters are generated from the Q3D and PEEC $S_{12\times12}$ parameters, with the aim to make the time-domain simulations with the PEEC and Q3D S-parameters equivalent to the time-domain simulations with the S parameters calculated using the HFSS. By importing $S_{6\times6}$ matrix into the ANSYS Nexxim circuit simulator, the P$i-$ ports of the components, i.e., an inductor L_{load}, a diode D_{DUT}, a power MOSFET M_{DUT}, and two voltage sources V_{test} and V_{Gdrive}, are connected between the corresponding ports P$i+$ and the circuit GND, i.e., reference 0 [41]. In this way, the potential of port P$i+$ in the transient simulation with $S_{6\times6}$ parameters represents the voltage between P$i+$ and P$i-$ in the transient simulation with $S_{12\times12}$ matrix.

3.3.2. Mesh

The highest frequency of interest for the EM modeling of fast switching transients is set to 1 GHz. The frequency range below 1 MHz is referred to as LF, whereas the HF range is defined as $f \geq 1$ MHz.

The HFSS 3-D FEM solver has a powerful meshing procedure implementing an adaptive mesh refinement technique. This allows achieving an optimized mesh for the solution frequency f_0 and the maximum variation of S-parameters ΔS in two consecutive iterations, as initially specified by the user, automatically. In the HFSS simulations, f_0 is set to 1 GHz and ΔS to 0.1%. This leads to 314 K tetrahedra.

The Q3D mesh is determined by the RL and the CG solvers, and each solver uses a different mesh. The RL solver uses a LF-optimized FEM to calculate the dc solution (R_{dc}, L_{dc}), modeling the current distribution across the whole cross section of conductors. A MoM solver based on the surface approximation is used to calculate the ac solution (R_{ac}, L_{ac}), assuming that the skin effect is fully developed so that the current is distributed only on the surface of the conductor. The transition between the dc and ac solution is estimated based on the modeling geometry and the frequency-dependent skin depth. In the Q3D simulations, an adaptive mesh strategy is used similar to the one used in the HFSS 3-D FEM solver. A relative convergence error for iterative RLCG solvers in Q3D is set to 0.1%. This results in the numbers of mesh elements equal to 10.5 K triangles, 35.5 K triangles, and 158 K tetrhedra for the CG, ACRL, DCR solvers, respectively.

The mesher implemented for the PEEC method first cuts the modeling structure in sub-domains and then runs the mesh with a specified maximum size of PEEC elements. The PEEC solver developed in-house can implement two types of mesh: uniform mesh and non-uniform mesh. The uniform mesh can capture the broadband EM behavior of a modeling structure more accurately but it requires a higher number of unknowns for larger structures. The PEEC non-uniform mesh has to be carefully implemented to capture both LF and HF effects with the same accuracy. A more accurate modeling with the non-uniform PEEC mesh would currently imply a higher number of unknowns. The in-house hexahedral PEEC mesher works as follows: firstly, the modeling structure shown in Figure 2 is cut into sub-regions in order to remove hanging nodes, cf. Figure 3a. This operation leads to 813 regions. Then, each region is locally meshed with an uniform criterion leading to 20,860 branch currents and 4867 node potentials, cf. Figure 3b. Finally, this PEEC mesh leads to the minimum and maximum PEEC elements size of 0.5 μm and 1.5 mm, respectively.

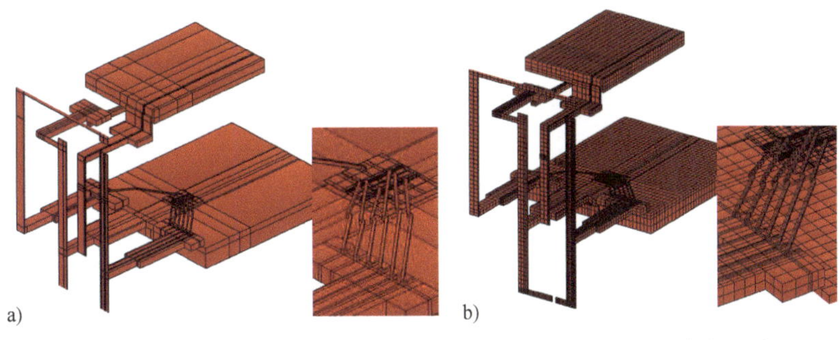

Figure 3. PEEC meshing procedure: (**a**) cutting the 3-D modeling domain, and (**b**) meshing.

3.3.3. Resulting S-Parameters

The S-parameters were first calculated by the in-house developed PEEC solver, the HFSS 3-D FEM solver, the Q3D RLCG-solver and Q3D RL-solver. They are shown in Figure 4. In the Q3D, S-parameters can be calculated either by an equivalent TLM or a lumped RLGC model. The TLM distributes the lumped resistances (R), inductances (L), conductances (G) and capacitances (C) uniformly across a system of coupled transmission lines, which corresponds to adding more internal nodes in between the input modeling nodes. The S-parameters, S_{12}, S_{14}, S_{34}, and S_{24}, are selected to show the differences between the modeling approaches. S_{14} and S_{24} contribute to the gate-power loops coupling, while S_{12} and S_{34} are associated to the power loop impedance and the gate loop impedance, respectively. In the HF range, the PEEC S-parameters are physically more closer to the HFSS S-parameters than the Q3D RLCG S-parameters. To fully match the PEEC and HFSS S-parameters is challenging mainly due to the PEEC mesh requirements.

As explained in Section 3.2, the TLM can increase accuracy for certain modeling structures in power electronics. Using the Q3D RLCG model, the S_{12} and S_{34} show a higher deviation from the S-parameters calculated by the HFSS 3-D FEM solver and the PEEC solver, as it can be seen in Figure 4a,b, respectively. In particular, the accuracy of S_{12} and S_{34} is improved by using the Q3D RLCG-TLM, while the S_{14} and S_{24}, shown in Figure 4c,d, calculated by the Q3D RLCG-TLM show still a visible deviation from the corresponding HFSS and PEEC S-parameters in the HF range above 100 MHz. The effects of electric-field induced couplings can be identified by comparing the Q3D-RLCG and Q3D-RL S-parameters.

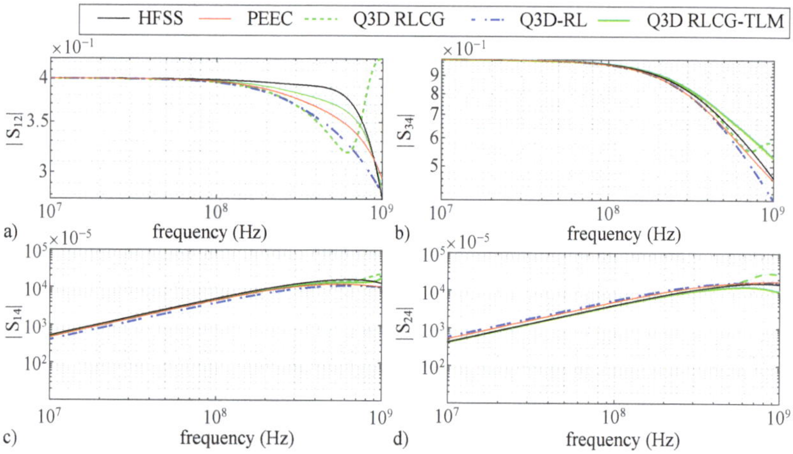

Figure 4. The magnitude of selected S parameters calculated in the HF range by the PEEC solver developed in-house, the HFSS 3-D FEM solver, the Q3D RLCG-solver, the Q3D RLCG-solver with TLM, and the Q3D RL-solver: (**a**) S_{12}, (**b**) S_{14}, (**c**) S_{34} and (**d**) S_{24}.

The importance of these differences observed in the S-parameters calculated by five different methods is evaluated by modeling the switching transients of power MOSFETs as shown in the next section.

4. Device-Circuit Layout Coupled Simulations

The aim of this section is to demonstrate the main differences between three numerical techniques with respect to predicting fast switching transients of power semiconductor devices, such as Si Super Junction (SJ) and SiC power MOSFETs. The potential differences between switching transient simulations are demonstrated on two examples, implementing the EM macromodels, i.e., $S_{6\times6}$ parameters, as extracted by the PEEC solver developed in-house, and the commercial tools ANSYS Q3D and HFSS (see above). The current commutation between a SiC diode D_{DUT} (IDWD40G120C5) [42] and a power MOSFET M_{DUT} is simulated in ANSYS Nexxim using the DPT circuit shown in Figure 2. As layout optimization is crucial for achieving the best performance of fast switching power devices, M_{DUT} is in the first example modeled as a fast switching high efficiency Si SJ power MOSFET (IPP60R180C7), M_{DUT1}, Ref. [43] and in the second example as a SiC trench power MOSFET (IMZ120R030M1H) [44] (M_{DUT2}), using the vendor-provided Spice models.

4.1. Example 1

The first example demonstrates the simulation of HF oscillations in the power loop amplified from the gate loop oscillations via the high transconductance of the MOSFET, i.e., $i_d \approx g_m V_{gs}$ modeling M_{DUT1} as a 600 V, 180 mΩ fast switching Si SJ power MOSFET. The current waveforms simulated in ANSYS Nexxim using the DPT circuit from Figure 2a with the PEEC-, Q3D- and HFSS- $S_{6\times6}$ parameters are shown in Figure 5 for M_{DUT1} switching the current $I_L = 40$ A at $V_{test} = 400$ V for $R_{gate} = 2\ \Omega$ and $C_{gd,ext} = 10$ pF.

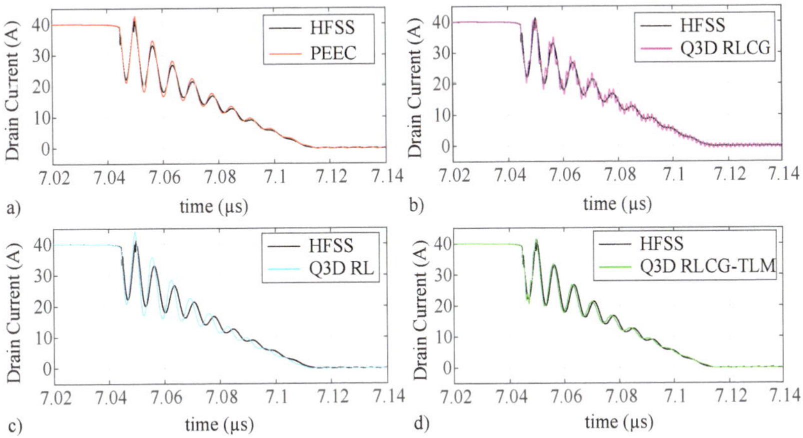

Figure 5. Simulations of HF oscillations observed in the drain current waveform of a fast switching Si SJ power MOSFET M_{DUT1} for $I_L = 40$ A, $V_{test} = 400$ V, $R_{gate} = 2\ \Omega$, $C_{gd,ext} = 10$ pF using the extracted $S_{6\times6}$ parameters: (**a**) HFSS vs. PEEC, (**b**) HFSS vs. Q3D RLCG, (**c**) HFSS vs. Q3D RL, and (**d**) HFSS vs. Q3D RLCG-TLM.

The detailed comparison of the drain current, i_d, waveforms in Figure 6a demonstrates the main differences of three numerical techniques.

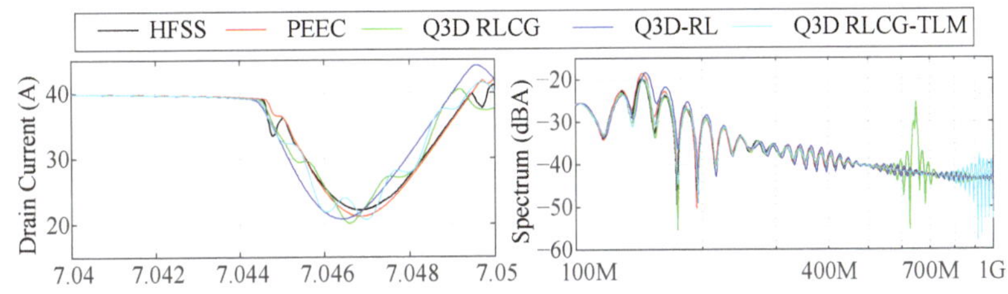

Figure 6. Detailed comparison of current waveforms shown on Figure 5: (**a**) the zoom of all waveforms at the beginning of the switching transition, and (**b**) the spectrum of turn-off current waveforms.

A good matching between the transient simulations with the PEEC- and HFSS- 6 × 6 S-parameters, cf. Figures 5a and 6a, points out that the QS field assumption used by the PEEC solver is still valid. The $S_{6\times6}$ parameters extracted by the Q3D RLCG solvers with and without the TLM introduce HF ringing of the drain current, which is not observed in the transient simulations using the PEEC- and HFSS- $S_{6\times6}$ parameters, see Figures 5b,d and 6a. By using the $S_{6\times6}$ parameters from Q3D-RL Solver, the i_d waveform does not contain the HF ringing, however, a mismatch to the i_d waveform simulated with the HFSS $S_{6\times6}$ parameters is visible, cf. Figure 5c. Comparing the i_d waveforms at the beginning of the current switching transient as shown in Figure 7a, it can be observed that the $S_{6\times6}$ parameters from the Q3D-RL solver, not modeling electric field effects, do not capture the first current valleys modeled by the PEEC- and HFSS-$S_{6\times6}$ parameters, while the Q3D RLCG- and Q3D RLCG-TLM-$S_{6\times6}$ parameters introduce additional HF oscillations. The differences of the EM modeling approaches can quantitatively evaluated by comparing the spectra of the turn-off current waveforms as shown in Figure 6b. The Q3D RLCG model leads to higher amplitude harmonics around 600 MHz–700 MHz, which is shifted towards 800 MHz–1 GHz and reduced by using the TLM approach. Accordingly, the TLM improves the accuracy of the Q3D RLCG modeling in a wider frequency range, however the distributed RLCG coupled transmission line system set automatically by the Q3D, does not eliminate all HF inaccuracies. Taking the current spectrum simulated by the HFSS-$S_{6\times6}$ parameters as reference, the maximum errors of the spectra simulated by the 3-D PEEC-, the Q3D RLCG-, the Q3D RLCG-TLM- and the Q3D RL-$S_{6\times6}$ parameters are 14.1%, 32.6%, 33.5% and 18.3%, respectively.

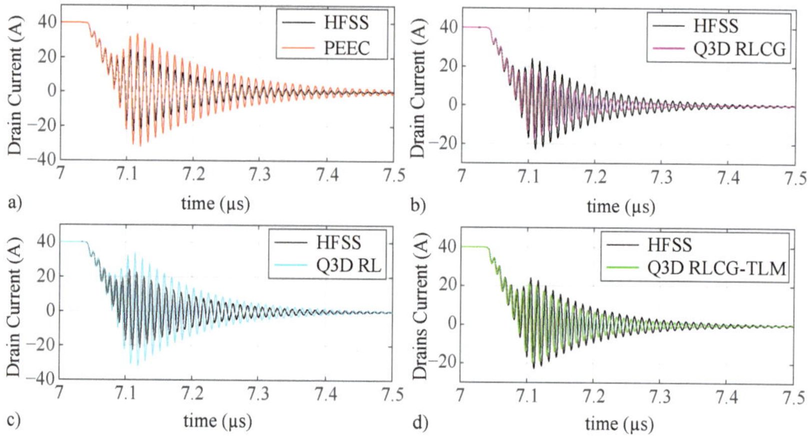

Figure 7. Simulations of HF oscillations observed in the drain current waveform of a SiC trench power MOSFET M_{DUT2} for $I_L = 40$ A, $V_{test} = 600$ V, $R_{gate} = 1\ \Omega$, $C_{gd,ext} = 1$ pF using the extracted 6×6 S-parameters: (**a**) HFSS vs. PEEC, (**b**) HFSS vs. Q3D RLCG, (**c**) HFSS vs. Q3D RL, and (**d**) HFSS vs. Q3D RLCG-TLM.

4.2. Example 2

The second example presents the current switching waveforms of a 1.2 kV, 30 mΩ SiC trench power MOSFET containing HF ringing mainly caused by the layout parasitics. The transient current switching waveforms simulated using different $S_{6\times6}$ parameters and their spectra are shown in Figures 7 and 8. The comparison in Figures 7a and 8a shows that the transient simulations with the PEEC- and HFSS-$S_{6\times6}$ parameters capture the same resonance frequency of HF oscillations of ≈98 MHz but with different amplitudes. This leads to higher amplitude of HF harmonics in the spectrum calculated by the 3-D PEEC-$S_{6\times6}$ parameters in comparison to the spectrum of the HFSS-$S_{6\times6}$ parameters. To better match the 3-D PEEC modeling with HFSS 3-D FEM modeling, an optimized non-uniform meshing is required. On the other hand, the transient simulations with the $S_{6\times6}$ parameters extracted by the Q3D RLCG and RL solvers contain the HF oscillations shifted by ≈2 MHz in the resonant frequency, cf. Figure 7b,c. The $S_{6\times6}$ parameters extracted by the Q3D RLCG solver, cf. Figure 7b, additionally contribute to a higher damping that can be attributed to higher capacitive effects modeled by the Q3D-RLCG S-parameters. The TLM improves the Q3D RLCG modeling, however, similarly to the previous example, the spectra calculated by the Q3D RLCG with and without TLM both show high amplitude harmonics in the HF range not observed in the spectra calculated by HFSS 3-D FEM, Q3D RL, and 3-D PEEC methods. Taking the current spectrum simulated by the HFSS-$S_{6\times6}$ parameters as reference, the maximum errors of the spectra simulated by the 3-D PEEC-, the Q3D RLCG-, the Q3D RLCG-TLM- and the Q3D RL-$S_{6\times6}$ parameters are 39%, 74%, 75% and 51%, respectively.

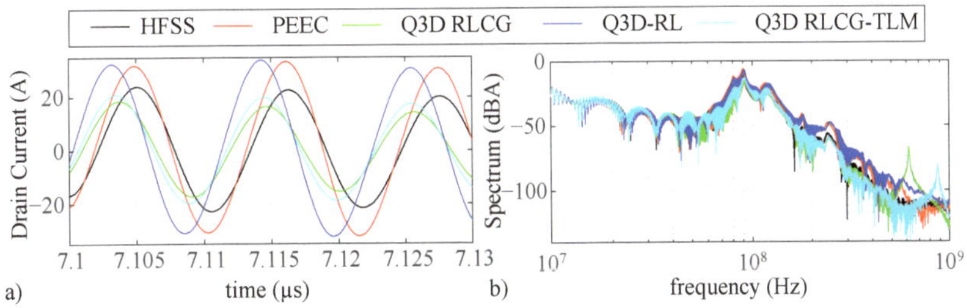

Figure 8. Detailed comparison of current waveforms shown on Figure 7: (**a**) the zoom of all waveforms at the beginning of the switching transition, and (**b**) the spectrum of current switching waveforms.

4.3. Results Analysis

The differences observed in the transient simulations of the MOSFET drain current using the two examples described above demonstrate the capabilities of three numerical techniques, the FEM, the equivalent-RLCG-circuit method, and the PEEC method, used in computational electromagnetics for PE applications. Similarly to [22], the performed analysis also points out that consideration of the broadband frequency range is necessary for accurately capturing the fast switching transitions of modern power devices. Namely, under the QS field assumption, which can be assumed for a large number of PE applications, it is shown that the PEEC-based EM modeling can be as accurate as the well-established FEM-based EM modeling. The frequency range up to hundreds of MHz is important for accurate simulations of PE circuits and their layouts. The RLCG-equivalent-circuit-based methods have to be used carefully since here the modeling of electric-field induced coupling effects is approximated in comparison to the 3-D PEEC modeling, leading to pronounced artificial oscillations and/or overestimation of capacitive effects. As long as a layout features dominantly inductive behavior, the modeling using the Q3D RL-solver, the PEEC solver, and the HFSS 3-D FEM solver is behaving more similar. To accurately capture HF ringing behavior originating from both magnetic and electric field coupling effects often occuring in the switching transients of power semiconductor device due to layout parasitics, the FEM-based or the PEEC-based EM solvers represent a better alternative than the RLCG-equivalent circuit EM solvers.

5. Conclusions

This paper shows the capabilities of three numerical techniques, the PEEC method, the HFSS 3-D FEM, and the Q3D RLCG-equivalent-circuit method typically used for broadband EM modeling of PES. Special attention is given to the differences between the QS 3-D PEEC- and the RLCG-equivalent-circuit-based EM modeling, which have been frequently overlooked so far. The HFSS 3-D FEM solver has a high potential for modeling broadband EM behavior of power converters. However, the FEM-based EM solvers require that the ports are small in size and referenced to a common return current path. Here, in general, the PEEC method offers a higher flexibility in setting-up of the ports. In addition, the commercial FEM-based EM solvers are typically optimized for either LF or HF EM simulations, whereas the required accuracy of broadband EM modeling depends on the application. The broadband capabilities and accuracy of the PEEC method are shown on a modeling example of a 3-D structure rather simplified in comparison to circuit layout geometries in real applications. This is mainly do to the present difficulties with automatized PEEC meshing of more complex structures. Consequently, the current on-going research on the PEEC method is focusing on the development of a PEEC mesher which can automatically handle more complex geometries, and a broadband surface-based PEEC formulation for capturing HF effects at lower computational costs.

Author Contributions: Conceptualization, I.K.-B., D.R., G.A., J.E., and U.G.; methodology, I.K.-B., D.R., G.A., J.E., and U.G.; software, I.K.-B., and D.R.; validation, I.K.-B., and D.R.; formal analysis, I.K.-B., D.R., G.A., J.E., and U.G.; investigation, I.K.-B., D.R., G.A.; resources, I.K.-B., D.R., G.A., J.E., and U.G.; data curation, I.K.-B., and D.R.; writing–original draft preparation, I.K.-B.; writing–review and editing, I.K.-B., D.R., G.A., J.E., and U.G.; visualization, I.K.-B.; supervision, I.K.-B.; project administration, I.K.-B.; funding acquisition, U.G. All authors have read and agreed to the published version of the manuscript.

Funding: This research received no external funding.

Conflicts of Interest: The authors declare no conflict of interest.

Abbreviations

The following abbreviations are used in this manuscript:

MDPI	Multidisciplinary Digital Publishing Institute
DOAJ	Directory of open access journals
EM	Electromagnetic
PE	Power electronics
PES	Power electronic systems
HF	High Frequency
LF	Low Frequency
SiC	Silicon Carbide
PEEC	Partial Element Equivalent Circuit
FEM	Finite Element Method
3-D	three-dimensional

References

1. Liu, T.; Wong, T.T.; Shen, Z.J. A new characterization technique for extracting parasitic inductances of SiC power MOSFETs in discrete and module packages based on two-port S-parameters measurement. *IEEE Tran. Power Electron.* **2018**, *33*, 9819–9833. [CrossRef]
2. Lemmon, A.; Graves, R. Parasitic extraction procedure for silicon carbide power modules. In Proceedings of the 2015 IEEE International Workshop on Integrated Power Packaging (IWIPP), Chicago, IL, USA, 3–6 May 2015; pp. 91–94.
3. Shahabi, A.; Lemmon, A.N. Multibranch Inductance Extraction Procedure for Multichip Power Modules. *IEEE J. Emerg. Sel. Top. Power Electron.* **2020**, *8*, 272–285. [CrossRef]
4. Mukunoki, Y.; Nakamura, Y.; Konno, K.; Horiguchi, T.; Nakayama, Y.; Nishizawa, A.; Kuzumoto, M.; Akagi, H. Modeling of a SiC MOSFET with focus on internal stray capacitances and inductances, and its verification. *IEEE Tran. Ind. Appl.* **2018**, *54*, 2588–2597. [CrossRef]
5. Pace, L.; Defrance, N.; Videt, A.; Idir, N.; De Jaeger, J.C.; Avramovic, V. Extraction of Packaged GaN Power Transistors Parasitics Using S-Parameters. *IEEE Tran. Electron. Devices* **2019**, *66*, 2583–2588. [CrossRef]
6. Iida, H.; Hasegawa, K.; Omura, I. Mutual inductance influence to switching speed and TDR measurements for separating self-and mutual inductances in the package. In Proceedings of the 31st International Symposium on Power Semiconductor Devices and ICs (ISPSD), Shanghai, China, 19–23 May 2019; pp. 503–506.
7. Kovacevic-Badstuebner, I.; Grossner, U.; Popescu, D. Tools for Broadband Electromagnetic Modeling of Power Semiconductor Packages and External Circuit Layouts. In Proceedings of the 32nd International Symposium on Power Semiconductor Devices and ICs (ISPSD), Vienna, Austria, 13–18 September 2020; pp. 388–391. [CrossRef]
8. Popescu, D.; Treiber, M. Broadband TCAD mixed-mode simulation framework for predictive modeling of fast dynamic switching events. In Proceedings of the 31st Int. Sym. on Power Semiconductor Devices and ICs (ISPSD), Shanghai, China, 19–23 May 2019; pp. 327–330.
9. Le, Q.; Evans, T.; Peng, Y.; Mantooth, H.A. PEEC Method and Hierarchical Approach Towards 3D Multichip Power Module (MCPM) Layout Optimization. In Proceedings of the IEEE International Workshop on Integrated Power Packaging (IWIPP), Toulouse, France, 24–26 April 2019; pp. 131–136. [CrossRef]
10. Che, C.; Zhao, H.; Guo, Y.; Hu, J.; Kim, H. Investigation of Segmentation Method for Enhancing High Frequency Simulation Accuracy of Q3D Extractor. In Proceedings of the IEEE International Conference on Computational Electromagnetics (ICCEM), Shanghai, China, 20–22 March 2019; pp. 1–3. [CrossRef]
11. Jorgensen, A.B.; Munk-Nielsen, S.; Uhrenfeldt, C. Overview of Digital Design and Finite-Element Analysis in Modern Power Electronic Packaging. *IEEE Trans. Power Electron.* **2020**, *35*, 10892–10905. [CrossRef]
12. Zeng, Y.; Yi, Y.; Liu, P. An Improved Investigation into the Effects of the Temperature-Dependent Parasitic Elements on the Losses of SiC MOSFETs. *Appl. Sci.* **2020**, *10*, 7192. [CrossRef]

13. Gabriadze, G.; Chiqovani, G.; Gheonjian, A.; Oganezova, I.; Demurov, A.; Kutchadze, Z.; Bunlon, X.; Ajebbar, F.; Jobava, R. Enhanced PEEC Model Based on Automatic Voronoi Decomposition of Triangular Meshes. *IEEE Trans. Electromagn. Compat.* **2020**, *62*, 2196–2208. [CrossRef]
14. Liu, H.; Huang, X.; Lin, F.; Yang, Z. Loss Model and Efficiency Analysis of Tram Auxiliary Converter Based on a SiC Device. *Energies* **2017**, *10*, 2018. [CrossRef]
15. Alimeling, J.H.; Hammer, W.P. PLECS-piece-wise linear electrical circuit simulation for Simulink. In Proceedings of the IEEE 1999 International Conference on Power Electronics and Drive Systems. PEDS'99 (Cat. No.99TH8475), Hong Kong, China, 27–29 July 1999; Volume 1, pp. 355–360. [CrossRef]
16. Muesing, A.; Kolar, J.W. Successful online education—GeckoCIRCUITS as open-source simulation platform. In Proceedings of the 2014 International Power Electronics Conference (IPEC-Hiroshima 2014—ECCE ASIA), Hiroshima, Japan, 18–21 May 2014; pp. 821–828. [CrossRef]
17. Stark, R.; Tsibizov, A.; Nain, N.; Grossner, U.; Kovacevic-Badstuebner, I. Accuracy of Three Inter-Terminal Capacitance Models for SiC Power MOSFETs under Fast Switching. *IEEE Trans. Power Electron.* **2021**. [CrossRef]
18. Triverio, P.; Grivet-Talocia, S.; Nakhla, M.S.; Canavero, F.G.; Achar, R. Stability, Causality, and Passivity in Electrical Interconnect Models. *IEEE Trans. Adv. Packag.* **2007**, *30*, 795–808. [CrossRef]
19. Romano, D.; Antonini, G.; Grossner, U.; Kovacvic-Badstuebner, I. Circuit synthesis techniques of rational models of electromagnetic systems: A tutorial paper. *Int. J. Numer. Model. Electron. Netw. Devices Fields* **2019**, *32*. [CrossRef]
20. Gustavsen, B.; Semlyen, A. Rational approximation of frequency domain responses by vector fitting. *IEEE Trans. Power Deliv.* **1999**, *14*, 1052–1061. [CrossRef]
21. Ekman, J. Electromagnetic Modeling Using the Partial Element Equivalent Circuit Method. Ph.D. Thesis, Lulea University of Technology, Lulea, Sweden, 2003.
22. Schroeder, A.; Wunsch, B.; Kicin, S. Wideband Macro-Modelling of Power Modules for Transient Electromagnetic Analysis. In Proceedings of the 11th International Conference on Integrated Power Electronics Systems (CIPS), Berlin, Germany, 24–26 March 2020; pp. 1–5.
23. Ruehli, A.E.; Antonini, G.; Jiang, L. *Circuit Oriented Electromagnetic Modeling Using the PEEC Techniques*; John Wiley & Sons, Inc.: Hoboken, NJ, USA, 2017.
24. Ho, A. Ruehli, P. Brennan, C. The modified nodal approach to network analysis. *IEEE Trans. Circuits Syst.* **1975**, *22*, 504–509.
25. Hartman, A.; Romano, D.; Antonini, G.; Ekman, J. Partial Element Equivalent Circuit Models of Three-Dimensional Geometries Including Anisotropic Dielectrics. *IEEE Trans. Electromagn. Compat.* **2018**, *60*, 696–704. [CrossRef]
26. Kovacevic-Badstuebner, I.; Romano, D.; Antonini, G.; Lombardi, L.; Grossner, U. Full-Wave Computation of the Electric Field in the Partial Element Equivalent Circuit Method Using Taylor Series Expansion of the Retarded Green's Function. *IEEE Trans. Microw. Theory Tech.* **2020**, *68*, 3242–3254. [CrossRef]
27. Nguyen, T.S.; Duc, T.L.; Tran, T.S.; Guichon, J.M.; Chadebec, O.; Meunier, G. Adaptive Multipoint Model Order Reduction Scheme for Large-Scale Inductive PEEC Circuits. *IEEE Trans. Electromagn. Compat.* **2017**, *59*, 1143–1151. [CrossRef]
28. Romano, D.; Antonini, G. Partitioned Model Order Reduction of Partial Element Equivalent Circuit Models. *IEEE Trans. Components, Packag. Manuf. Technol.* **2014**, *4*, 1503–1514. [CrossRef]
29. Kamon, M.; Tsuk, M.; White, J. FASTHENRY: A multipole-accelerated 3-D inductance extraction program. *IEEE Trans. Microw. Theory Tech.* **1994**, *42*, 1750–1758. [CrossRef]
30. Vialardi, E.; Clavel, E.; Chadebec, O.; Guichon, J.M.; Lionet, M. Electromagnetic simulation of power modules via adapted modelling tools. In Proceedings of the 14th International Power Electronics and Motion Control Conference (EPE-PEMC), Ohrid, Macedonia, 6–8 September 2010; pp. T2-78–T2-83. [CrossRef]
31. Nabors, K.; White, J. FastCap: A multipole accelerated 3-D capacitance extraction program. *IEEE Trans. Comput. Aided Des. Integr. Circuits Syst.* **1991**, *10*, 1447–1459. [CrossRef]
32. Cottet, D.; Paakkinen, M. Scalable PEEC-SPICE modelling for EMI analysis of power electronic packages and subsystems. In Proceedings of the 8th Electronics Packaging Technology Conference, Singapore, 6–8 December 2006; pp. 871–878. [CrossRef]
33. Ardon, V.; Aime, J.; Chadebec, O.; Clavel, E.; Guichon, J.M.; Vialardi, E. EMC Modeling of an Industrial Variable Speed Drive With an Adapted PEEC Method. *IEEE Trans. Magn.* **2010**, *46*, 2892–2898. [CrossRef]
34. Rondon-Pinilla, E.; Morel, F.; Vollaire, C.; Schanen, J.L. Modeling of a Buck Converter With a SiC JFET to Predict EMC Conducted Emissions. *IEEE Trans. Power Electron.* **2014**, *29*, 2246–2260. [CrossRef]
35. Musing, A.; Heldwein, M.L.; Friedli, T.; Kolar, J.W. Steps Towards Prediction of Conducted Emission Levels of an RB-IGBT Indirect Matrix Converter. In Proceedings of the Power Conversion Conference (PCC), Nagoya, Japan, 2–5 April 2007; pp. 1181–1188. [CrossRef]
36. Evans, P.L.; Castellazzi, A.; Johnson, C.M. Design Tools for Rapid Multidomain Virtual Prototyping of Power Electronic Systems. *IEEE Trans. Power Electron.* **2016**, *31*, 2443–2455. [CrossRef]
37. Kovacevic-Badstuebner, I.; Grossner, U.; Romano, D.; Antonini, G.; Ekman, J. A more accurate electromagnetic modeling of WBG power modules. In Proceedings of the IEEE 30th International Symposium on Power Semiconductor Devices and ICs (ISPSD), Chicago, IL, USA, 13–17 May 2018; pp. 260–263. [CrossRef]
38. Kovacevic-Badstuebner, I.; Romano, D.; Lombardi, L.; Grossner, U.; Ekman, J.; Antonini, G. Accurate Calculation of Partial Inductances for the Orthogonal PEEC Formulation. *IEEE Trans. Electromagn. Compat.* **2021**, *63*, 82–92. [CrossRef]

39. Ruehli, A.E.; Antonini, G.; Esch, J.; Ekman, J.; Mayo, A.; Orlandi, A. Nonorthogonal PEEC formulation for time- and frequency-domain EM and circuit modeling. *IEEE Trans. Electromagn. Compat.* **2003**, *45*, 167–176. [CrossRef]
40. Kovacevic-Badstuebner, I.; Stark, R.; Grossner, U.; Guacci, M.; Kolar, J.W. Parasitic Extraction Procedures for SiC Power Modules. In Proceedings of the 10th International Conference on Integrated Power Electronics Systems (CIPS), Stuttgart, Germany, 20–22 March 2018; pp. 1–6.
41. Sun, W. Accurate EM simulation of SMT components in RF designs. In Proceedings of the IEEE Radio Frequency Integrated Circuits Symposium (RFIC), Honolulu, HI, USA, 4–6 June 2017; pp. 140–143. [CrossRef]
42. Infineon. 5th Generation CoolSiC 1200V Schottky Diode (IDWD40G120C5). Available online: https://www.infineon.com/dgdl/Infineon-IDWD40G120C5-DataSheet-v02_00-EN.pdf?fileId=5546d462689a790c016933d56ffd548f (accessed on 14 May 2021).
43. Infineon. CoolMOS C7 600 V PowerTransistor (IPP60R180C7). Available online: https://www.infineon.com/dgdl/Infineon-IPP60R180C7-DS-v02_00-EN.pdf?fileId=5546d4624cb7f111014d483fe4ba707b (accessed on 14 May 2021).
44. Infineon. CoolSiC 1200V SiC Trench MOSFET Silicon Carbide MOSFET (IMZ120R030M1H). Available online: https://www.infineon.com/dgdl/Infineon-IMZ120R030M1H-DataSheet-v02_02-EN.pdf?fileId=5546d46269e1c019016a92fdcc776696 (accessed on 14 May 2021).

energies

MDPI

Article

Application of High-Frequency Leakage Current Model for Characterizing Failure Modes in Digital Logic Gates

Zahra Abedi [1], Sameer Hemmady [1], Thomas Antonsen [2], Edl Schamiloglu [1] and Payman Zarkesh-Ha [1,*]

[1] Department of Electrical & Computer Engineering, University of New Mexico, Albuquerque, NM 87131, USA; zabedi@unm.edu (Z.A.); shemmady@unm.edu (S.H.); edls@unm.edu (E.S.)

[2] Departments of Electrical and Computer Engineering, Physics, University of Maryland, College Park, MD 20742, USA; antonsen@umd.edu

* Correspondence: pzarkesh@unm.edu

Abstract: In this paper, a predictive model is developed to characterize the impact of high-frequency electromagnetic interference (EMI) on the leakage current of CMOS integrated circuits. It is shown that the frequency dependence can be easily described by a transfer function that depends only on a few dominant parasitic elements. The developed analytical model is successfully compared against measurement data from devices fabricated using 180 nm, 130 nm, and 65 nm standard CMOS processes through TSMC. Based on the predictive model, the impact of EMI on leakage current in a CMOS inverter is reduced by increasing the frequency from 10 MHz to 4 GHz.

Keywords: electromagnetic interference; leakage current; predictive modeling of CMOS inverters; VLSI systems

check for updates

Citation: Abedi, Z.; Hemmady, S.; Antonsen, T.; Schamiloglu, E.; Zarkesh-Ha, P. Application of High-Frequency Leakage Current Model for Characterizing Failure Modes in Digital Logic Gates. *Energies* **2021**, *14*, 2906. https://doi.org/10.3390/en14102906

Academic Editors: Ivica Stevanovic and Bernhard Wunsch

Received: 26 February 2021
Accepted: 12 May 2021
Published: 18 May 2021

1. Introduction

Power dissipation in electronics has always been one of the major concerns of the massive growth in portable computing and wireless communication technologies. The exponential relationship between power consumption and the square of the power supply voltage has led to the scaling down of CMOS transistors to reduce dynamic power draw and offset reliability issues [1,2]. Transistor threshold voltages are also reduced to maintain a high drive current and achieve performance improvements [3]. However, these measures result in an exponential increase in leakage current when the transistor's size is in the nanometer range. Consequently, leakage current becomes a significant portion of CMOS circuit's total power consumption [4]. Although enormous efforts have been made to mitigate leakage current, such as MTCMOS Power Gating, Super Cutoff CMOS Circuitry, Forced Transistor Stacking, and Sleepy Stack techniques [5], the design of such leakage-free nanoscale CMOS circuits turns out to be very challenging.

Moreover, as the device size enters the nanometer scale and the bias voltage is reduced, the integrated circuits and systems become more vulnerable to EMI [6,7]. A low level of electromagnetic coupling from an external EMI source can still disrupt the operation of a circuit by causing soft, reversible upset, or hard, permanent damage, even when numerous techniques for reducing the effect of EMI are implemented, such as shielding and filtering of the digital modules and power supplies [8,9]. Although the modeling of the EMI susceptibility of digital circuits related to delay time and power supply current [10], or upset of CMOS inverters in static operation [11], is recently attracting attention, there is no model to predict the impact of EMI on the leakage current.

Leakage or static power dissipation occurs due to the presence of various leakage currents, such as subthreshold leakage, reverse-bias source/drain junction leakages, gate oxide tunneling leakage, and Gate-Induced Drain Leakage (GIDL). Because subthreshold leakage current is the dominant leakage in sub-100 nm circuits [12], we focus on evaluating the said current's changes when an RF signal is injected onto the gates of CMOS inverters, which are the fundamental units in electronic ICs and systems.

Subthreshold leakage is the drain-source current of a transistor during operation in the weak inversion regime. Unlike the strong inversion regime, in which drift current dominates, subthreshold conduction is due to the diffusion current of the minority carriers in the channel for a metal oxide semiconductor (MOS) device. The magnitude of the subthreshold current is a function of temperature, supply voltage, device size, and the process parameters [13,14]. A CMOS device comprises both PMOS and NMOS transistors; therefore, its leakage current will be determined based on which MOSFET is in the OFF state.

In our previous work, we reported on a predictive model to estimate the impact of EMI on the leakage current of a digital integrated circuit at the low frequency of 10 MHz [15]. Based on our developed predictive model, we found that the rate of increase in leakage current, as a function of injected RF power, is independent of transistor scaling, technology node, or power supply voltage.

This paper's main contribution is to extend our predictive, analytical model to higher frequencies, focusing on elemental CMOS inverters. Our predictive model for the device's leakage current under such high frequency injected interference is validated with experimental results.

The remainder of this paper is organized as follows: In Section 2, the high-frequency model of the leakage current under EMI is established and the main factors that affect the leakage current are discussed. Section 3 presents the experimental setup and the test chip prototype. The measured results and comparisons are shown in Section 4. Finally, we conclude by summarizing the key results uncovered through this research.

2. Predictive Model for Leakage Current under EMI

In this section, we develop an analytic model to predict the inverter's leakage susceptibility as a function of the frequency and power of the injected EMI. Our initial experiments and research have focused on large-signal injection at a low frequency of 10 MHz, where the impact of parasitic device capacitances was negligible [16,17]. Figure 1a depicts the leakage current of a CMOS inverter with and without RF injection at its input terminal. In an inverter, the leakage current will be determined with the transistor in the OFF state. For example, if the applied voltage to the gate is zero and the NMOS is OFF, the leakage current is due to the NMOS; however, if the applied voltage to the gate is set to V_{DD}, then the PMOS is OFF, and the leakage current is due to the PMOS. In both cases, leakage current appears due to diffusion between the source and the drain of MOSFET and can be calculated using the following expression [18]:

$$I_{Subthreshold} = I_{off} e^{\frac{V_{GS}}{nV_T}} \left(1 - e^{-\frac{V_{DS}}{V_T}}\right), \quad (1)$$

where I_{off} is the OFF state current, V_{GS} is the gate-source voltage, V_{DS} is the drain-source voltage, $V_T = KT/q$ is the thermal potential (≈25 mV at room temperature), and n is the subthreshold swing coefficient.

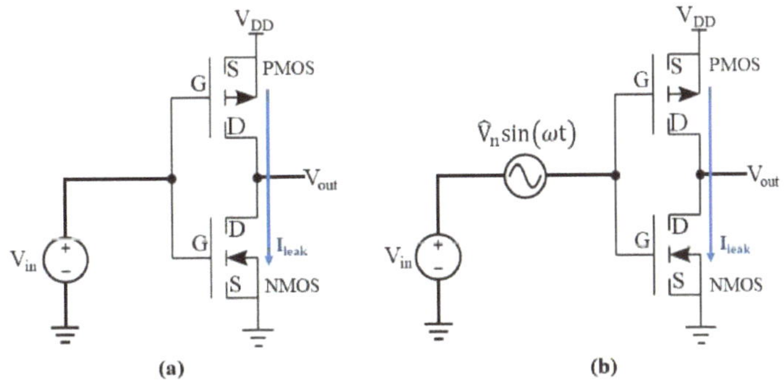

Figure 1. Leakage current of a CMOS inverter in standby mode (**a**) with (**b**) without RF injection.

Since V_{GS} voltage is zero, V_{DS} can be estimated by the power supply voltage, V_{DD}. Therefore, by substituting $V_{GS} = \hat{V}_n \sin(\omega t)$ and $V_{DS} \approx V_{DD}$ in Equation (1), the leakage current under EMI over one cycle of the RF injection ($T = \frac{2\pi}{\omega}$), and knowing $V_{DD} \gg V_T$ will yield (see [15])

$$\overline{I_{Leak}} = I_{off} \cdot I_0\left(\frac{\hat{V}_n}{nV_T}\right), \tag{2}$$

where I_0 is the modified Bessel function of order zero.

Our previous study revealed that leakage current is very sensitive to EMI at low frequency, where only a few hundred millivolts of noise can increase the leakage current by a factor of 1000. However, here we must consider the effect of parasitic loading and capacitance of the test circuit by extraneous instrumentation devices because of high RF frequency injection. Therefore, to further investigate the correlation between RF injection and leakage current, a deeper study of the "transfer function" is required. Here, we propose modeling the frequency response of the device under test (DUT) using assumptions and approximations that will be substantiated by comparing to experimental data.

2.1. Assumptions

One of the basic assumptions of experiments with RF injection is that the characteristic impedance of the system's transmission line is 50 Ω. Although the injected EMI can be applied to any inverter terminal, we assume the excitation is fed to the input, which is typically the most sensitive terminal of the device and, in many practical applications, the most likely input of the RF signal [19].

2.2. Predictive Model

The basic building block in CMOS digital logic devices is the inverter. Although the inverter in different series and parallel configurations is used in logic topologies, a single inverter stage is analyzed here for ease of analysis. Moreover, to achieve precise control of the coupled voltage characteristic at the device input, we apply an RF signal directly onto inverter pins using microwave probes. Because the parasitic impedances have a more significant influence on device performance, the model developed here explores high-frequency injection, where the parasitic impedances arise in the inverter, transmission line, and source.

2.3. Transfer Function

The cascaded combination of frequency-varying transfer functions of the system, coupling to cables, and component susceptibility often means that the total transfer function from the incident field to the malfunction of an internal component shows a significant variation as a function of frequency. As shown in Figure 2, the transfer function is responsible for defining the peak injection received by the input of the inverter. The transfer function is given by:

$$H(s) = \frac{\hat{V}_{out}}{\hat{V}_{in}}, \quad (3)$$

where $\hat{V}_{out}$ is the phasor showing the level of RF signal coupling into input of device and $\hat{V}_{in}$ is the voltage of the RF source, which can take several forms depending on the physical property of the cables, connectors, probes, or DUT. The effective voltage then is

$$\hat{V}_{eff} = H(s)\hat{V}_n \quad (4)$$

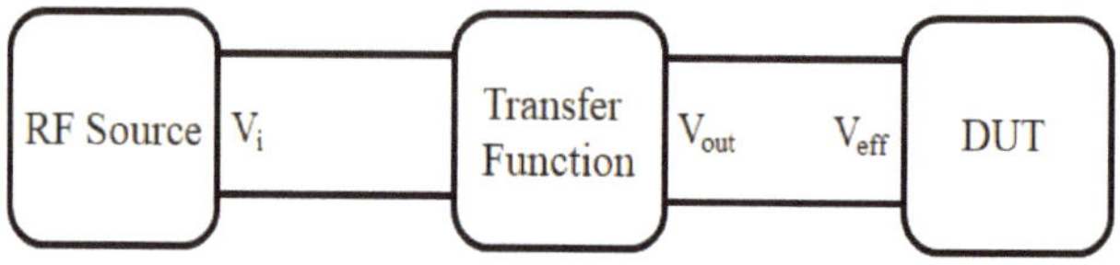

Figure 2. Schematic of the voltage transfer function.

Here, because we use the precision ground-signal-ground (G-S-G) probes for RF signal injection, the dominant elements of the equivalent circuit will be the parasitic resistance of the RF source, the transmission line (Z_0), and substrate (R_{sub}), and the parasitic capacitance of the pad (C_{pad}), as shown in Figure 3.

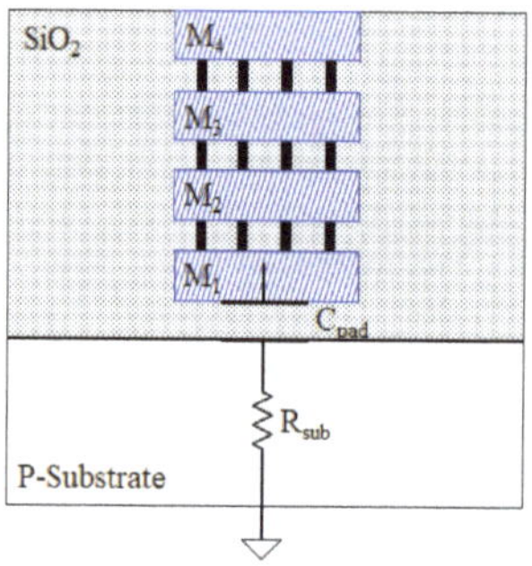

Figure 3. Cross-section of the pad capacitance and substrate resistance.

2.4. Pad Capacitance

Although the device size has reduced substantially in advanced CMOS technology, the contact metal size still needs to be relatively large to ensure bonding reliability. As a result, the pad capacitance cannot be decreased by scaling the CMOS technology. Therefore, in most RF applications, the pad capacitance is considered part of the matching network. The cross-sectional view of the pad capacitance is depicted in Figure 3. The first plate is formed using M_4, M_3, M_2, and M_1, which are Ohmic contacts and connected through a via. It is worth mentioning that different metals (or alloys) are used to decrease the Schottky barrier height and, therefore, the contact resistance. The oxide layer SiO_2 between plates 1 and 2, which is a p-type substrate (P-Substrate), can be used as an insulating dielectric between the parallel plates. Therefore, using a metallic parallel-plate model, the pad capacitance is given by:

$$C_{pad} = \epsilon_0 \epsilon_r \frac{A}{d}, \quad (5)$$

where ϵ_r, and d are the relative permittivity and thickness of the insulator, respectively, ϵ_0 is the permittivity of free space, and A is the area that will be determined from the smaller areas of plates 1 and 2, which in this case is the area of the M_1. The area of M_1 on a fabricated test chip is 150 μm × 360 μm, and the relative permittivity of SiO_2 is 3.9. Therefore, the value of pad capacitance, which is determined by the thickness of the oxide layer, will be in the range of 2 pF to 8 pF.

2.5. Substrate Resistance

The bottom plate of the pad capacitor is the silicon substrate, which introduces some Ohmic resistance to the ground terminal depending on its doping concentration. At high frequency, this substrate resistance, which is typically around 20 to 30 Ω, must be taken into consideration. In this circuit model, the substrate resistance (R_{sub}) is the resistance between the pad capacitor and the reference ground terminal.

2.6. Frequency Response of Equivalent Circuit

According to the equivalent circuit (Figure 4), the peak injection received by the input of the inverter can be calculated as:

$$\hat{V}_{eff} = \frac{X_c + R_{sub}}{Z_0 + X_c + R_{sub}} \hat{V}_n, \tag{6}$$

where X_C is the pad capacitor impedance is $X_c = 1/(j\omega C_{pad})$, and jw = s. Therefore:

$$\hat{V}_{eff} = \frac{1 + j\omega C_{pad} R_{sub}}{1 + j\omega C_{pad}(Z_0 + R_{sub})} \hat{V}_n \tag{7}$$

and:

$$\left|\hat{V}_{eff}\right| = \frac{\sqrt{1 + \left(2\pi f C_{pad} R_{sub}\right)^2}}{\sqrt{1 + \left(2\pi f C_{pad}(R_{sub} + Z_0)\right)^2}} \left|\hat{V}_n\right| \tag{8}$$

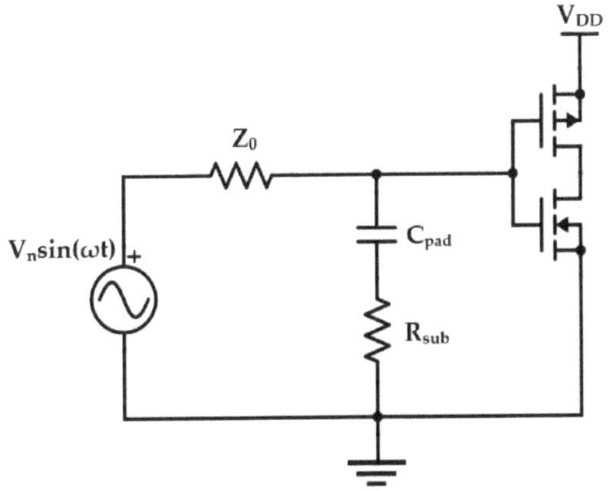

Figure 4. Simplified schematic of the experimental test circuit showing Z_0, C_{pad}, and R_{sub}.

The zeros and poles of the system are defined as the roots of the numerator and the denominator of the transfer function, respectively. According to Equation (8), the zeros and poles of the system will be at frequencies f_0 and f_p, respectively, given by:

$$f_0 = j\frac{1}{2\pi C R_{sub}} \text{ and} \tag{9}$$

$$f_p = j\frac{1}{2\pi C(Z_0 + R_{sub})} \tag{10}$$

The voltage transfer functions of 180 nm, 130 nm, and 65 nm technologies relative to frequency are calculated and plotted in Figure 5. This is analogous to the insertion loss typically determined from the S11, but easier to experimentally obtain given the high impedance of the gate inputs, which is not conducive for S-parameter measurements with a Vector Network Analyzer. Table 1 shows the physical properties and cutoff (−3 dBm) frequencies of the fabricated devices.

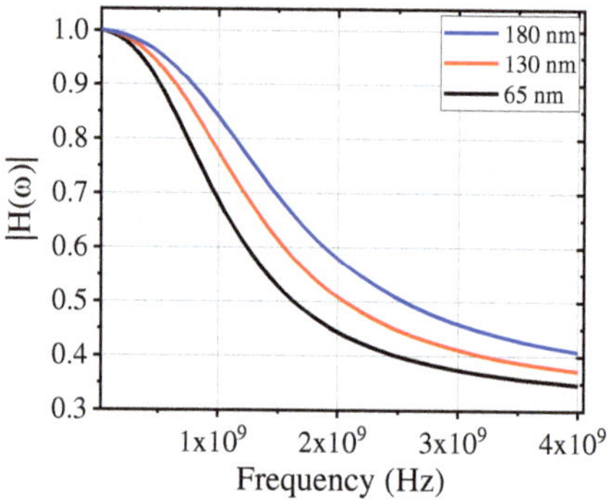

Figure 5. Voltage transfer versus frequency.

Table 1. Cutoff frequencies of 180 nm, 130 nm, and 65 nm technology node.

Technology Node	C_{pad} (pF)	R_{sub} (Ω)	Cutoff Frequency (GHz)
180 nm	3.27	26	1.5
130 nm	4	24	1.2
65 nm	5	26	0.95

This model's advantage is its ability to calculate the frequency response of leakage current using a transfer function for the circuit. The transfer function can be varied according to the measurement setup, fabrication process, and dominant elements at high frequency. The leakage current is given by:

$$\overline{I_{Leak}} = I_{off} \cdot I_0(\frac{\hat{V}_{eff}}{nV_T}). \tag{11}$$

Note that the transfer function can be developed for any other applications depending on the test setup. Typically, 50 Ohms matching termination cannot be implemented in digital applications, where the gate input is high impedance. Therefore, in practical digital circuits, the transfer function will be dependent on the interconnect capacitance/resistance as well as the substrate resistance. The presented methodology is a guide for designers to generate their own transfer function depending on their test setup.

3. Experimental Setup and Methodology

To understand the impact of EMI on leakage current and compare the experimental data with the developed predictive model, several test chips were designed and fabricated using various TSMC's standard CMOS technology nodes, including 180 nm, 130 nm, and 65 nm. To investigate the effects of EMI on leakage current, measurements both with and without microwave injection are performed on a test chip at the input of 1X INV using G_S_G probes having a 350-μm pitch on a coplanar probe station. A photograph of the test chip under the probes is shown in Figure 6. The three RF probes provide V_{in}, V_{DD}, and GND connections for the 1× inverter on the test chip, respectively. The RF probes have three pins with ground-signal-ground configuration, as shown in Figure 6.

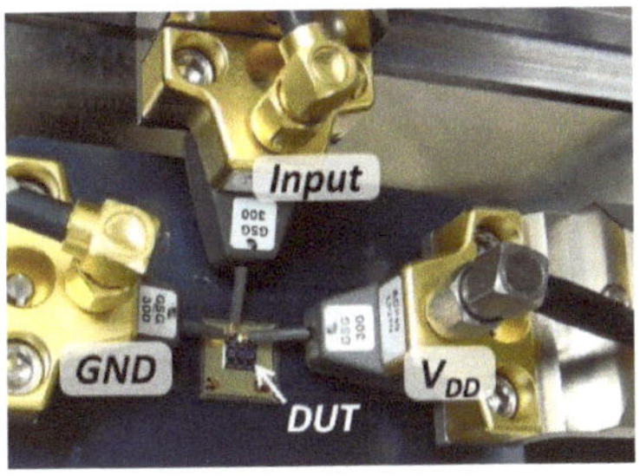

Figure 6. Photograph of the test chip and the RF probes setup used in the measurements [15].

A RF signal generator (Keysight, Santa Rosa, CA, USA) is used to excite the input of the 1× inverter. To measure the leakage current due to NMOS, the inverter input is not biased (no dc gate voltage); however, for leakage current of PMOS, DC voltage and RF signal are injected onto the input using a bias tee. The carrier frequency ranges from 10 MHz to 4 GHz, and the input signal amplitude is set to less than the threshold voltage to satisfy leakage current condition. The leakage current is measured using a Keithley semiconductor analyzer with fixed V_{DD}, and a controlled RF signal injected onto the device's input.

The experimental setup is shown in Figure 7. The entire system in our experimental setup is controlled via LabView (National Instruments, Austin, TX, USA), where the voltages and RF injection power are parametrically swept, and the device leakage current is measured and stored. To avoid exciting photocurrents, all measurements are performed in a darkened room. Figures 8 and 9 show the dependence of leakage current on the RF power and frequency based on measurement data of 180 nm, 130 nm, and 65 nm technology. As expected, the leakage current of the inverter increases as the amplitude of RF signal increases; however, this change is smaller at high frequencies.

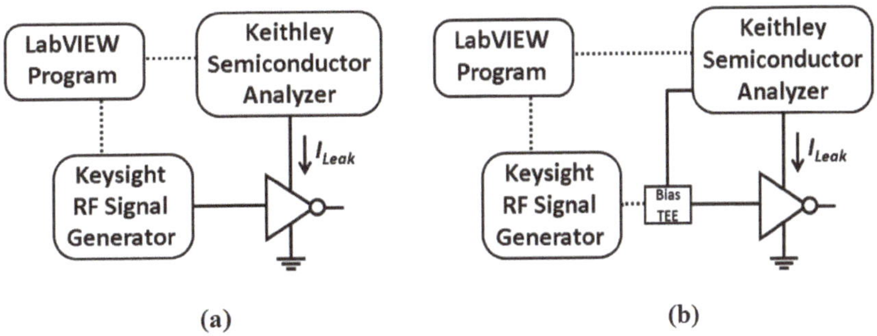

Figure 7. The block diagram of the experimental setup with the LabVIEW control program, measuring leakage current of (**a**) NMOS; (**b**) PMOS.

Since power is an important priority in the design of submicron technology in an integrated circuit, the increase of leakage current of the inverter due to EMI could result in serious disruption in the total power budget distribution of the entire system [20]. Therefore, the entire integrated circuit may experience logic failure. In addition, at higher power injection, leakage current is more sensitive to the change in the peak power of the interference. At fixed RF power, leakage current decreases by increasing RF frequency from

10 MHz to 4 GHz. RF power effects are strongly suppressed at frequencies above 1.2 GHz. This indicates that the inverter is more susceptible to EMI at the low-frequency range.

After the RF injection event, the devices are measured again in order to determine whether there were permanent changes in their operating characteristics. No permanent damage was observed. Therefore, these effects should be considered as "soft" error effects where the device can return to its normal operation without permanent damage; however, they are able to decrease the battery life of the device.

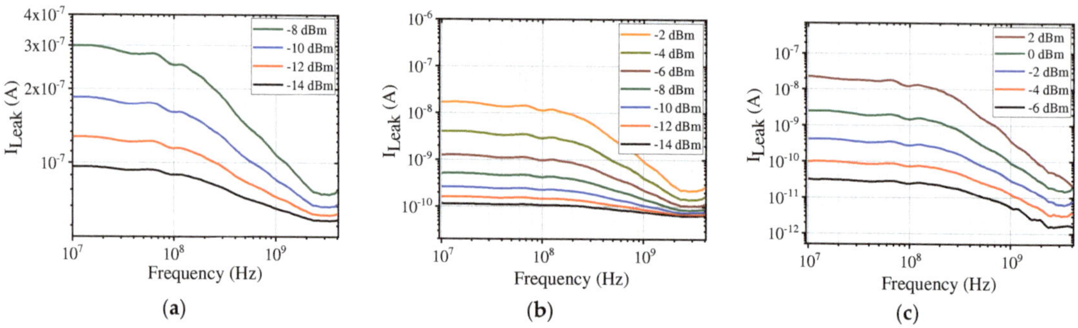

Figure 8. The measured leakage current due to a NMOS for a 1× inverter fabricated in the (**a**) 65nm, (**b**) 130 nm, and (**c**) 180 nm standard CMOS technology node.

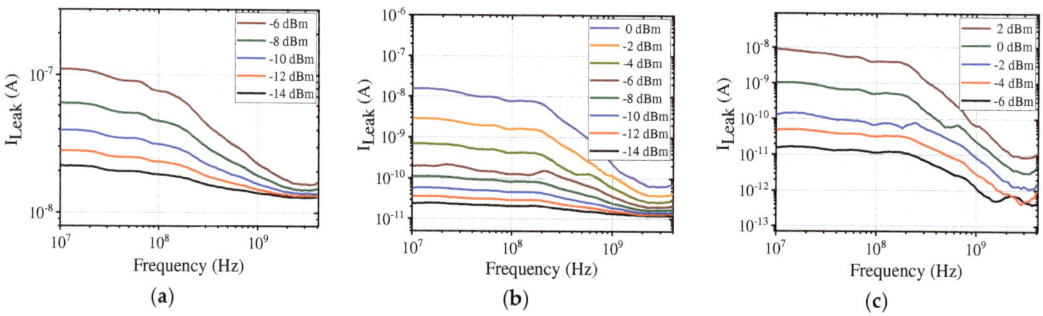

Figure 9. The measured leakage current due to a PMOS for a 1× inverter fabricated in the (**a**) 65 nm, (**b**) 130 nm, and (**c**) 180 nm standard CMOS technology node.

4. Comparison of Predictive Model with Measured Data

To validate the accuracy of the proposed predictive model, the analytical model is compared against measured data from the fabricated test chips using TSMC's 180 nm, 130 nm, and 65 nm standard CMOS processes. Figures 10 and 11 depict the calculated effective voltage $\hat{V}_{eff}$ that gets injected into the input of the inverter using Equation (8). Figures 12 and 13 illustrate the predictive model in solid line and the normalized leakage currents (I_{Leak}/I_{off}) of PMOS and NMOS.

The leakage current of the devices under investigation is measured based on the electrical system shown in Figure 7. Then the measured leakage current at different powers is fitted to the model with fit parameters of I_{off}, n, C_{pad}, and R_{sub}. I_{off} and n are responsible only for defining the initial point. C_{pad} and R_{sub} are the parameters of the transfer function that are dominant in determining the trend of model at high frequency. C_{pad} is the dominant fitting parameter for the middle range of frequency, while R_{sub} adds a zero to the transfer function and plays the main role in frequencies higher than 1 GHz. The pertinent device parameters are extracted by comparing the measurements to the model and given in Table 2.

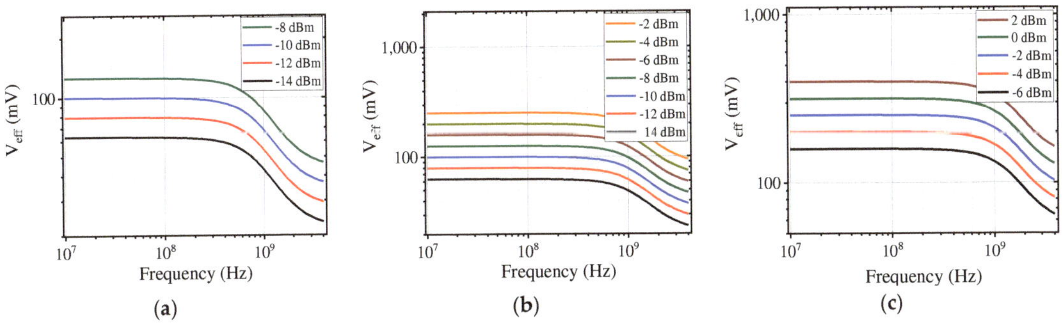

Figure 10. Calculated Veff for NMOS in the inverter (**a**) 65 nm, (**b**) 130 nm, and (**c**) 180 nm technology node.

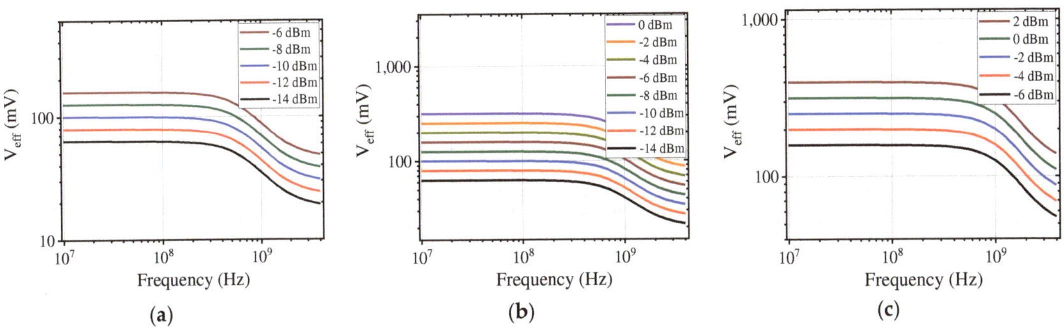

Figure 11. Calculated Veff for PMOS in the inverter (**a**) 65 nm, (**b**) 130 nm, and (**c**) 180 nm technology node.

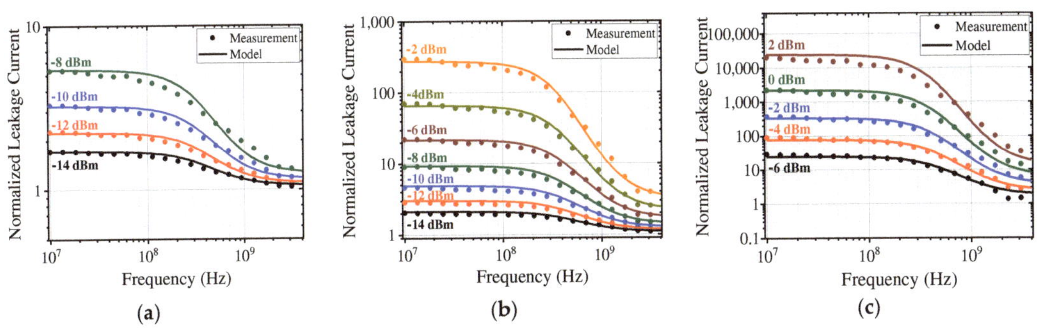

Figure 12. Comparison between the analytical model and the measured data for leakage current of NMOS for (**a**) 65 nm, (**b**) 130 nm, and (**c**) 180 nm technology node.

In general, the model, to first-order, predicts the values and behavior of the measured leakage current of devices. At low frequencies around 10 MHz, the model predicts a small slope in the change of leakage current, which measurements also confirm that leakage current is a weak function of applied RF frequency in the low frequency regime. In the high frequency range, the model predicts the change in the slope of leakage current, while there is a mismatch in the absolute values of I_{leak} at some frequencies and powers.

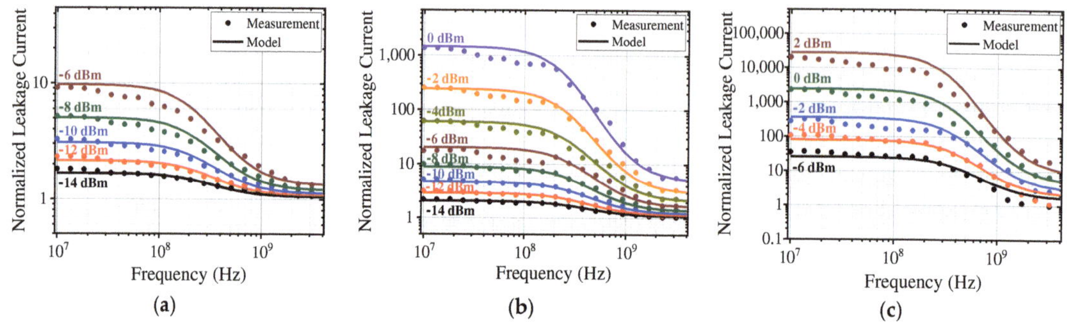

Figure 13. Comparison between the analytical model and the measured data for leakage current of PMOS for (**a**) 65 nm, (**b**) 130 nm, and (**c**) 180 nm technology node.

Table 2. Transistor sizes and parameters.

	Device Type	Width (nm)	Length (nm)	V_{th} (V)	V_{DD} (V)	I_{off} (A)	n	C_{pad} (pF)	R (Ω)
180 nm	PMOS	900	180	0.59	1.8	4.5×10^{-13}	1.25	4	20
	NMOS	450	180	0.44	1.8	1.1×10^{-12}	1.26	3.27	26
130 nm	PMOS	650	130	0.39	1.2	1.12×10^{-11}	1.32	5.4	23
	NMOS	330	130	0.32	1.2	5.7×10^{-11}	1.3	4	24
65 nm	PMOS	325	65	0.13	1	1.2×10^{-8}	1.61	7	20
	NMOS	173	65	0.24	1	5.6×10^{-8}	1.57	5	26

The purpose of our research is to generate simple rule-of-thumb guidelines for MOS design engineers who have to design their products to operate in extreme EM environments. While we acknowledge the "quantitative" mismatch between the model and data, for reasons cited above, these parasitic aspects can be characterized and can lead to system-specific EM susceptibility curves. However, the general trend in the data is still captured by our simple and elegant model thereby paving the way for a quick assessment of damage thresholds in MOS circuits, where there are billions of transistors in a single chip.

5. Conclusions

To the best of our knowledge there has not been any previous predictive model for leakage current under EMI. Our previously published model captures the leakage model only at low frequencies [15]. The new predictive model presented in this paper extends our previous model by taking frequency into consideration. The model uses only a few primitive device and system parameters to successfully predict the behavior of EMI injection on the leakage current in a CMOS inverter over a wide range of frequencies up to 4 GHz. Although the model is developed for a CMOS inverter, it can also be applied to any other complex logic gates using the leakage current of the equivalent transistor in the logic gate. The developed analytical model is successfully compared against measurement data for devices fabricated using 180 nm, 130 nm, and 65 nm standard CMOS process through TSMC. This simple model can be utilized by EMI/EMC engineers to quickly ascertain the soft and hard error decision limits for their circuits under external RF interference.

Author Contributions: Conceptualization, T.A., P.Z.-H.; developing the model, P.Z.-H., Z.A.; collecting and analyzing measured data, Z.A.; writing—original draft preparation Z.A.; writing—review and editing, S.H., E.S., P.Z.-H.; supervision, S.H., E.S., P.Z.-H. All authors have read and agreed to the published version of the manuscript.

Funding: This research is funded by United States Air Force Research Laboratory (AFRL) and Air Force Office of Scientific Research (AFSOR) with AFOSR/AFRL COE Grant #FA9550-15-1-0171 and by DURIP Grant #FA9550-15-1-0379.

Data Availability Statement: Not applicable.

Conflicts of Interest: The authors declare no conflict of interest.

References

1. Lakshmi Narayanan, S.; Korah, R.; Swarnalatha, A. Novel 10-T Write Driver SRAM Design Using 45 nm CMOS Technology with Leakage Current Reduction Scheme for FPGA Routing Switch Architecture. *J. Nanoelectron. Optoelectron.* **2017**, *12*, 1–13.
2. Friedman, E.G.; Kursun, V. Node Voltage Dependent Subthreshold Leakage Current Characteristics of Dynamic Circuits. In Proceedings of the International Symposium on Signals, Circuits and Systems, San Jose, CA, USA, 22–24 March 2003.
3. Roy, K.; Mukhopadhyay, S.; Mahmoodi-Meimand, H. Leakage Current Mechanisms and Leakage Reduction Techniques in Deep-Submicrometer CMOS Circuits. *Proc. IEEE* **2003**, *91*, 305–327. [CrossRef]
4. Ho, Y.; Hsu, C. Standby Power Reduction Using Dynamic Standby Control with Voltage Keeper. *IEICE Electron.* **2017**, 1–7. [CrossRef]
5. Priya, M.G.; Baskaran, K.; Krishnaveni, D. Leakage Power Reduction Techniques in Deep Submicron Technologies for VLSI Applications. *Procedia Eng.* **2012**, *30*, 1163–1170. [CrossRef]
6. Kim, K.; Iliadis, A.A. Impact of Microwave Interference on Dynamic Operation and Power Dissipation of CMOS Inverters. *IEEE Trans Electromagn Compat.* **2007**, *49*, 329–338. [CrossRef]
7. Deepaksubramanyan, B.S.; Nunez, A. Analysis of Subthreshold Leakage Reduction in CMOS Digital Circuits. In Proceedings of the 50th Midwest Symposium on Circuits and Systems, Montreal, QC, Canada, 5–8 August 2007.
8. Ferchau, J.U.; Kotyuk, K.A.; Diaz, R.J. Electronic Assembly with Improved Grounding and EMI Shielding. U.S. Patent No. 5,311,408, 10 May 1994.
9. Lee, S.; Jeong, S.; Hong, S.; Sim, B.; Kim, J. Design and Analysis of EMI Shielding Method using Intermediate Coil for Train WPT System. In Proceedings of the Wireless Power Transfer Conference (WPTC), Montreal, QC, Canada, 3–7 June 2018.
10. Hwang, S.; Hong, J.; Han, S.; Huh, C.; Huh, U.; Choi, J. Delay time and power supply current characteristics of CMOS inverter broken by intentional high power microwave. In Proceedings of the Microwave Conference, Bangkok, Thailand, 11–14 December 2007.
11. Kim, K.; Iliadis, A. Critical Upsets of CMOS Inverters in Static Operation Due to High-Power Microwave Interference. *IEEE T Electromagn C* **2007**, *49*, 876–885. [CrossRef]
12. Abraham, D.; George, A. Effect of Fin Shape on GIDL and Subthreshold Leakage Currents. *Int. J. Sci. Eng. Technol.* **2015**, *1*, 135–145.
13. Nandyala, V.R.; Mahapatra, K.K. A Circuit Technique for Leakage Power Reduction in CMOS VLSI Circuits. In Proceedings of the International Conference on VLSI Systems, Bengaluru, India, 10–12 January 2016.
14. Roy, K.; Mukhopadhyay, S.; Mahmoodi-Meimand, H. Leakage Current in Deep-Submicron CMOS Circuits. *J. Circuits Syst. Comp.* **2002**, *11*, 575–600. [CrossRef]
15. Abedi, Z.; Hemmady, S.; Antonsen, T.; Schamiloglu, E.; Zarkesh-Ha, P. Electromagnetic Compatibility in Leakage Current of CMOS Integrated Circuits. In Proceedings of the International Symposium on Electromagnetic Compatibility (EMC EUROPE), Barcelona, Spain, 2–6 September 2019.
16. Sule, N.H.; Abedi, Z.; Hemmady, S.; Schamiloglu, E.; Zarkesh-Ha, P. Predictive Modeling for Estimating the Limits for Nonpersistent Effects in MOSFET Response Under Large Signal Gate Injection. *IEEE T Electromagn C* **2020**, *62*, 1763–1770. [CrossRef]
17. Powell, T.; Sule, N.; Hemmady, S.; Zarkesh-Ha, P. Predictive Model for Extreme Electromagnetic Interference on CMOS Inverters. In Proceedings of the International Symposium on Electromagnetic Compatibility (EMC EUROPE), Amsterdam, The Netherlands, 27–30 August 2018.
18. Sheu, B.J.; Scharfetter, D.L.; Ko, P.; Jeng, M.C. BSIM: Berkeley Short-Channel IGFET Model for MOS Transistors. *IEEE J. Solid State Circuits* **1987**, *22*, 558–566. [CrossRef]
19. Tadano, H.; Hattori, Y. A Study of DC Operating Point Shifts in MOSFETs with Large RF Signals. *Electron. Commun. Jpn. I Commun.* **2001**, *84*, 18–26.
20. Sengupta, D.; Saleh, R. Generalized Power-Delay Metrics in Deep Submicron CMOS Designs. *IEEE Trans. CAD* **2007**, *26*, 183–189. [CrossRef]

energies

MDPI

Article

Black-Box Modelling of Low-Switching-Frequency Power Inverters for EMC Analyses in Renewable Power Systems

Lu Wan [1], Abduselam Hamid Beshir [1], Xinglong Wu [1], Xiaokang Liu [1], Flavia Grassi [1,*], Giordano Spadacini [1], Sergio Amedeo Pignari [1], Michele Zanoni [2], Liliana Tenti [2] and Riccardo Chiumeo [2]

[1] Department of Electronics, Information and Bioengineering (DEIB), Politecnico di Milano, 20133 Milan, Italy; lu.wan@polimi.it (L.W.); abduselamhamid.beshir@polimi.it (A.H.B.); xinglong.wu@polimi.it (X.W.); Xiaokang.liu@polimi.it (X.L.); giordano.spadacini@polimi.it (G.S.); sergio.pignari@polimi.it (S.A.P.)

[2] Ricerca sul Sistema Energetico-RSE S.p.A., 20134 Milan, Italy; michele.zanoni@rse-web.it (M.Z.); liliana.tenti@rse-web.it (L.T.); riccardo.chiumeo@rse-web.it (R.C.)

* Correspondence: flavia.grassi@polimi.it

Abstract: Electromagnetic interference (EMI) from renewable power systems to the grid attracts more attention especially in the low-frequency range, due to the low switching frequency of high-power inverters. It is significantly important to derive EMI models of power inverters as well as to develop strategies to suppress the related conducted emissions. In this work, black-box modelling is applied to a three-phase inverter system, by implementing an alternative procedure to identify the parameters describing the active part of the model. Besides, two limitations of black-box modelling are investigated. The first regards the need for the system to satisfy the linear and time-invariant (LTI) assumption. The influence of this assumption on prediction accuracy is analysed with reference to the zero, positive and negative sequence decomposition. It is showing that predictions for the positive/negative sequence are highly influenced by this assumption, unlike those for the zero sequence. The second limitation is related to the possible variation of the mains impedance which is not satisfactorily stabilized at a low frequency outside the operating frequency range of standard line impedance stabilization networks.

Keywords: black-box modelling; conducted emissions; power inverter

Citation: Wan, L.; Beshir, A.H.; Wu, X.; Liu, X.; Grassi, F.; Spadacini, G.; Pignari, S.A.; Zanoni, M.; Tenti, L.; Chiumeo, R. Black-Box Modelling of Low-Switching-Frequency Power Inverters for EMC Analyses in Renewable Power Systems. *Energies* **2021**, *14*, 3413. https://doi.org/10.3390/en14123413

Academic Editors: Ivica Stevanovic and Bernhard Wunsch

Received: 21 April 2021
Accepted: 7 June 2021
Published: 9 June 2021

1. Introduction

With the widespread of renewable energy technologies, such as photovoltaic (PV) panels and wind farms, the development of a suitable tool to predict and reduce the conducted emission (CE) generated by renewable power systems has gained increasing attention from the Electromagnetic Compatibility (EMC) community due to the severe electromagnetic interference (EMI) threats these noise currents can cause when propagating through the power grid [1]. Indeed, they can be responsible for interference with consequent malfunctioning in smart communications and metering systems [2], thus reducing the reliability and interoperability of the whole infrastructure. Particularly, since the switching frequencies of most high-power converters are still limited to a few kHz only [3,4], these applications have brought to the attention of the community the partial lack of regulations in the frequency interval between 2 kHz (maximum frequency foreseen by International Power Quality Standards) and 150 kHz (minimum frequency foreseen by International EMC Standards) [5–7], which is covered (often only partially) by a few of Standards for specific applications only [8]. In this frequency interval, also the availability of devices suitable to run the required measurement and tests is limited. An explicative example is the Line Impedance Stabilization Network (*LISN*), whose use in EMC test setups is fundamental to assure that the CEs exiting the device under test were measured on a stable impedance, despite possible variations of the mains impedance. Topology and values of *LISN* components depend on the specific Standards and are selected to ensure the correct operation of the *LISN* in the frequency interval foreseen by the Standards. For instance, the *LISN* layout specified in the International Special Committee on Radio Interference (CISPR) 22 Standard is different from the one recommended by the CISPR 16 Standard, as they are expected to cover two different frequency intervals: The former from 150 kHz to 30 MHz, the latter from 9 kHz to 30 MHz. Using the *LISN* outside its recommended frequency range may jeopardize test repeatability leading to unpredictably different CE levels.

In this respect, developing effective strategies to model power converters is relevant not only in view of predicting the generated CE and of deriving guidelines to design proper countermeasures (e.g., design of suitable passive, active, or hybrid EMI filters) but also in view of developing new test setups, procedures, and measurement devices to supplement the regulations currently available in the Standards and to extend their applicability starting from lower frequencies (i.e., 2 kHz or lower).

The challenge is to derive power converter models able to represent not only the functional part of the system (i.e., the switching activity of the power modules along with their control system) but also all those undesired mechanisms and parasitic paths leading to the generation and propagation of the CE exiting the converter. In the literature, either circuit models for time-domain analysis and behaviour models for frequency-domain analysis can be found [9,10]. The former approach is definitely more flexible since it involves an explicit circuit representation of the functional and parasitic behaviour of the converter. However, it usually requires detailed knowledge of the internal structure of the converter along with a huge effort to identify suitable setups and procedures to extract from the measurement data (or from the manufacturer data-sheets) proper values for the involved circuit components. Moreover, this approach may be scarcely effective when time-domain simulation of complex systems is the target, since the complexity of the obtained network may rise to convergence issues [11].

Behavioural modelling strategies can help to overcome most of the aforesaid limitations. This latter approach foresees to provide the converter with a black-box representation at the output ports in terms of an equivalent Thevenin/Norton circuit. The obtained "black-box" model involves a minimum number of active and passive components, whose frequency response is extracted from measurements carried out at the output ports of the converter. Although such a black-box approach does not offer any flexibility in the modelling of the converter control system, it is definitely beneficial for EMC analyses in complex systems. Indeed, it does not require any knowledge about the internal structure of the converter, since measurements at the external ports only are required to derive the frequency response of the involved model parameters [10].

The theoretical assumption behind this approach is that the device under analysis can be treated (at least approximately) as a linear and time-invariant (LTI) system. However, since switching modules exhibit an inherently time-variant and non-linear behaviour, a thorough investigation aimed at identifying the conditions for applicability and possible limitations of black-box modelling is required.

For this purpose, the black-box modelling technique in [12], which was originally adopted for modelling DC-DC converters in a satellite power system, is extended in this work. The model aims to predict the CE peaks related to the switching frequency and its harmonics exiting the AC side of a three-phase inverter connected to a PV panel. For the proposed modelling procedure, accuracy, applicability, and possible limitations of black-box modelling are systematically investigated, by paying particular attention to the low-frequency part of the spectrum down to 2 kHz (since the switching frequency of the inverter here considered is 5 kHz). To this end, an explicit model of the whole system (i.e., a model involving the switching modules, the control system as well as parasitic components) is preliminarily implemented in SPICE (it stands for Simulation Program with Integrated Circuit Emphasis) and used as a virtual environment to emulate the steps of the proposed experimental procedure and to obtain a prediction of the generated CE to be used as a reference to validate the proposed modelling strategy.

The accuracy of the obtained CE prediction is analysed not only in terms of physical voltages measured at the *LISN* output but also in terms of modal quantities. To this end, the decomposition of physical quantities into positive, negative and zero-sequence components is adopted, which is the one usually exploited in power systems analyses. Particular attention is devoted to investigating the influence on model effectiveness of the presence of a so-called mask impedance (here, the output filter installed at the ac-side of the inverter) as well as the bandwidth of the *LISN* exploited in the setup for model parameter extraction (active part).

The rest of the manuscript is organized as follows. Section 2 surveys the black-box modelling strategies available in the literature and discusses the LTI assumption. Section 3 describes the inverter system under analysis and introduces the test setups and procedures proposed for model-parameter extraction from measurement data. The accuracy of the proposed modelling technique is investigated in Section 4, by comparing the prediction obtained by the black-box model with those obtained by SPICE simulation of the whole system. Also, in this section possible limitations of the proposed approach are investigated and discussed. Conclusions are eventually drawn in Section 5.

2. Black-Box Modelling Procedures and LTI Assumption

2.1. Black-Box Modelling Procedures

In the literature, there are mainly two methods to extract black-box models of power converters. In one approach, a set of independent tests are carried out by connecting suitable external networks of known impedance to write and solve suitable equations involving the unknown parameters of the black-box model. The number of terminals of the system under analysis determines the topology of the obtained model (i.e., number of required impedances and sources) as well as the number of configurations [13]. For example, the black-box model of a DC/DC converter with three terminals (phase, neutral and ground) can be represented by five unknowns (i.e., two sources and three impedances). This means that at least three different measurement configurations are needed to write five linearly independent equations for the voltages/currents measured in each of the three setups [9]. The use of three setups leads to an overdetermined system of six equations in five unknowns, which can be solved by adopting, e.g., least-square algorithms [14].

An alternative approach is to separately characterize the active part (noise sources) and the passive part (impedance/admittance matrices) of the model. For the characterization of the passive part, most of the methods available in the literature foresee impedance measurement with the converter switched off through Impedance or Vector Network Analyzer (VNA) [15]. For instance, the method in [12] involves the use of a VNA to measure the scattering parameters then converted into admittances with the converter off, (see Figure 1a). Besides, the current sources are obtained starting from the measurement of output currents of the converter by connecting the converter to the power network through a *LISN* required by the standard measurement setups of conducted emissions. Moreover, alternative indirect measurement such as the so-called 'insertion loss' method [16] and analytical calculation method [17] are also investigated in some papers.

Approaches based on separate evaluation of the active and passive part (Figure 1a) seem to be more promising because (1) they require only two measurement setups, (2) the extraction of the unknown parameters from measurement data does not involve the solution of a system of overdetermined equations. This is a great advantage in terms of accuracy of the results as post-processing of the measurement data can often introduce significant numerical errors which degrade the accuracy of the extracted model.

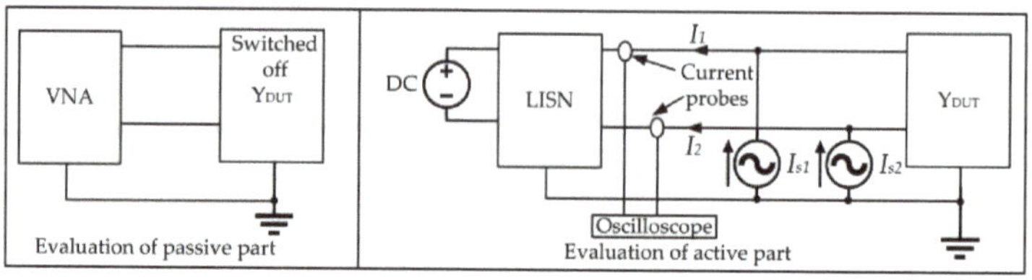

(a)

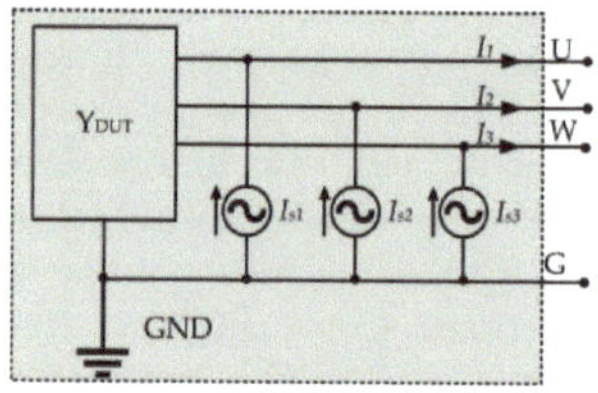

(b)

Figure 1. (**a**) Principle drawing of the black-box modelling technique in [12]: Identification of passive (left panel) and active (right panel) parameters separately. (**b**) Black-box model of the three-phase inverter system.

For these reasons, black-box modelling procedures evaluating the parameters of the active and passive parts of the model separately (Figure 1a) are adopted in this work and extended to a three-phase grid-connected inverter system. The extracted black-box model is shown in Figure 1b (Norton equivalent circuit) and it involves an admittance matrix and three current sources. The passive part of the model is evaluated by VNA measurement. The evaluation of the noise current sources is carried out with a *LISN* (instead of current probes) to avoid measurement uncertainty due to the presence of a fundamental 50 Hz component on the AC side of the inverter.

2.2. LTI Assumption

Black-box modelling provides an equivalent frequency-domain representation (in terms of the Thevenin/Norton circuit) at the external ports of the device to be modelled. Consequently, the applicability and effectiveness of this strategy are limited to the modelling of linear and time-invariant devices only. Since power converters are intrinsically non-linear and time-variant networks, black-box modelling requires preliminary verification that the device can be treated, at least approximately, as an LTI system. Indeed, if the system does not satisfy the LTI assumption, the use of different operating conditions to identify model parameters could lead to different black-box models, which makes the models unsuited to assure accurate predictions with other operating conditions.

In this regard, further investigations are needed to verify when and under what conditions it is possible to apply black-box modelling approaches for modelling power converters. In several works [18–20], such an assumption was assumed to be satisfied, since the converter under analysis was equipped with an EMI filter or decoupling capacitors/functional inductors, which can mask the inherently non-linear and time-varying behaviour of the switching modules.

However, this conclusion cannot be a priori extended to all power converters. For instance, in [21] it was experimentally observed that, unlike the common-mode (CM) impedance (whose frequency response is dominated by parasitic effects), the differential model (DM) impedance can exhibit a different frequency response depending on the on or off status of the converter. In these cases, extracting the parameters of the black-box model with the converter switched off may significantly impair prediction effectiveness and accuracy.

3. Identification of Model Parameters in a Virtual Test Bench

3.1. Inverter System under Analysis

Without loss of generality, a three-phase inverter connecting a photovoltaic panel to the power grid, along with an LCL (i.e., inductor-capacitor-inductor) filter is considered as a system under analysis. The principle drawing of this system is illustrated in Figure 2. Though a single-stage inverter system is investigated in this paper, it is worth mentioning that DC/DC converters are frequently added between the PV array and the inverter for DC-link voltage control and maximum power point tracking. The presence of this additional converter is disregarded here for the sake of simplicity as its possible presence is not expected to impact the accuracy of the proposed modelling procedure as long as the LCL filter is installed at the AC side of the inverter.

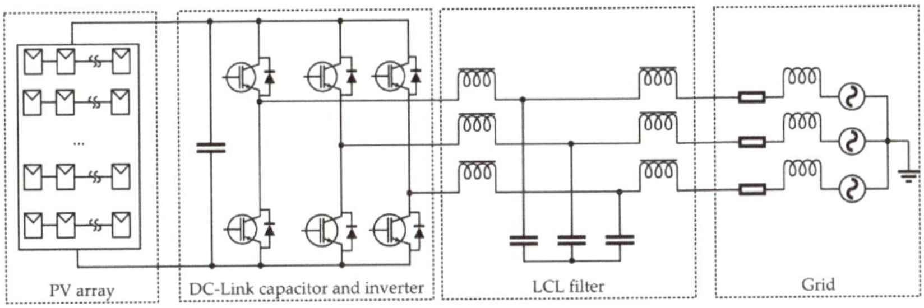

Figure 2. Principle drawing of a grid-connected three-phase inverter system.

The complete circuit model of this system, including functional parts and parasitic components, will be implemented in SPICE and used as the reference circuits. The circuit models of each component in the system will be discussed in the following paragraphs.

The solar panel Photowatt Technologies (Isère, France) PW 1650-24 V, whose parasitic parameters are adopted from [22], is used to construct the PV array for the system. In this work, the PV array contains 32 PV panels in series to provide the proper DC power supply.

The complete circuit model of one Insulated-Gate Bipolar Transistor (IGBT) switching leg (SKM200GB122D) and the capacitors at the DC side are shown in Figure 3a [23]. One 1.5 mF electrolytic capacitor and one 0.33 μF ceramic capacitor are connected at the input of each inverter leg, respectively—The functional model is augmented by including the following parasitic components: Stray inductances (L_{wa}, L_{w}) and resistance (R_{wa}) of the connecting wires, parasitic inductances (L_{EL}, L_{s}) and resistances (R_{d}, R_{EL}, R_{s}) of the DC capacitors, the stray inductances (L_{c}, L_{E}), the gate resistance (R_{G}) and internal capacitances (C_{gc}, C_{ge} and C_{ce}) of the IGBT/Diode devices (which are included in the SPICE models). Numerical values of the circuit elements in Figure 3a are collected in Appendix A (Table A1). The conventional Pulse-Width Modulation (PWM) strategy is implemented in SPICE with a 5 kHz switching frequency.

Grid-connected converters usually require an L or LCL filter attached at the output to reduce the harmonic currents in compliance with IEEE Standard 519–1992 and P1547-2003 requirements [24]. The LCL filter here adopted (see Figure 3b) was designed following the procedures in [25]. Relevant parameters are listed in Table A2 of Appendix A. Specifically, the equivalent circuit in [26] is adopted for the inductors, and datasheets are used to infer suitable values for the equivalent series resistance of the capacitors.

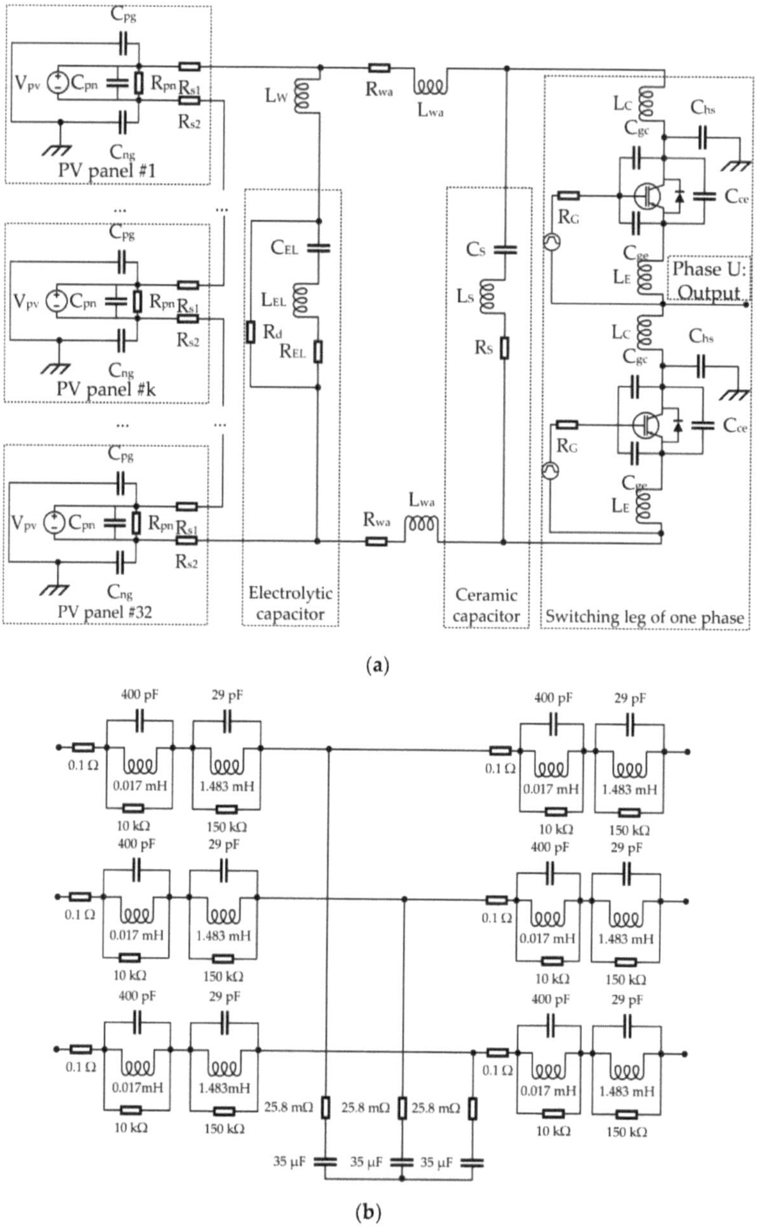

Figure 3. Circuit model of (**a**) the inverter with input capacitors and (**b**) the LCL filter.

It is worth mentioning that from the viewpoint of black-box modelling, such an LCL filter also plays a fundamental role in masking the nonlinear and time-varying behaviour of the inverter, thus making the LTI assumption generally satisfied. This property will be investigated in the following, by comparing the CEs predicted in the presence and absence of the LCL filter.

Eventually, the power grid is modelled by the use of three sinusoidal voltage sources (properly shifted in phase) connected in series with three impedances, obtained as the series connection of a 1 Ω resistor and a 0.2 mH inductor to mimic typical (average) mains impedance values.

3.2. Black-Box Modelling

The procedure here exploited to identify black-box model parameters involves two different setups. The former is used to evaluate the entries of the admittance matrix (passive part of the model), the latter to extract the frequency response of the noise (current) sources (the active part of the model).

3.2.1. Passive Part of the Black-Box Model

As the first step, the entries of the admittance matrix $\boldsymbol{Y}_{\text{dut}}$ are characterized through Vector Network Analyzer (VNA) measurement of the scattering parameters (S-parameter) at the output port of the LCL filter, as shown in Figure 4. For measurement, the gate signals and the DC-link voltage supply should be disconnected from the inverter for safety reason. To mimic VNA measurement conditions by the SPICE model, both the DC voltage sources and the gate signals sources are replaced by short circuits. The obtained 3 × 3 S-parameter matrix ($\boldsymbol{S}_{\text{dut}}$) is then converted into the corresponding 3 × 3 admittance matrix $\boldsymbol{Y}_{\text{dut}}$ by post-processing of measurement data [27].

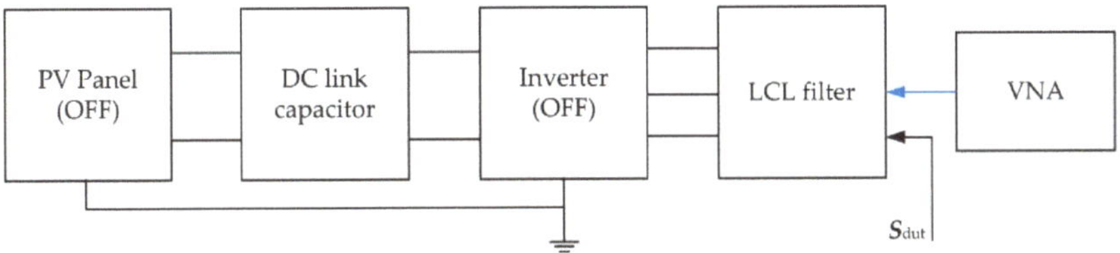

Figure 4. Measurement setup exploited to evaluate the entries of the admittance matrix (passive part of the black-box model).

To implement this step of the identification procedure in SPICE [28], VNA operation is mimicked through the blocks shown in Figure 5. In the specific configuration shown in the figure, the active VNA port is connected between phase 1 and ground, the other phases being connected to 50 Ω loads, and the entries of the first column of the S-parameter matrix $\boldsymbol{S}_{\text{dut}}$ (i.e., parameters S_{11}, S_{21} and S_{31}) are evaluated. Subsequently, exchanging the position of such an active port to phase 2 and phase 3 allows measuring the entries of the second (S_{12}, S_{22} and S_{32}) and the third column (S_{13}, S_{23} and S_{33}), respectively.

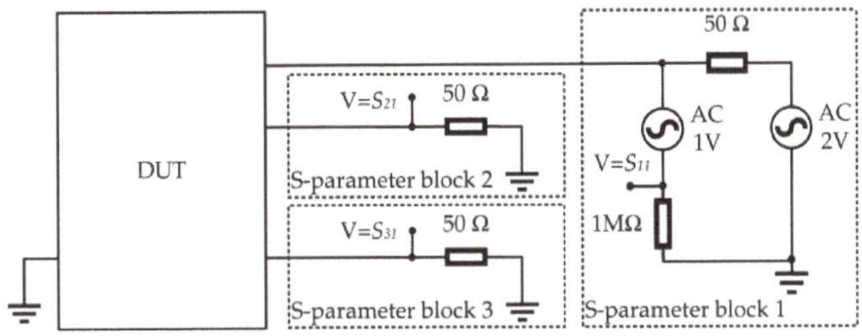

Figure 5. SPICE implementation of the measurement setup used to evaluate the passive part of the model: The three blocks connected at the *DUT* ports are used to emulate the three ports of a VNA.

3.2.2. Active Part of the Black-Box Model

To extract from measurements the frequency response of the current sources I_{s1}, I_{s2}, and I_{s3} (active part of the black-box model), the test setup in Figure 6 is exploited. The voltages V_4, V_5, and V_6 at the *LISN* output are measured in the time domain by the three channels of an oscilloscope (to obtain simultaneous measurements on the three phases), and then converted into the frequency domain by the Fourier Transform (FT).

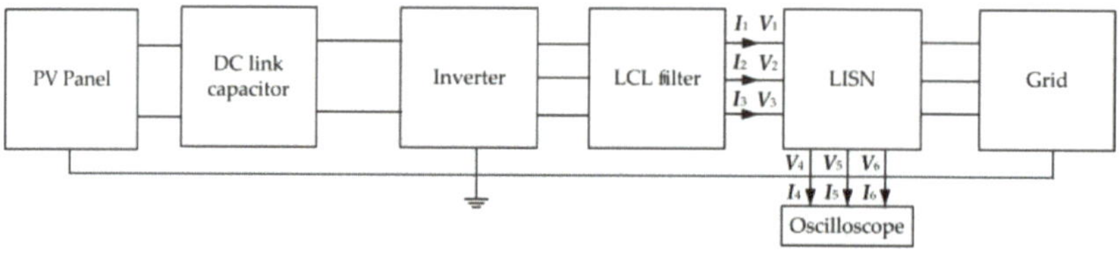

Figure 6. Measurement setup exploited to evaluate the active part of the black-box model, i.e., the three current sources I_{S1}, I_{S2} and I_{S3}.

When the black-box modelling approach is applied to the DC side to assess CEs to PV panels, current probes can be used to measure the currents directly (see Figure 1a). This approach was carried out in [12] to build the black-box model of a DC/DC converter and was validated by experiment.

With respect to [12], where the noise sources were directly extracted from the currents measured at the converter input (Figure 1a), the procedure proposed here requires additional efforts in terms of post-processing, since the *LISN* should be preliminarily characterized as a six-port network by VNA measurement (yielding a 6 × 6 S-parameter matrix, $\mathbf{S}_{LISN}$), and its effect should be de-embedded from the measurement data.

However, this solution is preferable when time-domain measurements are to be carried out at the AC side of the inverter, as it allows filtering out the fundamental 50 Hz component, whose presence could cause inaccurate sampling of higher spectral components.

In SPICE, the aforesaid setup was implemented by connecting 50 Ω resistors at the *LISN* output ports to emulate the oscilloscope channels. Additionally, to preliminarily characterize the *LISN* in terms of six-port network, the blocks already introduced in Figure 5 are connected on both sides of the *LISN*, with the active block connected to one of the six ports at a time, to evaluate the entries of the 6 × 6 S-parameter matrix, $\mathbf{S}_{LISN}$. In this step, it has to be carefully considered that the impedance connected at the output port of the *LISN* may influence the results for frequencies outside the *LISN* bandwidth. As it will be shown in Section 4, this effect is mostly visible in the low-frequency part of the spectrum, since several *LISN*s are designed to comply with EMC standards foreseeing CE measurement starting from 150 kHz.

Once this matrix is known, voltages (V_1, V_2, V_3) and currents (I_1, I_2, I_3) at the *DUT* side of the *LISN* are computed as

$$\begin{bmatrix} V_1 & V_2 & V_3 & I_1 & I_2 & I_3 \end{bmatrix}^T = ABCD_{LISN}\begin{bmatrix} V_4 & V_5 & V_6 & I_4 & I_5 & I_6 \end{bmatrix}^T \quad (1)$$

where $ABCD_{LISN}$ is the 6 × 6 *ABCD* matrix of the *LISN*, obtained starting from the corresponding S-parameters matrix $\mathbf{S}_{LISN}$ [27]. Eventually, the noise currents in the black-box model are retrieved as:

$$\begin{bmatrix} I_{s1} & I_{s2} & I_{s3} \end{bmatrix}^T = \begin{bmatrix} I_1 & I_2 & I_3 \end{bmatrix}^T + Y_{DUT}\begin{bmatrix} V_1 & V_2 & V_3 \end{bmatrix}^T \quad (2)$$

The magnitude of the extracted black-box-model parameters, i.e., the entries of the Y matrix and the three current sources, is plotted in Figure 7. The first plot, Figure 7a, shows the frequency response of *DUT* admittances: Due to the symmetry, diagonal entries are the same ($Y_{11} = Y_{22} = Y_{33}$) as well as off-diagonal entries ($Y_{12} = Y_{13} = Y_{23}$). Due to symmetry, also the frequency response of the three current sources is coincident, as shown in Figure 7b.

It is worth mentioning that *LISN* effectiveness in stabilizing the impedance seen from the *DUT* outlets in the whole frequency interval of interest is fundamental for model-parameter extraction. Indeed, since the *LISN* is connected with the mains for CE measurement, its effectiveness in providing a stable impedance, despite possible variation of the mains impedance, determines the accuracy of the noise-source extraction. Hence, to better investigate this specific aspect, in this work the two *LISN*s shown in Figure 8 will be considered, which are designed according to the requirements specified in the standard CISPR22, Figure 8a, and CISPR16, Figure 8b, respectively.

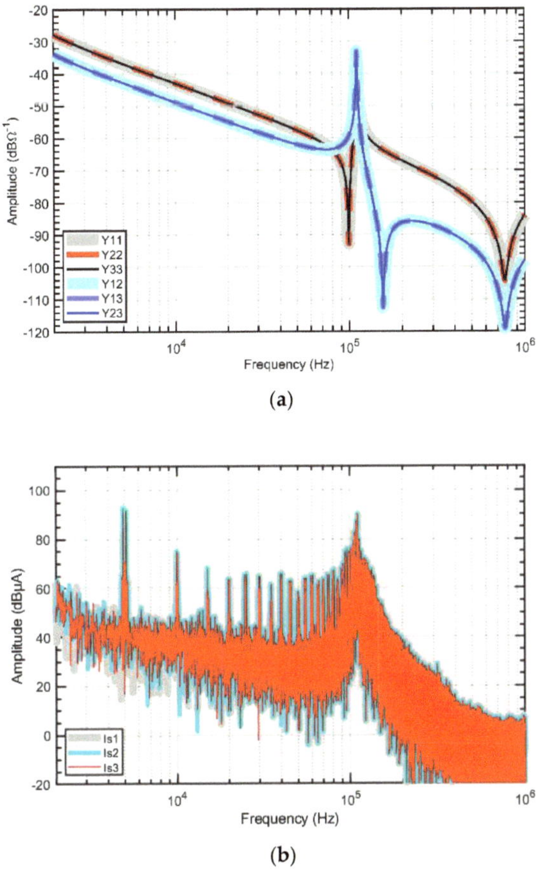

Figure 7. Extracted black-box model: (**a**) Selected six Y-parameter of *DUT*; (**b**) Three current sources.

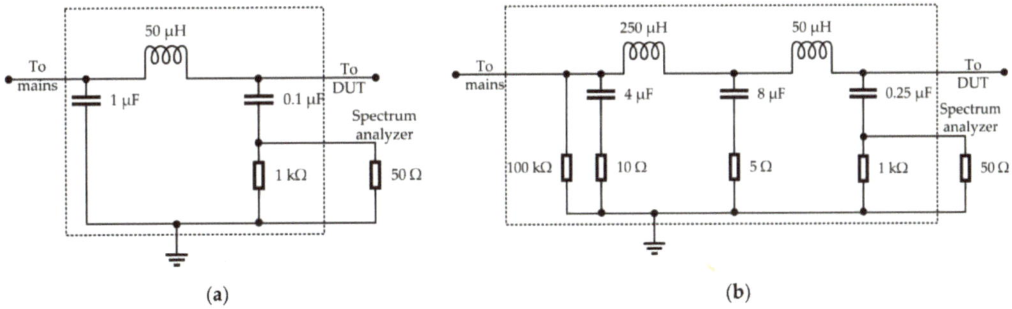

Figure 8. Schematics of the two *LISN*s considered in this work: (**a**) CISPR22 and (**b**) CISPR16-1-2.

3.3. Setup for Model Validation

To effectively validate the accuracy of the obtained black-box model, the conducted emissions exiting the converter under loading conditions sufficiently different from those exploited for model-parameter identification, Figure 6, are to be predicted and compared versus those obtained by exploiting the explicit model of the converter (here used as reference quantities). To this end, an additional EMI filter, Figure 9a, is included between the system and the *LISN* to appreciably modify the converter working conditions despite the presence of the *LISN*. The obtained measurement setup is shown in Figure 9b.

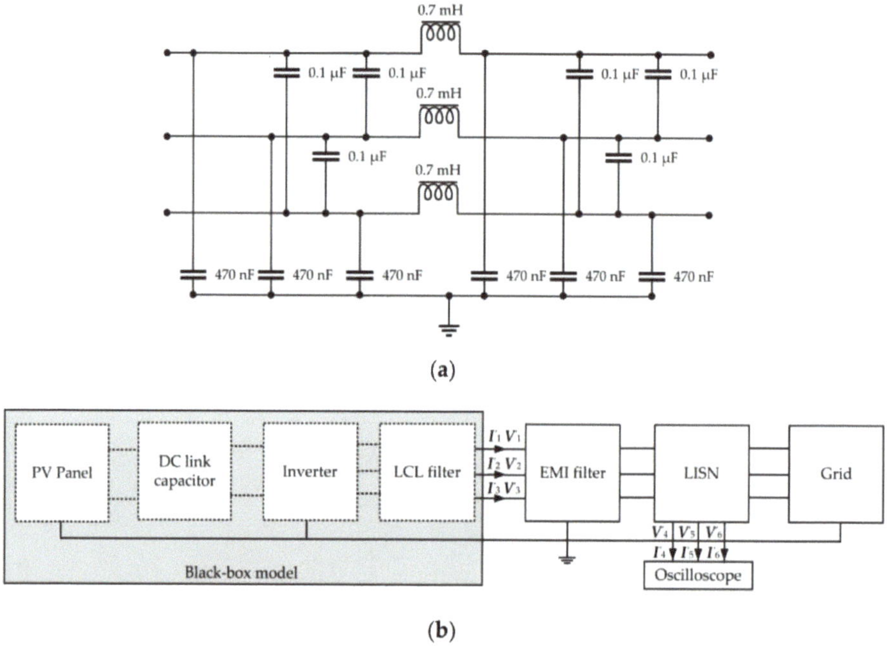

Figure 9. Validation of the black-box model: an additional EMI filter (**a**) is included in the measurement setup (**b**) to modify the working conditions of the converter.

4. Prediction of Conducted Emissions

4.1. Conducted Emissions Analysis

As the first step, the conducted emissions exiting the inverter are evaluated in the presence and the absence of the EMI filter in terms of voltages measured at the output (i.e., at the receiver side) of a *LISN* compliant with CISPR16 specifications. To this end, time-domain simulations are carried out in SPICE, and the frequency spectra of the voltages at the *LISN* output are obtained afterwards through FT.

The conducted emissions simulated in the absence of the EMI filter are then used as input data to extract the black-box model of the inverter. Conversely, the conducted emissions simulated in the presence of the filter are used to assess the effectiveness of the extracted black-box model in predicting the emissions exiting the inverter under analysis, in combination with a different set of working conditions.

To verify the suitability of the aforesaid two sets of working conditions in assessing the prediction effectiveness of the extracted black-box model, the conducted emissions from the converter in the presence and in the absence of the EMI filter are preliminary compared in Figure 10 (where the voltage V_4 on a specific phase is plotted). The significant difference between the two spectra confirms the effectiveness of introducing the EMI filter to generate a new set of working conditions for validation purposes. Apart from the first two spectrum lines at 5 kHz and 10 kHz, the EMI filter provides appreciable attenuation to all the other harmonics.

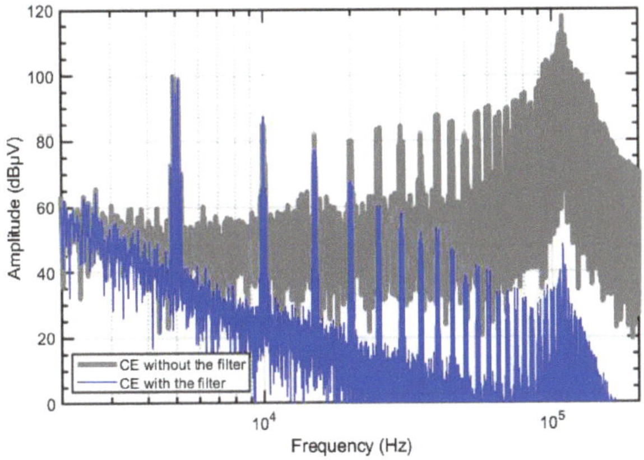

Figure 10. Conducted emissions at the AC side of the inverter in the presence and in the absence of the EMI filter.

Moreover, modal decomposition is applied to get a better understanding of the spectral content of the emissions exiting the converter. To this end, the three-phase voltages V_4, V_5, V_6 at the *LISN* output are decomposed into the corresponding positive-sequence, negative-sequence, and zero-sequence voltages V_+, V_-, V_0 by the similarity transformation, [29]:

$$\begin{bmatrix} V_+ \\ V_- \\ V_0 \end{bmatrix} = \frac{1}{3}\begin{bmatrix} 1 & e^{j2\pi/3} & e^{j4\pi/3} \\ 1 & e^{j4\pi/3} & e^{j2\pi/3} \\ 1 & 1 & 1 \end{bmatrix}\begin{bmatrix} V_4 \\ V_5 \\ V_6 \end{bmatrix} \tag{3}$$

In this regard, it is worth mentioning that for EMC analysis decomposing phase quantities (voltages and currents) into CM and DM components is a standard practice, which allows the designer to identify the noise component that is dominant in a specific frequency interval, so to properly design the required EMI filter. For a three-phase system, the transformation into positive/negative/zero sequences in (3) provides equivalent information in terms of CM (zero sequences) and DMs (positive and negative sequences) noise components, and it is here adopted since it is widely adopted also for power system analysis. Since the similarity transformation matrix in (3) is not singular, phase voltage (V_4, V_5, V_6) can be also obtained from modal quantities (V_+, V_-, V_0), by reversing (3).

One of the phase voltage (V_4) and the obtained modal voltages evaluated in the absence and the presence of the EMI filter are plotted in Figure 11a,b, respectively. The comparison versus the corresponding phase voltages (see the grey spectrum in Figure 11) reveals that the emissions from the converter are dominated by the positive and negative-sequence contributions up to nearly 35 kHz. Conversely, above this frequency the zero-sequence contribution is dominant. This result proves that the observed reduction of the converter CEs above 35 kHz is mainly to be ascribed to the effectiveness of the exploited EMI filter in damping the zero-sequence component of the noise currents exiting the converter.

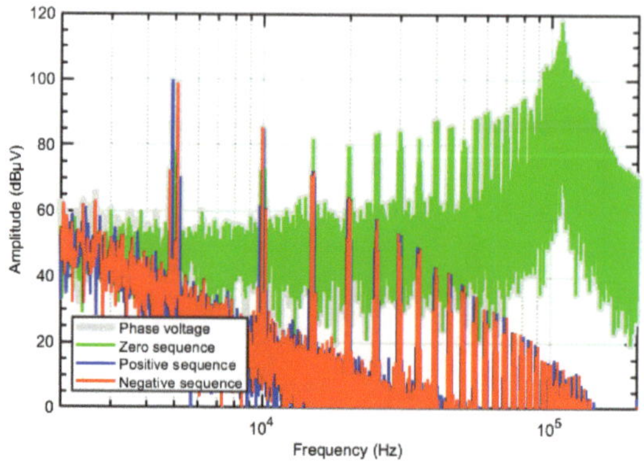

(**a**)

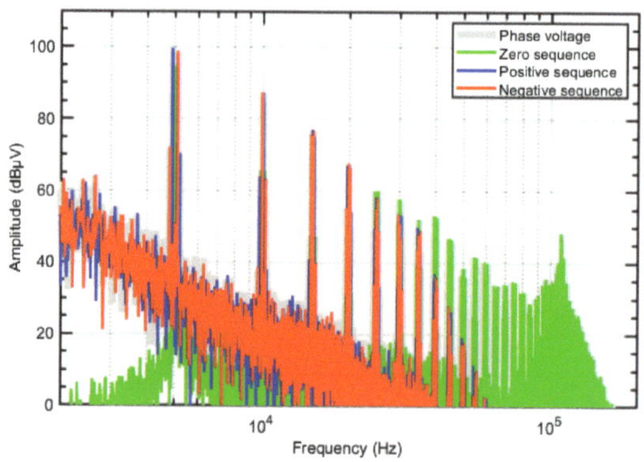

(**b**)

Figure 11. Decomposition of the inverter CEs into positive, negative and zero-sequence components: (**a**) without the EMI filter; (**b**) with the EMI filter.

4.2. Validation of the Modelling Procedure

The conducted emissions obtained by the SPICE model of the whole test setup described in Section 3 and those predicted by the black-box model are compared in Figure 12a in the frequency range from 2 kHz to 200 kHz. The CEs above 200 kHz are not plotted, as they are significantly attenuated (below −10 dBμV). The comparison proves that the black-box model can effectively predict the emissions exiting the converter, with maximum deviations within 3 dB around 110 kHz, where the system exhibits a resonance.

The corresponding comparison in terms of zero sequence and positive/negative sequence is shown in Figure 12b–d, respectively, and confirms the effectiveness of the proposed black-box model also in predicting modal quantities. The zero-sequence and the positive/negative sequences dominate above and below 35 kHz, respectively. The discrepancies observed above 100 kHz are mainly owing to numerical processing of data, due to the extremely low CE levels at high frequency.

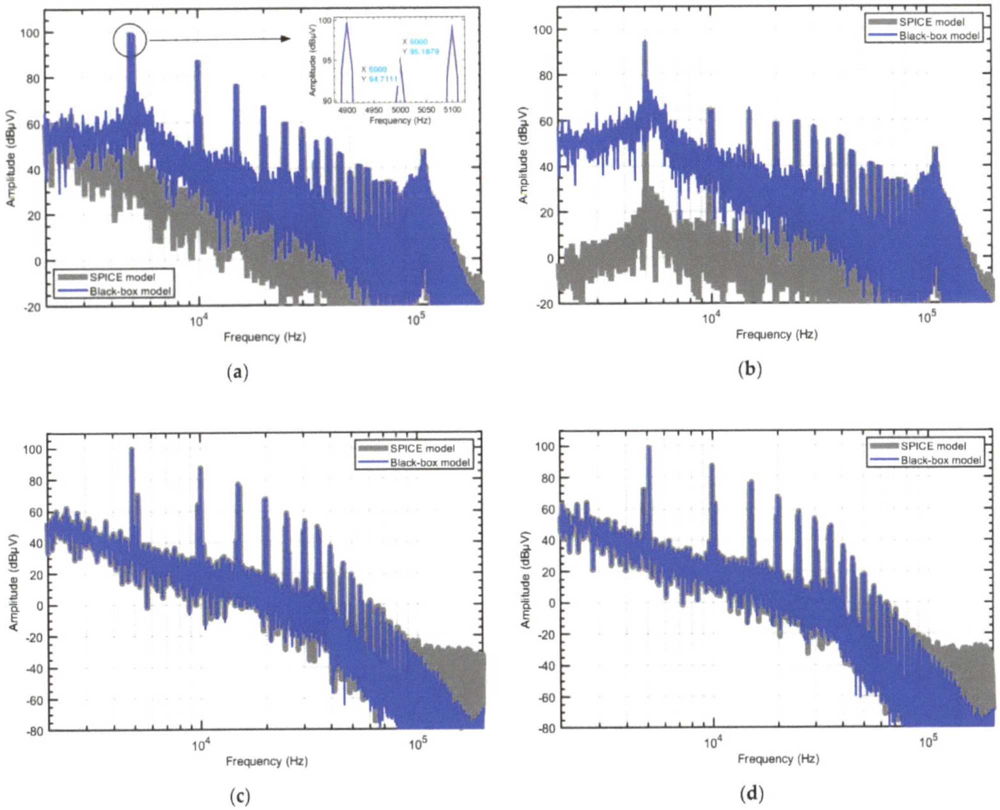

Figure 12. Conducted emissions obtained by the SPICE model in Section 3 and predicted by the black-box model (**a**) phase voltage (**b**) zero-sequence voltage (**c**) positive sequence voltage and (**d**) negative sequence voltage.

4.3. Limitations of Black-Box Modelling

4.3.1. Influence of the Mask Impedance

To investigate the role of the masking effect due to the LCL filter on model effectiveness, the LCL filter was removed, and the procedure for model parameter extraction was repeated. The obtained predictions are compared versus SPICE simulations in Figure 13a.

Specifically, for the system without the LCL filter, obvious mismatches are observed in predictions of CEs obtained by the black-box model from 100 to 400 kHz with respect to the accurate circuit simulation. In Figure 13b–d, three sequences of emissions are compared. It is observed that the discrepancies in this frequency interval are mainly from the positive and negative sequences instead of the zero sequence. With respect to Figure 12, the discrepancies observed in Figure 13 are to be ascribed to the inaccuracy of black-box modelling in the absence of the LCL filter, rather than to the noise floor of data-processing. As a matter of fact, in Figure 13 CE levels from 100 kHz to 400 kHz are still significantly larger (above 40 dBμV) than the noise floor.

This investigation puts in evidence that in the absence of an LCL filter (and its masking effect), the nonlinear and time-variant characteristics reflect the prediction of positive/negative sequence components. In this case, the black-box modelling approach is less accurate in the prediction of DM emissions, especially in the high frequency range.

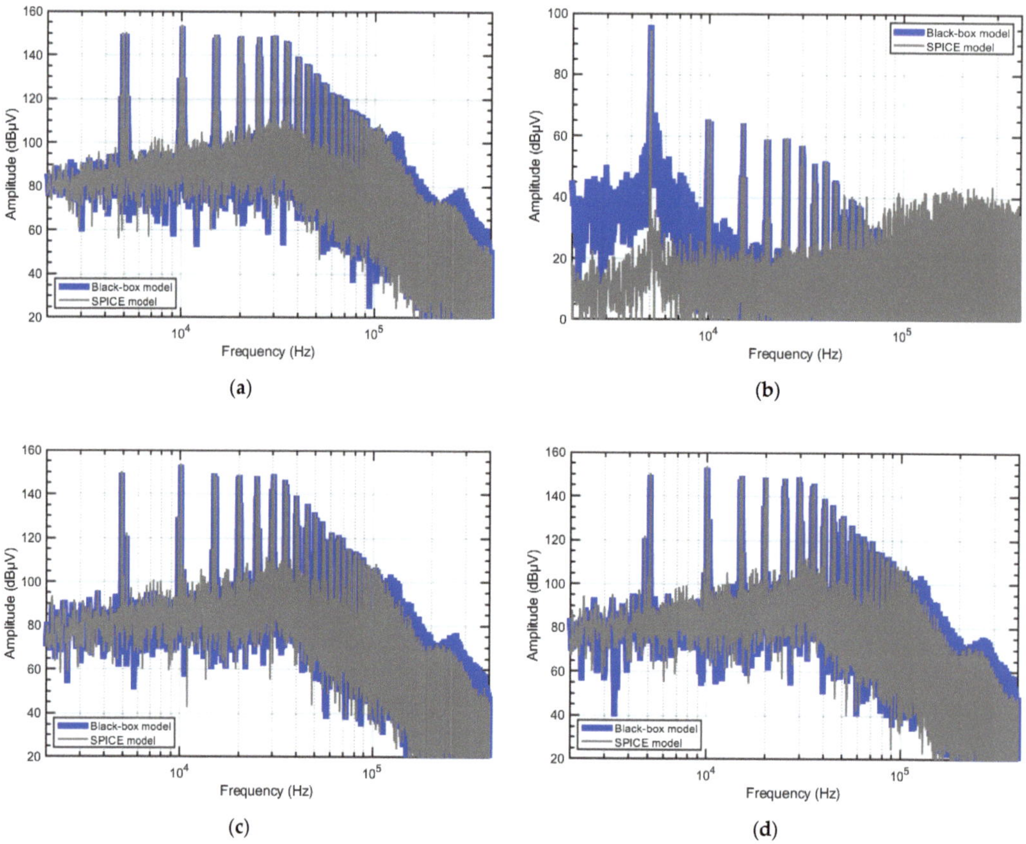

Figure 13. Conducted emissions in the absence of LCL filter obtained by SPICE model in Section 3 and predicted by the black-box model (**a**) phase voltage (**b**) zero-sequence voltage (**c**) positive sequence voltage and (**d**) negative sequence voltage.

4.3.2. Influence of the Power Network Impedance

For model-parameter evaluation, the ability of the exploited *LISN* in providing a stable impedance despite the possible variation of the impedance of the power grid plays a fundamental role. To extract the frequency response of the noise sources, the exploited *LISN* has to be disconnected from the power grid and preliminarily characterized by VNA measurements. At this stage, the impedance at the output port of the *LISN* (open-ended) is different from the actual impedance of the power grid (here mimicked by the series connection of a 0.2 mH inductor and a 1 Ω resistor), which is connected at the *LISN* output when noise currents are measured. Hence, if the *LISN* is used outside its bandwidth, the input impedance seen from its input port might be significantly different in the two test setups.

To investigate this specific aspect, different impedance values are assumed for the power grid, and the impedance seen at the input of the CISPR16 and CISPR22 *LISN*s are plotted in Figure 14a,b, respectively. The plots show that for frequencies below the *LISN* operation frequency range (i.e., for frequencies below 150 kHz for the CISPR22 *LISN*, and below 9 kHz for the CISPR16 *LISN*) non-negligible variations are observed in the impedance seen at the *LISN* input. This significantly impacts on CE measurement in the lower part of the spectrum, as shown in Figure 15, where the CEs measured by the two *LISN*s under analysis are compared (for these simulations the presence of the power grid was mimicked by the series connection of a 0.2 mH inductor and a 1 Ω resistor). The comparison shows significant differences in the measured CE levels for frequencies below 40 kHz, with maximum deviation in the order of 30 dB around 25 kHz.

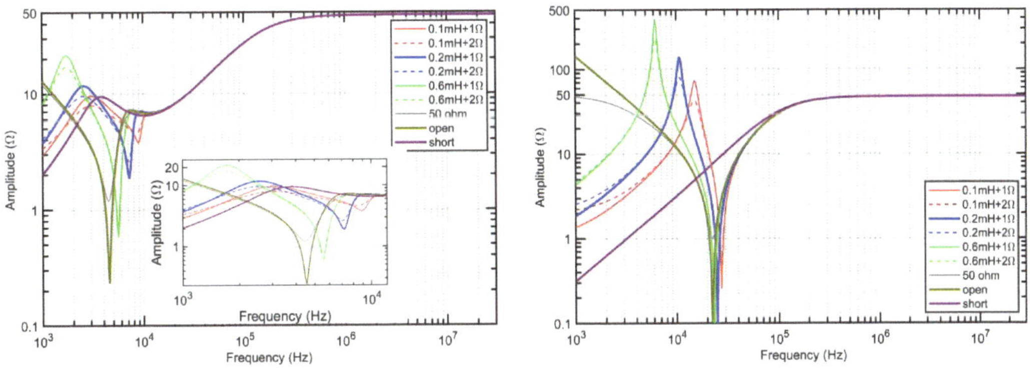

(a) (b)

Figure 14. Input impedance of the *LISN* for different impedances of the power grid attached at the output port: (**a**) CISPR16; (**b**) CISPR22 *LISN*.

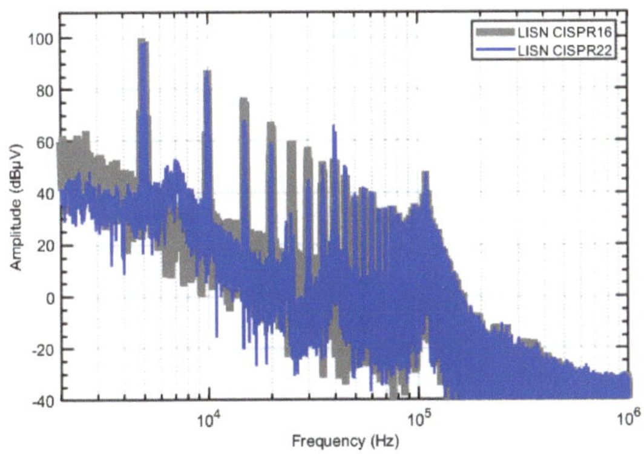

Figure 15. Conducted emissions (SPICE model) measured by using two different *LISN*s, i.e., CISPR16 (grey spectrum) and CISPR22 (blue spectrum).

To investigate the influence of the exploited *LISN*s on the effectiveness of the proposed black-box modelling procedure, the two *LISN*s under analysis were preliminarily characterized in terms of scattering parameters with the output port left open-ended. As previously observed, this configuration reproduces the experimental setup involving the VNA and the *LISN* disconnected by the power grid.

After that, the CEs exiting the systems were virtually measured by exploiting the two *LISN*s connected to the power grid (emulated by the series connection of a 0.2 mH inductor and a 1 Ω resistor), and the obtained CE levels were used as input data to determine the frequency response of the noise sources involved in the black-box model.

The CEs predicted by the obtained black-box models are compared versus those predicted by the SPICE model in Figure 16. The comparison shows that the predictions obtained by exploiting the CISPR16 *LISN* exhibit a deviation (less than 3 dB) at the switching frequency (5 kHz) only. Conversely, if the black-box model is extracted by exploiting the CISPR22 *LISN*, the predictions reveal significant deviations with respect to SPICE simulation below 150 kHz, since such a *LISN* is designed to assure a stable impedance starting from 150 kHz only.

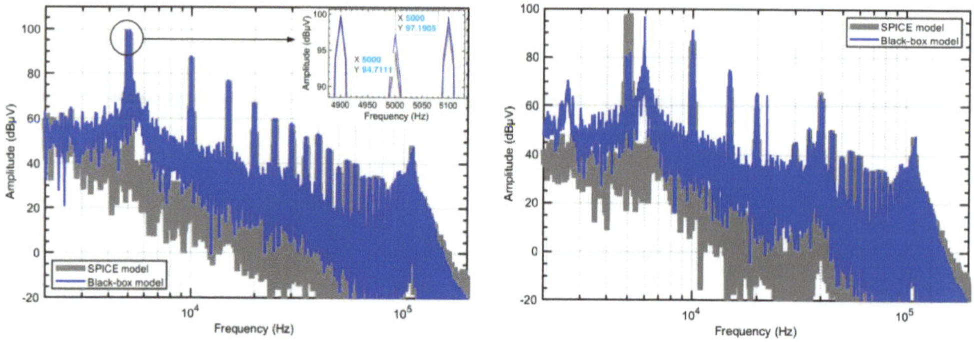

(a) (b)

Figure 16. CE levels predicted by exploiting two different *LISN*s, i.e., (**a**) CISPR16 and (**b**) CISPR22 *LISN*: SPICE versus black-box model.

5. Conclusions

In this work, the effectiveness and possible limitations of a black-box modelling procedure tailored to the representation of a three-phase inverter connected with a PV panel have been investigated. The procedure foresees separate characterization of the active and passive part of the model. Specifically, a suitable procedure is proposed to identify the noise sources (active part of the model), which resorts to the time-domain measurement of the voltages at the output of the *LISN*, instead of the currents exiting the converter, to filter out the 50 Hz fundamental component, whose contribution could impair the identification of higher frequency components. The comparison versus SPICE simulation proved the effectiveness of the proposed black-box modelling approach in predicting the CEs exiting the AC side of the grid-tied three-phase inverter system under analysis in the frequency range starting from 2 kHz.

The proposed analysis allowed a systematic investigation of two main limitations possibly degrading the prediction accuracy of the proposed black-box modelling technique during the measurement. The former limitation is related to the assumption that the system under analysis should be treated, at least approximately, as a linear and time-invariant system. In this regard, the fundamental role played by the LCL filter (installed at the converter output) in masking the inherent non-linear and time-invariant behaviour of the inverter switching modules has been proven. In general, this assumption needs to be confirmed prior to the application of the black-box modelling procedures. The latter limitation is related to the bandwidth of the *LISN* exploited to extract from measurement the frequency response of the noise sources (active part of the model). In this regard, it has been proven that using a *LISN* foreseen for CE measurement starting from 150 kHz, as foreseen by several EMC standards, may lead to significant degradation of prediction accuracy in the lower part of the spectrum (i.e., below 150 kHz), since the frequency response of the extracted noise sources is strongly influenced by the impedance connected at the output of the *LISN*. To eliminate the influence of power network impedance, it is necessary to develop a new type of *LISN* with stable impedance starting from 2 kHz, otherwise, the mains impedance must be measured or estimated to improve the accuracy of the black-box model predictions at low frequency.

Author Contributions: Conceptualization, M.Z., L.T., R.C. and S.A.P.; methodology, F.G. and G.S.; software, L.W. and A.H.B.; validation, X.W., X.L.; formal analysis, F.G., X.W. and L.W.; investigation, G.S., A.H.B. and L.W.; resources, M.Z., L.T. and F.G.; writing—original draft preparation, L.W.; writing—review and editing, F.G., A.H.B., X.W., X.L. and M.Z.; visualization, G.S.; supervision, F.G., S.A.P. and R.C.; project administration, M.Z., F.G. and S.A.P.; funding acquisition, F.G. and M.Z. All authors have read and agreed to the published version of the manuscript.

Funding: This project has received funding from the European Union's Horizon 2020 research and innovation programme under the Marie Skłodowska-Curie grant agreement No 812753. This work has been partially financed by the Research Fund for the Italian Electrical System in compliance with the Decree of 16 April 2018.

Conflicts of Interest: The authors declare no conflict of interest.

Appendix A

Table A1. Circuit model parameters of PV panel and inverter in Figure 3a.

Elements	Parameter	Value		Description
PV panel	V_{pv}	24	V	Nominal panel voltage
	C_{pg}, C_{ng}	4	pF	Parasitic capacitances between positive/negative DC bus and ground
	R_{s1}, R_{s2}	1.5	mΩ	DC bus panel feeder resistances
	C_{pn}	90	nF	Parasitic capacitance between positive and negative DC bus terminals
	R_{pn}	2.8	kΩ	Leakage resistance between positive and negative DC bus terminals
	L_w	10	nH	External wire inductance
Electrolytic capacitor	C_{EL}	1500	μF	Nominal capacitance
	L_{EL}	30	nH	Internal series inductance
	R_{EL}	40	mΩ	Internal series resistance
	R_d	100	kΩ	Discharging resistance
DC bus	R_{wa}	3.9	mΩ	Stray resistance
	L_{wa}	0.36	μH	Stray inductance
Ceramic capacitor	C_s	0.33	μF	Nominal capacitor
	L_s	30	nH	Internal series inductance
	R_s	30	mΩ	Internal series resistance
IGBT	L_C	40	nH	Collector stray inductance
	L_E	40	nH	Emitter stray inductance
	C_{gc}	700	pF	Stray capacitor between gate and collector
	C_{ce}	1	nF	Stray capacitor between collector and emitter
	C_{ge}	21.3	nF	Stray capacitor between gate and emitter
	C_{hs}	280	pF	Stray capacitor between IGBT and heatsink

Table A2. Design parameters of LCL filter circuit.

Parameter	Value		Description
S_B	7	kVA	Inverter rated power
V_B	380	V	Line to line output voltage (rms value)
V_{DC}	720	V	DC power supply
f_B	50	Hz	Output voltage frequency
I_B	10.64	A	Output current (rms value)
Z_B	20.63	Ω	Inverter base impedance
L_B	65.66	mH	Inverter base inductance
C_B	154.31	μF	Inverter base capacitance
r_g	0.003	/	Grid-side ripple current component
r_i	0.15	/	Inverter-side ripple current component
a	0.02	/	Capacitor voltage ripple attenuation
f_1	4.9	kHz	The First dominant harmonic frequency
x	0.316	/	Ratio = harmonic voltage at f_1/voltage at f_B
M	0.866	/	Amplitude modulation ratio
L_{ipu}	0.0228	/	PU inductance value of LCL filter at inverter side
L_{gpu}	0.0228	/	PU inductance value of LCL filter at grid side
C_{pu}	0.2191	/	PU capacitance value of LCL filter
L_i	1.4979	mH	Nominal inductance of LCL filter at inverter side
L_g	1.4979	mH	Nominal inductance of LCL filter at grid side
C	33.8	μF	Nominal capacitance of LCL filter

References

1. Araneo, R.; Lammens, S.; Grossi, M.; Bertone, S. EMC Issues in High-Power Grid-Connected Photovoltaic Plants. *IEEE Trans. Electromagn. Compat.* **2009**, *51*, 639–648. [CrossRef]
2. Rietveld, G.; Hoogenboom, D.; Acanski, M. Conducted EMI Causing Error Readings of Static Electricity Meters. In Proceedings of the 2018 Conference on Precision Electromagnetic Measurements (CPEM 2018), Paris, France, 8–13 July 2018; pp. 1–2.
3. Rönnberg, S.; Bollen, M.; Larsson, A. Emission from Small Scale PV-Installations on the Low Voltage Grid. *REPQJ* **2014**, 617–621. [CrossRef]
4. Darmawardana, D.; Perera, S.; Robinson, D.; Ciufo, P.; Meyer, J.; Klatt, M.; Jayatunga, U. Investigation of High Frequency Emissions (Supraharmonics) from Small, Grid-Tied, Photovoltaic Inverters of Different Topologies. In Proceedings of the 2018 18th International Conference on Harmonics and Quality of Power (ICHQP), Ljubljana, Slovenia, 13–16 May 2018; pp. 1–6.
5. Subhani, S.; Cuk, V.; Cobben, J.F.G. A Literature Survey on Power Quality Disturbances in the Frequency Range of 2–150 KHz. *REPQJ* **2017**, *1*, 405–410. [CrossRef]
6. Rönnberg, S.K.; Bollen, M.H.J.; Amaris, H.; Chang, G.W.; Gu, I.Y.H.; Kocewiak, Ł.H.; Meyer, J.; Olofsson, M.; Ribeiro, P.F.; Desmet, J. On Waveform Distortion in the Frequency Range of 2kHz–150kHz—Review and Research Challenges. *Electr. Power Syst. Res.* **2017**, *150*, 1–10. [CrossRef]
7. Meyer, J.; Khokhlov, V.; Klatt, M.; Blum, J.; Waniek, C.; Wohlfahrt, T.; Myrzik, J. Overview and Classification of Interferences in the Frequency Range 2–150 KHz (Supraharmonics). In Proceedings of the 2018 International Symposium on Power Electronics, Electrical Drives, Automation and Motion (SPEEDAM), Amalfi, Italy, 20–22 June 2018; pp. 165–170.
8. Noshahr, J.B.; Kalasar, B.M. Evaluating Emission and Immunity of Harmonics in the Frequency Range of 2–150 KHz Caused by Switching of Static Convertor in Solar Power Plants. *CIRED Open Access Proc. J.* **2017**, *2017*, 625–628. [CrossRef]
9. Frantz, G.; Frey, D.; Schanen, J.-L.; Revol, B. EMC Models of Power Electronics Converters for Network Analysis. In Proceedings of the 2013 15th European Conference on Power Electronics and Applications (EPE), Lille, France, 3–5 September 2013; pp. 1–10.
10. Kharanaq, F.A.; Emadi, A.; Bilgin, B. Modeling of Conducted Emissions for EMI Analysis of Power Converters: State-of-the-Art Review. *IEEE Access* **2020**, *8*, 189313–189325. [CrossRef]
11. Riba, J.-R.; Moreno-Eguilaz, M.; Bogarra, S.; Garcia, A. Parameter Identification of DC-DC Converters under Steady-State and Transient Conditions Based on White-Box Models. *Electronics* **2018**, *7*, 393. [CrossRef]
12. Spadacini, G.; Grassi, F.; Bellan, D.; Pignari, S.A.; Marliani, F. Prediction of Conducted Emissions in Satellite Power Buses. Available online: https://www.hindawi.com/journals/ijae/2015/601426/ (accessed on 15 February 2021).
13. Baisden, A.C.; Boroyevich, D.; Wang, F. Generalized Terminal Modeling of Electromagnetic Interference. *IEEE Trans. Ind. Appl.* **2010**, *46*, 2068–2079. [CrossRef]
14. Donnelly, T.J.; Pekarek, S.D.; Fudge, D.R.; Zarate, N. Thévenin Equivalent Circuits for Modeling Common-Mode Behavior in Power Electronic Systems. *IEEE Open Access J. Power Energy* **2020**, *7*, 163–172. [CrossRef]
15. Amara, M.; Vollaire, C.; Ali, M.; Costa, F. Black Box EMC Modeling of a Three Phase Inverter. In Proceedings of the 2018 International Symposium on Electromagnetic Compatibility (EMC EUROPE), Amsterdam, The Netherlands, 27–30 August 2018; pp. 642–647.
16. Liu, Y.; Jiang, S.; Wang, H.; Wang, G.; Yin, J.; Peng, J. EMI Filter Design of Single-Phase SiC MOSFET Inverter with Extracted Noise Source Impedance. *IEEE Electromagn. Compat. Mag.* **2019**, *8*, 45–53. [CrossRef]
17. Kerrouche, B.; Bensetti, M.; Zaoui, A. New EMI Model With the Same Input Impedances as Converter. *IEEE Trans. Electromagn. Compat.* **2019**, *61*, 1072–1081. [CrossRef]
18. Ales, A.; Schanen, J.; Moussaoui, D.; Roudet, J. Impedances Identification of DC/DC Converters for Network EMC Analysis. *IEEE Trans. Power Electron.* **2014**, *29*, 6445–6457. [CrossRef]
19. Bishnoi, H.; Baisden, A.C.; Mattavelli, P.; Boroyevich, D. Analysis of EMI Terminal Modeling of Switched Power Converters. *IEEE Trans. Power Electron.* **2012**, *27*, 3924–3933. [CrossRef]
20. Cheaito, H.; Diop, M.; Ali, M.; Clavel, E.; Vollaire, C.; Mutel, L. Virtual Bulk Current Injection: Modeling EUT for Several Setups and Quantification of CM-to-DM Conversion. *IEEE Trans. Electromagn. Compat.* **2017**, *59*, 835–844. [CrossRef]
21. Mazzola, E.; Grassi, F.; Amaducci, A. Novel Measurement Procedure for Switched-Mode Power Supply Modal Impedances. *IEEE Trans. Electromagn. Compat.* **2020**, *62*, 1349–1357. [CrossRef]
22. Rucinski, M.; Chrzan, P.J.; Musznicki, P.; Schanen, J.-L.; Labonne, A. Analysis of Electromagnetic Disturbances in DC Network of Grid Connected Building-Integrated Photovoltaic System. In Proceedings of the 2015 9th International Conference on Compatibility and Power Electronics (CPE), Costa da Caparica, Portugal, 24–26 June 2015; pp. 332–336.
23. Grandi, G.; Casadei, D.; Reggiani, U. Common- and Differential-Mode HF Current Components in AC Motors Supplied by Voltage Source Inverters. *IEEE Trans. Power Electron.* **2004**, *19*, 16–24. [CrossRef]
24. IEEE Power and Energy Society IEEE Recommended Practice and Requirements for Harmonic Control in Electric Power Systems. Available online: https://ieeexplore.ieee.org/document/6826459 (accessed on 8 June 2021).
25. Shahil, S. Step-by-Step Design of an LCL Filter for Three-Phase Grid Interactive Converter. Available online: http://rgdoi.net/10.13140/RG.2.1.3883.6964 (accessed on 8 June 2021).

26. Negri, S.; Wu, X.; Liu, X.; Grassi, F.; Spadacini, G.; Pignari, S.A. Mode Conversion in DC-DC Converters with Unbalanced Busbars. In Proceedings of the 2019 Joint International Symposium on Electromagnetic Compatibility, Sapporo and Asia-Pacific International Symposium on Electromagnetic Compatibility (EMC Sapporo/APEMC), Sapporo, Hokkaido, Japan, 3–7 June 2019; pp. 112–115.
27. Frickey, D.A. Conversions between S, Z, Y, H, ABCD, and T Parameters Which Are Valid for Complex Source and Load Impedances. *IEEE Trans. Microw. Theory Tech.* **1994**, *42*, 205–211. [CrossRef]
28. Robert, H. Extracting Scattering Parameters from SPICE. Available online: https://www.yumpu.com/en/document/read/40479516/create-s-parameter-subcircuits-for-microwave-and-rf-applications (accessed on 17 February 2021).
29. Paul, C.R. Decoupling the Multiconductor Transmission Line Equations. *IEEE Trans. Microw. Theory Tech.* **1996**, *44*, 1429–1440. [CrossRef]

energies

Article

Accurate Computation of Mutual Inductance of Non Coaxial Pancake Coils

Mauro Parise [1], Fabrizio Loreto [2], Daniele Romano [2], Giulio Antonini [2,*] and Jonas Ekman [3]

1 University Campus Bio-Medico of Rome, 00128 Rome, Italy; m.parise@unicampus.it
2 Department of Industrial and Information Engineering and Economy, University of L'Aquila, 67100 L'Aquila, Italy; fabrizio.loreto@graduate.univaq.it (F.L.); daniele.romano@univaq.it (D.R.)
3 Department of Computer Science, Electrical and Space Engineering, Luleå University of Technology, A2311 Luleå, Sweden; jonas.ekman@ltu.se
* Correspondence: giulio.antonini@univaq.it

Abstract: The computation of self and mutual inductances of coils is a classic problem of electrical engineering. The accurate modeling of coupled coils has received renewed interest with the spread of wireless power transfer systems. This problem has been quite well addressed for coplanar or perfectly coaxial coils but it is known that the misalignment conditions easily lead to a sharp decrease in the efficiency. Hence, it is crucial to take misalignment into account in order to properly design the overall wireless power transfer system. This work presents a study to compute analytically the mutual inductance of non-coaxial pancake coils with parallel axes. The accuracy of the proposed methodology is tested by comparison with the numerical results obtained using the tool Fast-Henry. Then, a wireless power transfer system, comprising a full bridge inverter is considered, showing the impact of the misalignment on the coupling between two pancake coils and, thus, between the source and the load.

Keywords: pancake coils; mutual inductance; wireless power transfer; planar coils

Citation: Parise, M.; Loreto, F.; Romano, D.; Antonini, G.; Ekman, J. Accurate Computation of Mutual Inductance of Non Coaxial Pancake Coils. *Energies* **2021**, *14*, 4907. https://doi.org/10.3390/en14164907

Academic Editors: Ivica Stevanovic and Bernhard Wunsch

Received: 25 June 2021
Accepted: 5 August 2021
Published: 11 August 2021

1. Introduction

The computation of mutual inductance between coaxial circular coils has been thoroughly addressed by many authors since the time of Maxwell who gave a formula for two circles whose axes intersect [1]. In [2,3] formulas for circular loops with parallel axes have been derived but, unfortunately, all these results can be applied only within a restricted range of parameters since they converge slowly in general [4].

Nowadays, numerical methods such as the finite element method (FEM) and boundary element method (BEM), allow calculation of the mutual inductance of realistic 3-D geometric arrangement of conductors in an accurate and fast way. However, analytic and semi-analytic methods are still of interest to address this problem as they considerably simplify the mathematical procedures, leading to a significant reduction in the computational effort [4]. General techniques have been developed over the years [5–10] which have been found to be useful in many different fields including eddy-current tomography [11], planar PCB inductors [12], coreless printed circuit board transformers [13], force and torque calculation [14,15], electromagnetic launchers [16], plasma science [17], and superconducting magnetic levitation [18]. Still, the methods presented in these works are finally based on a numerical computation [19–21].

The calculation of the mutual coupling between two coils with the spread of wireless power transfer (WPT) systems has received new interest. This problem has been quite well addressed for coplanar or perfectly coaxial coils [22,23] but it is known that the misalignment conditions easily lead to a sharp decrease in the efficiency. Misalignment can be either lateral or angular. In electrical vehicle charging systems, the most relevant is lateral misalignment. Hence, it becomes crucial to take this misalignment into account in order to properly design the overall WPT system [24–26]. In [27] the numerical solution proposed in [27] is applied to WPT system with many non-coaxial coils. While it is clear that the mutual inductance between non-coaxial coils can be computed by resorting to numerical integration [28], to authors' knowledge no analytical solution has been found to this problem. An accurate expression of mutual inductance of Archimedean spiral coils was presented and verified experimentally in [21] but, again, a numerical integration is required. By avoiding numerical integration, the mutual inductance can be calculated faster which can be used for the system optimization, i.e., evaluate faster the impact of misalignment on the WPT system performance e.g., [29]. This work presents a study to compute analytically the mutual inductance of non-coaxial pancake coils with parallel axes. The proposed formula in a series form is verified against numerical results obtained by means of the inductance extraction program Fast–Henry [30]. Finally, the case-study of a WPT system comprising two non-coaxial coils and a converter is presented, pointing out the effect of the misalignment on the performance.

2. Theory

Consider the two air-cored parallel flat coils shown in Figure 1. The turns of the coils have radii a_i $(i = 1, \ldots, N)$ and b_j $(j = 1, \ldots, M)$, while the radial displacement between the coil axes and the vertical spacing between the coil planes are indicated with ρ and h, respectively. The overall mutual inductance of the coils is given by

$$M_{ab} = \sum_{i=1}^{N} \sum_{j=1}^{M} \Phi(a_i, b_j, \rho), \tag{1}$$

where $\Phi(a, b, \rho)$ is the flux linkage per unit current between two generic turns with radii a and b.

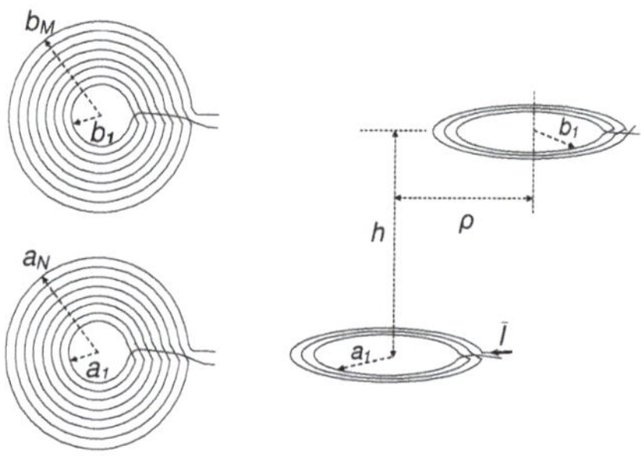

Figure 1. Sketch of two parallel pancake coils.

The goal of this section is to develop a rigorous procedure that allows analytical evaluation of the integral expression of $\Phi(a, b, \rho)$, that is [22]

$$\Phi(a, b, \rho) = \pi\mu_0 ab \int_0^\infty e^{-\lambda h} J_1(\lambda a)\, J_1(\lambda b)\, J_0(\lambda\rho) d\lambda, \tag{2}$$

where $J_m(\cdot)$ is the mth-order Bessel function, and μ_0 is the magnetic permeability of free space. To this end, we first replace $J_0(\lambda\rho)$ with its ascending power series expansion ([31] Eq. 9.1.12)

$$J_0(\lambda\rho) = \sum_{n=0}^{\infty} (-1)^n \frac{(\lambda\rho)^{2n}}{[(2n)!!]^2}, \tag{3}$$

with $(2n)!!$ being the double factorial $2 \cdot 4 \cdot 6 \cdot 8 \cdot \ldots \cdot (2n)$, and obtain

$$\Phi(a,b,\rho) = \pi\mu_0 a\, b \sum_{n=0}^{\infty} (-1)^n \frac{\rho^{2n}}{[(2n)!!]^2} \int_0^{\infty} e^{-\lambda h} J_1(\lambda a)\, J_1(\lambda b)\, \lambda^{2n} d\lambda. \tag{4}$$

Next, applying the identity

$$\lambda^{2n} e^{-\lambda h} = \frac{\partial^{2n} e^{-\lambda h}}{\partial h^{2n}}, \tag{5}$$

makes it possible to express the flux linkage between two arbitrary turns as

$$\Phi(a,b,\rho) = \sum_{n=0}^{\infty} (-1)^n \frac{\rho^{2n}}{[(2n)!!]^2} \frac{\partial^{2n}}{\partial h^{2n}} \Phi(a,b,0), \tag{6}$$

where $\Phi(a,b,0)$ is the flux corresponding to perfect alignment. An explicit representation for $\Phi(a,b,0)$ may be found starting from replacing the product of first-order Bessel functions with its finite integral representation according to Gegenbauer's addition theorem, namely [32] (Eq. 11.41.17)

$$J_1(\lambda a)\, J_1(\lambda b) = \frac{1}{\pi} \int_0^{\pi} J_0(\lambda c)\, \cos\phi\, d\phi, \tag{7}$$

where ϕ is the variable of integration, and

$$c = \sqrt{a^2 + b^2 - 2ab\cos\phi}. \tag{8}$$

This allows writing of the expression

$$\Phi(a,b,0) = \mu_0 ab \int_0^{\pi} \cos\phi\, d\phi \int_0^{\infty} e^{-\lambda h} J_0(\lambda c) d\lambda, \tag{9}$$

whose integral on the right-hand side is known and given by [33]

$$\int_0^{\infty} e^{-\lambda d} J_0(\lambda c) d\lambda = \frac{1}{\sqrt{d^2 + \varepsilon}}, \tag{10}$$

being $d^2 = a^2 + b^2 + h^2$, and $\varepsilon = -2ab\cos\phi$. Expanding (10) into a power series of ε, as follows

$$\frac{1}{\sqrt{d^2+\varepsilon}} = \frac{1}{d} \sum_{m=0}^{\infty} (-1)^m \frac{(2m-1)!!}{(2m)!!} \left(\frac{\varepsilon}{d^2}\right)^m \tag{11}$$

allows transformation of (9) into

$$\Phi(a,b,0) = \frac{\mu_0 ab}{d} \sum_{m=1}^{\infty} (-1)^m \frac{(2m-1)!!}{(2m)!!} \left(-\frac{2ab}{d^2}\right)^m \int_0^{\pi} \cos^{m+1}\phi\, d\phi, \tag{12}$$

where it has been taken into account that the finite integral over ϕ is non-null only for $m \neq 0$. Finally, since it holds [34]

$$\int_0^{\pi} \cos^{m+1}\phi = \begin{cases} \pi\, m!!\,/(m+1)!!, & \text{odd } m \\ 0, & \text{even } m \end{cases} \tag{13}$$

after setting $m = 2l + 1$, it is found that

$$\Phi(a,b,0) = \frac{\pi\mu_0 ab}{d}\sum_{l=0}^{\infty}\frac{(4l+1)!!}{(2l+2)!!(2l)!!}\left(\frac{ab}{d^2}\right)^{2l+1}. \tag{14}$$

Combining (14) with (6) provides an explicit expression for the mutual inductance of two misaligned turns with radii a and b, that is

$$\Phi(a,b,\rho) = \frac{\pi\mu_0 ab}{d}\sum_{n=0}^{\infty}(-1)^n\frac{\rho^{2n}}{[(2n)!!]^2}\sum_{l=0}^{\infty}\frac{(4l+1)!!}{(2l+2)!!(2l)!!}(ab)^{2l+1}\frac{\partial^{2n}}{\partial h^{2n}}\left[\frac{1}{(a^2+b^2+h^2)^{2l+1}}\right], \tag{15}$$

which, since it yields

$$\frac{\partial^{2n}}{\partial h^{2n}}\left[\frac{1}{(a^2+b^2+h^2)^{2l+1}}\right] = \frac{(2n)!}{(2l)!}\sum_{m=0}^{n}(-1)^{m+n}\frac{2^{2m}(2l+m+n)!}{(n-m)!(2m)!}\cdot\frac{h^{2m}}{d^{2(2l+m+n+1)}} \tag{16}$$

becomes

$$\Phi(a,b,\rho) = \frac{\pi\mu_0 ab}{d}\sum_{n=0}^{\infty}\frac{(2n-1)!!}{(2n)!!}\left(\frac{\rho}{d}\right)^{2n}\sum_{l=0}^{\infty}\frac{(4l+1)!!}{l!(l+1)!(2l)!}\left(\frac{ab}{2d^2}\right)^{2l+1}f_{ln}\left(\frac{2h}{d}\right), \tag{17}$$

where

$$f_{ln}(x) = \sum_{m=0}^{n}(-1)^m\frac{(2l+m+n)!}{(n-m)!(2m)!}x^{2m}. \tag{18}$$

Use of (17) and (18) in conjunction with (1) provides the mutual inductance of two parallel non-coaxial pancake coils. Moreover, expression (14) alone may be also applied to the computation of the overall self-inductance of each coil. For instance, for the coil at the bottom of Figure 1, it yields [22]

$$L_a = \sum_{i=1}^{N}L_t(a_i) + 2\sum_{i=1}^{N}\sum_{j=i+1}^{N}\Phi(a_i,a_j,0), \tag{19}$$

with $L_t(a)$ being the self-inductance of a thin-wire circular loop with radius a, given by [22,35]

$$L_t(a) = \mu_0 a\left[\log\left(\frac{8a}{r_w}\right) - 2\right], \tag{20}$$

where r_w is the wire radius. It should be observed that both the integral expression of $\Phi(a,b,\rho)$ and formula (20) are valid subject to the thin-wire assumption, which holds when r_w is negligible if compared to the radii of the turns.

3. Results and Discussion

To test the developed theory, the derived Formula (17) is first applied to the computation of the flux linkage between two identical coils made up of 10 turns. For each coil, the radius of the inner turn is taken to be equal to 3 cm, while the spacing between the edges of two adjacent turns is 2 cm. The mutual inductance is calculated against the radial distance ρ, and the obtained results, illustrated in Figures 2 and 3, are compared with the outcomes from the multipole-accelerated three-dimensional inductance extraction program Fast-Henry. Figure 2 depicts $\rho-$profiles of the mutual inductance corresponding to $h = 10$ cm, with the truncation index N of the outer sum in (17) taken as a parameter. Thus, the figure illustrates the effect of adding a higher-order term to the truncated power series expansion of the flux, seen as a function of ρ. To ensure highly accurate computation of the nth term of the expansion, the inner sequence of partial sums in (17) is terminated when the relative difference between the two last partial sums is smaller than a specified tolerance, which is assumed to be 10^{-12}. As is evident from the analysis of the plotted curves, the data arising from using the partial sum in (17) with $N = 3$ agree well with those produced by the Fast-Henry solver. As a consequence, the sequence of partial sums converges to the exact solution as N is increased. Figure 2 also points out that, for a given value of the truncation index N, the accuracy of the outcomes from (17) is affected by the misalignment between the coils, and that, in particular, it worsens as the misalignment increases. However, further increasing N makes it possible to achieve a high degree of accuracy of the results of the computation regardless of the value of the misalignment. For instance, in the $0 < \rho < 15$ cm range, the choice of $N = 8$ ensures that the relative difference between the last two elements of the outer sequence of partial sums in (17) is always smaller than 10^{-8}. This means that setting the desired tolerance of 10^{-8} for stopping the outer sequence of partial sums always leads to truncate the series at $N = 8$.

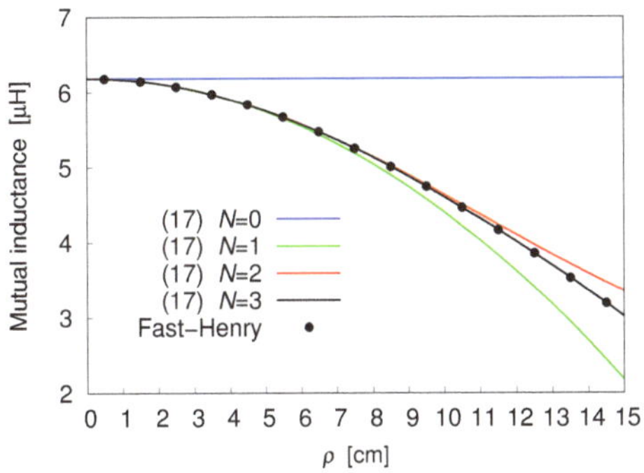

Figure 2. Mutual inductance of two 10-turn coils spaced 10 cm apart, as a function of ρ.

Convergence of (17) is also confirmed by the $\rho-$profiles plotted in Figure 3. Here, the same coils as in the previous example are considered, while their planes are now separated by the distance $h = 20$ cm. As can be seen, even if the distance has been doubled, perfect matching is still observed between the outcomes of Fast-Henry solver and the trend arising from the proposed series-form solution truncated at $N = 3$. This suggests that the rate of convergence of (17) is not affected by a variation of h.

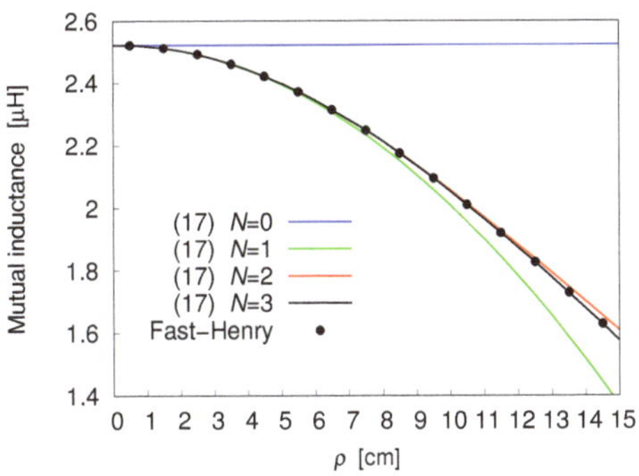

Figure 3. Mutual inductance of two 10-turn coils spaced 20 cm apart, as a function of ρ.

4. WPT System with a Full Bridge Inverter

A possible application of the coupled coils system presented is in the Wireless Power Transfer (WPT) area. The purpose of this paragraph is to provide an example where a couple of coils are employed in order to transfer an amount of power from a source to a distant load. The WPT coupling between two pancake coils is studied performing several simulations using the Plexim software "PLECS"; the target is to compute the load voltage over a load resistance R_L due to the exchange of power between two distant coupled coils [36]. The schematic of the configuration under investigation is shown in Figure 4. The central element of the circuit represents the coupling between a primary coil having a self inductance L_1 and the secondary coil with a self inductance L_2; the effect of the mutual coupling is described by the mutual inductance parameter L_m that, besides depending on the geometry and the materials employed for each single coil, depends certainly on the distance and the misalignment between the coils.

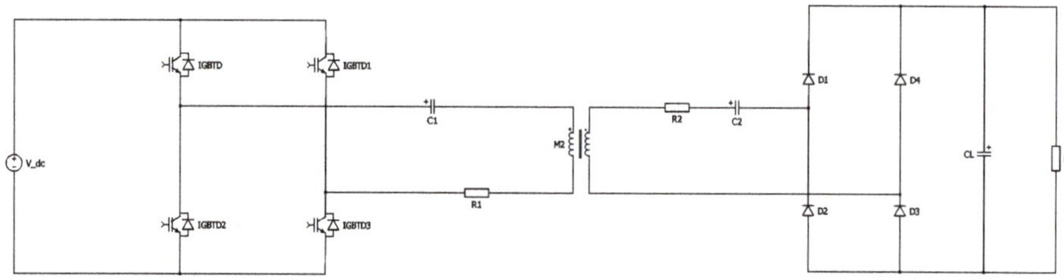

Figure 4. Circuit scheme of the inductive power transfer system which includes an inverter for the primary coil and a rectifier for the secondary coil.

The primary coil and its series resonant capacitor C_1 are connected to an ideal full-bridge inverter. The secondary coil and its series resonant capacitor C_2 are connected to an ideal full-bridge rectifier which converts the induced AC voltage of the secondary coil to DC voltage. The components R_1 and R_2 represent primary and secondary effective series DC resistances, respectively, which in this case are equal because the geometry is the same for both the coils. The load capacitor C_L is a filter capacitor. The full bridge inverter is supposed to be operated by a switching frequency $f_s = 20$ kHz [36]; hence, a bipolar voltage is generated by the inverter on the primary side. The global voltage applied to the primary side is sketched in Figure 5. In order to guarantee the zero voltage switching (ZVS) operation of the inverter, the primary side resonant frequency f_{r1}, determined by C_1 and L_1, is chosen slightly smaller than the switching frequency [36].

$$f_s = 1.05 f_{r1} = \frac{1.05}{2\pi\sqrt{L_1 C_1}} \tag{21}$$

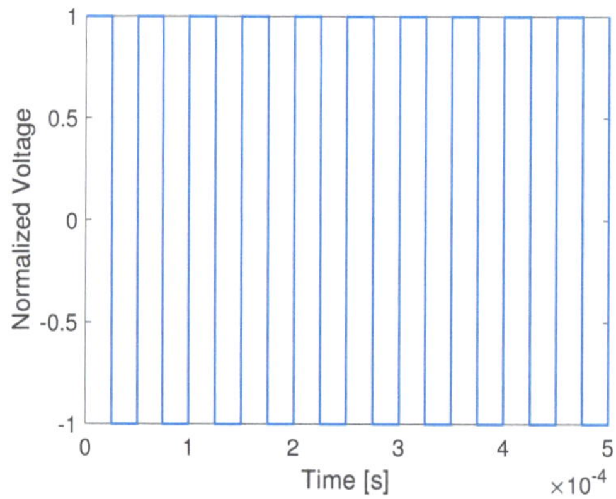

Figure 5. Voltage applied to the WPT system.

The secondary side resonant frequency f_{r2}, determined by C_2 and L_2, is tuned exactly to the switching frequency:

$$f_{r2} = \frac{1}{2\pi\sqrt{L_2 C_2}} = f_s \tag{22}$$

It is clear that the capacitors values C_1 and C_2 can be easily determined by (21) and (22) knowing only the self inductances values and the inverter switching frequency. Twelve different configurations have been considered for the simulations, varying both the distances and the misalignments between the coils under observation. All the set-up parameters are reported in Table 1 and the different configurations are reported in Table 2.

Table 1. Circuit parameters.

Switching frequency	$f_s = 20$ kHz
Primary capacitor	$C_1 = 4.1$ µF
Primary resistance	$R_1 = 1.66$ mΩ
Secondary capacitor	$C_2 = 3.73$ µF
Secondary resistance	$R_2 = 1.66$ mΩ
Self inductances	$L_1 = L_2 = 17$ µH
Load capacitor	$C_L = 300$ µF
Load resistance	$R_L = 40$ Ω
Conductors radius	$r_w = 5$ mm

Table 2. Test configurations.

Distance h [cm]	Misalignment ρ [cm]	L_m [µH]
10	0.5	6.18
10	3.5	5.97
10	6.5	5.47
10	9	4.88
15	0	3.87
15	3.5	3.76
15	6.25	3.52
15	8.5	3.24
20	0	2.52
20	3.25	2.47
20	6	2.34
20	8.5	2.18

The load voltage on the resistance R_L, normalized to the DC source voltage, is computed and illustrated, for each value of the misalignment, in Figures 6–8. It is clearly seen that the misalignment has a significant impact on the rectified load voltage. Furthermore, for the case $h = 10$ cm, $\rho = 0.5$ cm and $L_m = 6.18$ µH, the efficiency for increasing values of the load resistance R_L has been evaluated; the efficiency is defined as the average power delivered to the load divided by the average power supplied by the full-bridge inverter. It is easy to verify from Figure 9 that the optimal load resistance for this specific configuration is around $R_L = 1.1$ Ω.

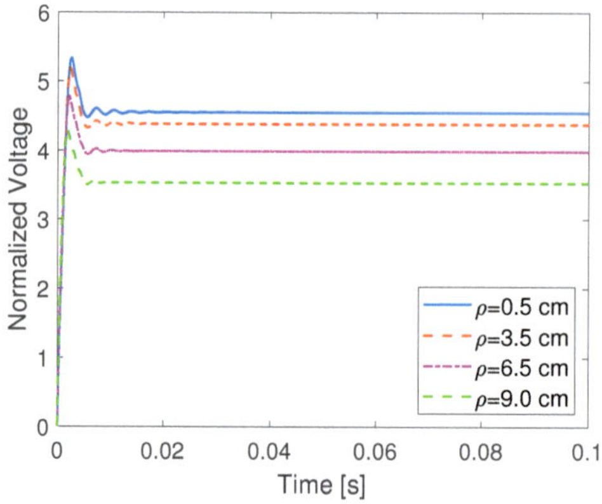

Figure 6. Load normalized voltages with $h = 10$ cm.

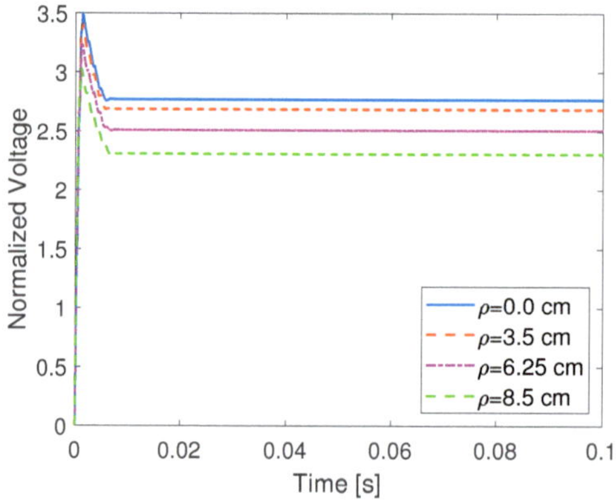

Figure 7. Load normalized voltages with $h = 15$ cm.

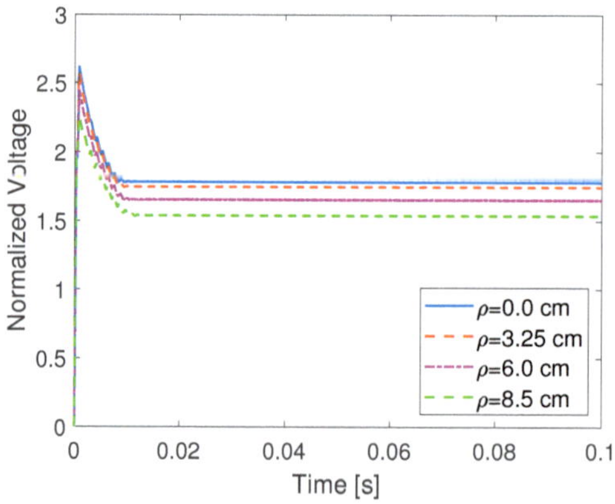

Figure 8. Load normalized voltages with $h = 20$ cm.

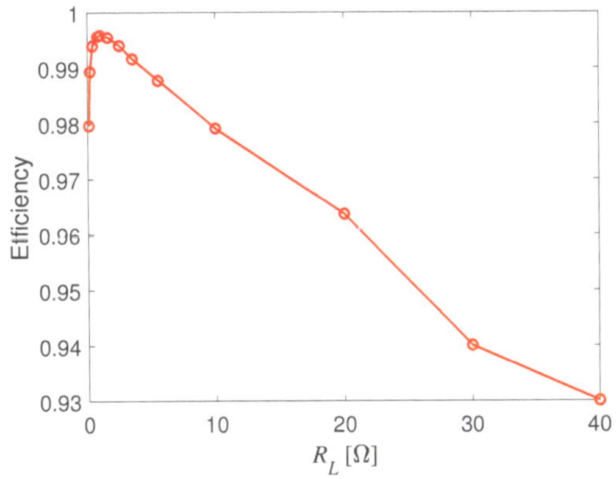

Figure 9. WPT efficiency varying the load resistance R_L and assuming $h = 10$ cm, $\rho = 0.5$ cm.

As a last test, the load voltage was computed considering four different load conditions: $R_L = [1, 10, 20, 40]\ \Omega$, when: $h = 10$ cm, $\rho = 0.5$ cm and $L_m = 6.18$ μH; the results are shown in Figure 10.

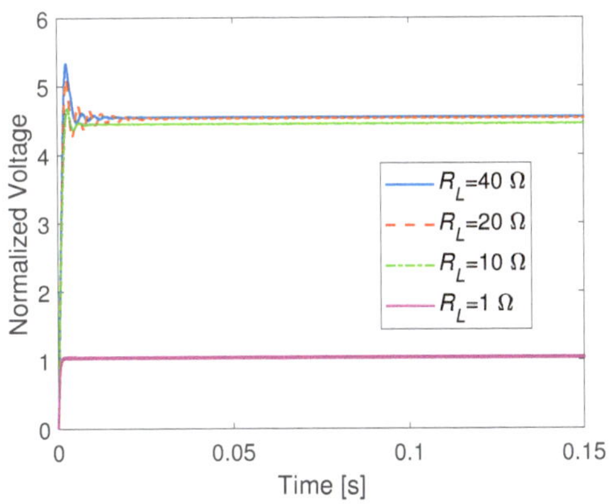

Figure 10. Load normalized voltages assuming $h = 10$ cm, $\rho = 0.5$ cm and varying the load resistance R_L.

5. Conclusions

This work has presented a study to compute analytically the mutual inductance of non-coaxial pancake coils with parallel axes. The proposed formula was tested by comparison with the numerical results obtained using the tool Fast-Henry. Then, the proposed formula was used to compute the mutual inductance between two pancake coils in a wireless power transfer system, comprising a full bridge inverter. The impact of the misalignment on the coupling between the source and the load was verified and quantified.

Author Contributions: Conceptualization, M.P., F.L., D.R., G.A. and J.E.; methodology, M.P., F.L., D.R., G.A. and J.E.; software, M.P., F.L. and D.R.; validation, M.P., F.L. and D.R.; formal analysis, M.P., F.L., D.R., G.A. and J.E.; investigation, M.P., F.L., D.R. and G.A.; resources, M.P., G.A. and J.E.; data curation, M.P., F.L., D.R. and G.A.; writing—original draft preparation, M.P., F.L. and G.A.; writing—review and editing, M.P., F.L., D.R., G.A. and J.E.; visualization, M.P., F.L. and G.A.; supervision, M.P. and G.A.; project administration, G.A.; funding acquisition, G.A. All authors have read and agreed to the published version of the manuscript.

Funding: This research received no external funding.

Acknowledgments: The authors are grateful to Ulrike Grossner and Ivana Kovacevic-Badstuebner for their hints which have allowed improvemnet of the quality of the paper.

Conflicts of Interest: The authors declare no conflict of interest.

Abbreviations

The following abbreviations are used in this manuscript:

MDPI	Multidisciplinary Digital Publishing Institute
DOAJ	Directory of open access journals
EM	Electromagnetic
PE	Power electronics
PES	Power electronic systems
HF	High Frequency
LF	Low Frequency
SiC	Silicon Carbide
PEEC	Partial Element Equivalent Circuit
FEM	Finite Element
3-D	Three-dimensional

References

1. Maxwell, J.C. *A Treatise on Electricity and Magnetism*; Dover: New York, NY, USA, 1954.
2. Butterworth, S. On the coefficients of mutual induction of eccentric coils. *Philos. Mag. J. Sci.* **1916**, *31*, 443–454. [CrossRef]
3. Snow, C. *Formulas for Computing Capacitance and Inductance*; Bureau of Standards Circular 544: Washington, DC, USA, 1954.
4. Babic, S.; Sirois, F.; Akyel, C.; Girardi, C. Mutual Inductance Calculation Between Circular Filaments Arbitrarily Positioned in Space: Alternative to Grover's Formula. *IEEE Trans. Magn.* **2010**, *46*, 3591–3600. [CrossRef]
5. Babic, S.; Akyel, C. New analytic-numerical solutions for the mutual inductance of two coaxial circular coils with rectangular cross section in air. *IEEE Trans. Magn.* **2006**, *42*, 1661–1669. [CrossRef]
6. Babic, S.I.; Akyel, C. Calculating Mutual Inductance Between Circular Coils With Inclined Axes in Air. *IEEE Trans. Magn.* **2008**, *44*, 1743–1750. [CrossRef]
7. Conway, J.T. Inductance calculations for noncoaxial coils using Bessel functions. *IEEE Trans. Magn.* **2007**, *43*, 1023–1034. [CrossRef]
8. Conway, J.T. Noncoaxial Inductance Calculations Without the Vector Potential for Axisymmetric Coils and Planar Coils. *IEEE Trans. Magn.* **2008**, *44*, 453–462. [CrossRef]
9. Engel, T.; Rohe, S. A comparison of single-layer coaxial coil mutual inductance calculations using finite-element and tabulated methods. *IEEE Trans. Magn.* **2006**, *42*, 2159–2163. [CrossRef]
10. Cirimele, V.; Torchio, R.; Virgillito, A.; Freschi, F.; Alotto, P. Challenges in the Electromagnetic Modeling of Road Embedded Wireless Power Transfer. *Energies* **2019**, *12*. [CrossRef]
11. Su, Y.P.; Liu, X.; Hui, S.Y.R. Mutual Inductance Calculation of Movable Planar Coils on Parallel Surfaces. *IEEE Trans. Power Electron.* **2009**, *24*, 1115–1123. [CrossRef]
12. Sonntag, C.L.W.; Lomonova, E.A.; Duarte, J.L. Implementation of the Neumann formula for calculating the mutual inductance between planar PCB inductors. In Proceedings of the 2008 18th International Conference on Electrical Machines, Vilamoura, Portugal, 6–9 September 2008; pp. 1–6. [CrossRef]
13. Tang, S.; Hui, S.; Chung, H. A low-profile wide-band three-port isolation amplifier with coreless printed-circuit-board (PCB) transformers. *IEEE Trans. Ind. Electron.* **2001**, *48*, 1180–1187. [CrossRef]
14. Ravaud, R.; Lemarquand, G.; Lemarquand, V. Force and Stiffness of Passive Magnetic Bearings Using Permanent Magnets. Part 1: Axial Magnetization. *IEEE Trans. Magn.* **2009**, *45*, 2996–3002. [CrossRef]
15. Ravaud, R.; Lemarquand, G.; Lemarquand, V. Force and Stiffness of Passive Magnetic Bearings Using Permanent Magnets. Part 2: Radial Magnetization. *IEEE Trans. Magn.* **2009**, *45*, 3334–3342. [CrossRef]
16. Keyi, Z.; Bin, L.; Zhiyuan, L.; Shukang, C.; Ruiping, Z. Inductance Computation Consideration of Induction Coil Launcher. *IEEE Trans. Magn.* **2009**, *45*, 336–340. [CrossRef]
17. Engel, T.G.; Mueller, D.W. High-Speed and High-Accuracy Method of Mutual-Inductance Calculations. *IEEE Trans. Plasma Sci.* **2009**, *37*, 683–692. [CrossRef]
18. Kajikawa, K.; Yokoo, R.; Tomachi, K.; Enpuku, K.; Funaki, K.; Hayashi, H.; Fujishiro, H. Numerical Evaluation of Pulsed Field Magnetization in a Bulk Superconductor Using Energy Minimization Technique. *IEEE Trans. Appl. Supercond.* **2008**, *18*, 1557–1560. [CrossRef]
19. Babic, S.; Martinez, J.; Akyel, C.; Babic, B. Mutual Inductance Calculation between Misalignment Coils for Wireless Power Transfer of Energy. *Prog. Electromagn. Res. M* **2014**, *38*, 91–102. [CrossRef]

20. Luo, Z.; Wei, X. Mutual Inductance Analysis of Planar Coils with Misalignment for Wireless Power Transfer Systems in Electric Vehicle. In Proceedings of the 2016 IEEE Vehicle Power and Propulsion Conference (VPPC), Hangzhou, China, 17–20 October 2016; pp. 1–6. [CrossRef]
21. Liu, S.; Su, J.; Lai, J. Accurate Expressions of Mutual Inductance and Their Calculation of Archimedean Spiral Coils. *Energies* **2019**, *12*, 2017. [CrossRef]
22. Zierhofer, C.M.; Hochmair, E.S. Geometric approach for coupling enhancement of magnetically coupled coils. *IEEE Trans. Biomed. Eng.* **1996**, *43*, 708–714. [CrossRef]
23. Parise, M.; Antonini, G.; Romano, D. On the Flux Linkage between Pancake Coils in Resonance-Type Wireless Power Transfer Systems. *Int. J. Antennas Propag.* **2020**, *2020*, 1–6. [CrossRef]
24. Simonazzi, M.; Sandrolini, L.; Zarri, L.; Reggiani, U.; Alberto, J. Model of Misalignment Tolerant Inductive Power Transfer System for EV Charging. In Proceedings of the 2020 IEEE 29th International Symposium on Industrial Electronics (ISIE), Delft, The Netherlands, 17–19 June 2020; pp. 1617–1622. [CrossRef]
25. Fontana, N.; Brizi, D.; Barmada, S.; Monorchio, A. A Methodology for Efficiency Recovering in Wireless Power Transfer Applications with Misalignment. In Proceedings of the 2020 XXXIIIrd General Assembly and Scientific Symposium of the International Union of Radio Science, Rome, Italy, 29 August–5 September 2020. [CrossRef]
26. ElGhanam, E.; Hassan, M.; Osman, A.; Kabalan, H. Design and Performance Analysis of Misalignment Tolerant Charging Coils for Wireless Electric Vehicle Charging Systems. *World Electr. Veh. J.* **2021**, *12*, 89. [CrossRef]
27. Steckiewicz, A.; Stankiewicz, J.M.; Choroszucho, A. Numerical and Circuit Modeling of the Low-Power Periodic WPT Systems. *Energies* **2020**, *13*, 2651. [CrossRef]
28. Tal, N.; Morag, Y.; Levron, Y. Magnetic Induction Antenna Arrays for MIMO and Multiple-Frequency Communication Systems. *Prog. Electromagn. Res. C* **2017**, *75*, 155–167. [CrossRef]
29. Miller, J.M.; Onar, O.C.; Chinthavali, M. Primary-Side Power Flow Control of Wireless Power Transfer for Electric Vehicle Charging. *IEEE J. Emerg. Sel. Top. Power Electron.* **2015**, *3*, 147–162. [CrossRef]
30. Kamon, M.; Tsuk, M.J.; White, J.K. FASTHENRY: A multipole-accelerated 3-D inductance extraction program. *IEEE Trans. Microw. Theory Tech.* **1994**, *42*, 1750–1758. [CrossRef]
31. Abramowitz, M.; Stegun, I.A. *Handbook of Mathematical Functions: With Formulas, Graphs, and Mathematical Tables*; Courier Corporation: Washington, DC, USA, 1964; Volume 55.
32. Watson, G.N. *A Treatise on the Theory of Bessel Functions*; Cambridge mathematical library, Cambridge University Press: Cambridge, UK, 1944.
33. Erdelyi, A. *Tables of Integral Transforms*; McGraw-Hill: New York, NY, USA, 1954; Volume 1.
34. Gradshteyn, I.S.; Ryzhik, I.M. *Table of Integrals, Series, and Products*; Academic Press: New York, NY, USA, 2007.
35. Paul, C.R. *Inductance: Loop and Partial*; John Wiley & Sons: Hoboken, NJ, USA, 2010.
36. Rim, C.T.; Mi, C. *Wireless Power Transfer for Electric Vehicles and Mobile Devices*; JohnWiley & Sons: Hoboken, NJ, USA, 2017.

MDPI
St. Alban-Anlage 66
4052 Basel
Switzerland
Tel. +41 61 683 77 34
Fax +41 61 302 89 18
www.mdpi.com

Energies Editorial Office
E-mail: energies@mdpi.com
www.mdpi.com/journal/energies

www.ingramcontent.com/pod-product-compliance
Lightning Source LLC
Chambersburg PA
CBHW041400210926
51850CB00013B/3134
* 9 7 8 3 0 3 6 5 1 9 8 7 6 *